D1807310

# Governance, Development, and Social Inclusion in Latin America

Series Editors
Rebecka Villanueva Ulfgard
International Studies
Instituto Mora
Mexico City, Mexico

César Villanueva
Department of International Studies
Universidad Iberoamericana
Mexico City, Mexico

This series seeks to go beyond a traditional focus on the virtues of intra-regional and inter-regional trade agreements, liberal economic policies, and a narrow security agenda in Latin America. Instead, titles deal with a broad range of topics related to international cooperation, global and regional governance, sustainable development and environmental cooperation, internal displacement, and social inclusion in the context of the Post-2015 Development Agenda—as well as their repercussions for public policy across the region. Moreover, the series principally focuses on new international cooperation dynamics such as South-South and triangular cooperation, knowledge sharing as a current practice, and the role of the private sector in financing international cooperation and development in Latin America. The series also includes topics that fall outside the traditional scope of studying cooperation and development, in this case, (in)security and forced internal displacement, cultural cooperation, and *Buen Vivir* among indigenous peoples and farmers in Latin America. Finally, this series welcomes titles which explore the tensions and dialogue around how to manage the imbalance between state, markets, and society with a view to re-articulating cooperation and governance dynamics in the 21st century.

More information about this series at
http://www.palgrave.com/gp/series/15135

Matilda Baraibar Norberg

# The Political Economy of Agrarian Change in Latin America

## Argentina, Paraguay and Uruguay

Matilda Baraibar Norberg
Department of Economic History
and International Relations
Stockholm University
Stockholm, Sweden

Governance, Development, and Social Inclusion in Latin America
ISBN 978-3-030-24585-6          ISBN 978-3-030-24586-3   (eBook)
https://doi.org/10.1007/978-3-030-24586-3

© The Editor(s) (if applicable) and The Author(s), under exclusive license to Springer
Nature Switzerland AG 2020
This work is subject to copyright. All rights are solely and exclusively licensed by the
Publisher, whether the whole or part of the material is concerned, specifically the rights
of translation, reprinting, reuse of illustrations, recitation, broadcasting, reproduction
on microfilms or in any other physical way, and transmission or information storage and
retrieval, electronic adaptation, computer software, or by similar or dissimilar methodology
now known or hereafter developed.
The use of general descriptive names, registered names, trademarks, service marks, etc. in this
publication does not imply, even in the absence of a specific statement, that such names are
exempt from the relevant protective laws and regulations and therefore free for general use.
The publisher, the authors and the editors are safe to assume that the advice and
information in this book are believed to be true and accurate at the date of publication.
Neither the publisher nor the authors or the editors give a warranty, expressed or implied,
with respect to the material contained herein or for any errors or omissions that may have
been made. The publisher remains neutral with regard to jurisdictional claims in published
maps and institutional affiliations.

Cover image: © Architectura/Alamy Stock Photo

This Palgrave Macmillan imprint is published by the registered company Springer Nature
Switzerland AG
The registered company address is: Gewerbestrasse 11, 6330 Cham, Switzerland

# SERIES EDITORS' PREFACE

When thinking about agriculture and economic growth in today's Latin America, it is unavoidable to consider the wide contrasts between technology-intensive, industrial-scale projects, on the one hand, and small farmers with limited productivity, on the other. This development is not unique to Latin America but is shared with many other developing regions in the world, in Africa and Southeast Asia, for example.[1] The massive expansion of eucalyptus, palm plantations, or soybeans on farming land is now commonplace in emerging economies, which has caused significant land-use and land-cover change (LULCC). However, despite protests from environmental and human rights activists, consumer organizations, and the scientific community as regards the adverse ecological, health-related, and social consequences of LULCC, political action with the objective of mitigation is not addressing fully the depth or scope of its effects on nature and human beings.

Take for example the soybean expansion. The dynamics of the global soybean chain (but it could also be coffee beans, maize, or wheat) in the first twenty years of this century have dramatically changed consumption patterns and affected both local communities and the environment throughout the world. Today, industrial-scale soybean producers are joining loggers and cattle ranchers in speeding up destruction and further fragmentation of immense, untouched territories. Now, clearing and tilling the land for soybean production—with the effect of turning forests into deserts—is only part of the challenge. Soybean cultivation puts an end to habitat for wildlife (including endangered or unknown species), destroys ecosystems, and increases greenhouse gases that contribute to global warming. Intense use of synthetic fertilizers, pesticides, and herbicides (for example, glyphosate, which is a likely human carcinogen) not only threatens the environment and animal life, but it also endangers human health. Moreover, the soybean global production is situated in a chain that includes labor, land, and capital, to extract as much gains as possible (often disguised as development) at the expense of specific cultures and their local habitats.

On October 1, 2018, the Director-General of the United Nations Food and Agriculture Organization (FAO), José Graziano da Silva, at the biennial meeting of FAO's Committee on Agriculture (COAG, 1–5 October) with representatives of governments, private sector and civil society organizations, expressed concern that "current farming practices have contributed to deforestation, water scarcity, soil depletion and high levels of greenhouse gas emissions". He called for a shift to move away from "high-input and resource-intensive farming and food systems" to production systems that help preserving biodiversity, ecosystems, and the environment. "To do so requires reducing the use of pesticides and chemicals, increasing crop diversification, and improving land conservation practices," the Director-General said.[2] This is a discourse that seems to be falling on deaf ears. In fact, it is in stark contrast with the current tendencies in many developing countries. Brazil under the leadership of Jair Bolsonaro will be a case in point. The case of Brazil has been well-researched, but what about its southern neighbors?

Drawing on several years of research, this exceptional study by Matilda Baraibar, *The Political Economy of Agrarian Change in Latin America—Argentina, Paraguay and Uruguay*, offers unique insights into the controversial and multifaceted issue of agropastoral expansion in Latin America in the context for increasing global demand for meat and dairy products. Through a cross-national and historical comparative approach, the reader is invited to rethink the challenges facing the state with regard to issues of autonomy and capacity in the soybean and beef complexes in Argentina, Paraguay, and Uruguay. Baraibar's study is a must read for anyone interested in capturing the historical context, modern development, and contemporary challenges surrounding the LULCC phenomenon in Latin America, which has become the world's largest provider of both "pasturelands for livestock grazing" and "croplands for high-protein vegetable feed grains, particularly soybeans" (all quotes in this

Preface are taken from Chapter 1). Argentina, Paraguay, and Uruguay display the "highest rate of export specialization in agrofood products, which makes their economies highly dependent on the development of the international agrofood market", in addition to varying state capacity for designing strategies with the view to tackle the demands from contemporary agrofood globalization. This state capacity has been conditioned by national institutions shaped over time and national regulations.

Mapping out several strands of research in the field of economic history, Baraibar extracts key concepts and theorized notions and then creates her distinctive conceptual-analytical grid with which she approaches a comprehensive empirical material (including several interviews with high-level officials in the three countries). She valuably identifies some key limitations in two broad fields: land-use and land-cover change (LULCC), on the one hand, concentrating on "distant drivers and world systemic explanations" and pressures arising from economic and political globalization, versus "myopically centering the analysis on formal state regulations as if recent agrarian change could be exclusively governed by the nation-states," on the other. The void she identifies is then replaced with a multifaceted reflection on the state's room for maneuver in the globalized economy, global markets, and global production chains. The result is a multilayered, intriguing study, which should leave a mark in the academic literature on this global phenomenon as it displays significant depth, originality, and critical analysis.

We find it particularly interesting the way Baraibar scrutinizes the role of the state in Latin America. She emphasizes it is important to go beyond the exaggerated supposition of Latin American states "either as powerless victims" (short of state autonomy) or "completely failed states or in perfect control". Her study speaks to one of our aims with this series, namely to explore the tensions and dialogue around how to manage the imbalance between states, markets, and societies. The two approaches: (1) an overemphasis on global forces with a mischaracterization of the role of the Latin American state and public regulations, as well as (2) a state-centrist approach that helps little to understand the pressures on the Latin American state arising from transnational agribusiness, have to be complemented with an analysis focused on bringing the state back into the picture, she claims. Besides, the sovereign power of the state is still in place, and should be used to regulate, protect and facilitate environmental, territorial and peoples' capacities and rights, she insists.

The design of Baraibar's study is innovative and revolves around the idea that the state is certainly important, but "it is not the only game in town." Therefore, she applies three distinct approaches to dissect changes in the soybean and beef complexes in Argentina, Paraguay, and Uruguay. First, to fixate the "political economy of agrarian change," she centers on the creation of institutions and regulations in a *historical approach* in which *food regimes* "functions as a chronological periodization." Second, she adopts a *world systemic approach*. Here, the soybeans and beef production in the three countries is approached through a *global commodity chain* analysis. Third, Baraibar applies a *regulatory approach* that enables the examination of regulatory frameworks in a comparative light (to address how Argentina, Paraguay, and Uruguay cope with the transformative pressures brought by agrofood globalization). She advocates that in order to understand the political economy of current agrarian change in Latin America, one has to keep a good historical record of many different changes that have taken place associated with regulations, institutions and actors that made key decisions. The ways in which agrarian change has been articulated in the three countries can only be understood by including characteristic features in specific national regulatory frameworks and then "connecting them to both macro (international) and micro (specific areas of territories) levels."

It should be pointed out that Baraibar engages in a notable discussion between the state-centric approach and the world systems approach to shed light on the conflict between the (in)capacity of states to enforce their mandate and rights, on the one hand, and agribusinesses' (self) interests and their predilection for neoliberal global market mechanisms, on the other. Baraibar skillfully penetrates the connections between the national, the regional and the international arenas through mapping out the genealogy of specific policies. This methodology enables her to track down the actors, their agendas and the ways in which the decision-making process took place, giving rise to regulations, institutionalization or else, a regime. Her study is enriched by a significant number of in-depth personal interviews with key stakeholders from a wide array of positions, who participated in the policy formulation and implementation of many of the policies discussed throughout the chapters.

Now, Baraibar does not seek to issue prescriptions for how to remedy some of the persistent imbalances in the soybean and beef complexes involving Argentina, Paraguay, and Uruguay. With acute precision, she identifies some of the challenges brought on by distant-driven agrarian change when it comes to governance of food production systems. Moreover, she indirectly highlights dilemmas of development and how to handle the social costs arising from this particular phenomenon. In the final chapter, she suggests that "stronger and more coordinated regulative efforts could lead to more sustainable pathways for agrarian change in Latin America." As this series seeks to go beyond a traditional focus on the virtues of intraregional and inter-regional trade agreements, and (neo-)liberal economic policies, Baraibar's thought-provoking work is a very welcome contribution to bring greater focus on Latin America in debates on contemporary global challenges, such as precisely agropastoral expansion.

One of the many outstanding qualities of Baraibar's exhaustive study is that she unveils the intertwined nature of LULCC change in the increasing expansion of the soybeans and beef complexes, which is also cross-nationally integrated. It is also remarkable how she examines some of the adverse social-ecological effects resulting from changes in the production process; massive losses of farming land and diversity of crops. In addition, she takes into consideration social aspects following LULCC change such as "displacement of family farmers" and "agriculture without farmers" (reliance on sophisticated technological solutions, automatization). Looking at the transformations in the agency–structure relationship in the changing production process, it is clear that the value added from processing soybeans has been transferred from the local farmers to multinational corporations. Hence, concentration of land and agribusiness are cornerstones in the dilemma of decreasing food sovereignty.

In contrast, we reason that the tensions inherent in such transformations could also be seen through the classical distinction made by Karl Polanyi in his seminal work *The Great Transformation* (1944), where the self-regulating market aims to commodify land, labor, and money, and the countermovement tries to regulate and protect the social strata, or the community. In our times, however, Polanyi's idea of a "double movement" does not function the same way as in the times of the emergence of market economy in England. First, the nation-state cannot longer be exclusively accountable for the regulation of social and market relations. A pattern of global governance is vested in international organizations, which in turn establish policies and provide funding for development. However, such supra-state authorities embrace the logic of capital accumulation, free market-oriented growth, and deregulations, which has become a generalized imperative throughout the world. Their regard for sustainability and social relations is secondary, whereas the state is called for to provide with policies and regulations defining such issues.

A few decades ago, Phillip McMichael cautioned policy-makers against the practices carried out by private interests in the name of globalization and prosperity. He claimed that "development" became identified, by many countries in the Global South, with the ideal of consolidating a growing national economy fueled by a domestic-global link and an exchange between agricultural and industrial sectors.[3] As we have witnessed these last two decades, this ideal has not delivered on its promise mainly because of the diminishing state power and the increasing private interests' capacity to mold policies accordingly. McMichael says that "The corporate food regime, constituted by the double movement of empire and food sovereignty thus reveals both the immanent, and the historical, conditions governing the politics of capitalist empire in the twenty-first century. In the crisis of the Doha Round [2004], the discourse of development is most clearly framed by the dialectic of 'food security' versus 'food sovereignty'."[4] Baraibar's study is a remarkable update of this debate that is relevant not only for South America's emerging economies or developing countries in other regions of the world, but also for the debate itself in academia, policy circles, and wider society.

We wish that Baraibar's inspiring study and future investigations will ring loud in research communities in many parts of the world. This, because the issue of exogenously driven demand for primary materials— be they soybean, copper, or something else—and the pressures on the state in the Global South for transformative change with regard to regulations and standards is a pressing challenge worth being debated seriously from, precisely, the point of view of constrained state autonomy and capacity under globalization.

Mexico City, Mexico                    Rebecka Villanueva Ulfgard
May 2019                               César Villanueva

# NOTES

1. See, for example, *BRICS and MICs: Implications for Global Agrarian Transformation*, Ben Cousins, Saturnino M. Borras Jr., Sérgio Sauer, and Jingzhong Ye (eds.), Routledge (Rethinking Globalizations Series), 2018.
2. n/a. "Transforming agriculture and food systems to nourish people and to nurture the planet", Food and Agriculture Organization of the United Nations, October 1, 2018. Available at http://www.fao.org/news/story/en/item/1155122/icode/.
3. Philip McMichael (ed.), 1994, *The Global Restructuring of Agro-Food Systems*, Cornell University Press; Philip McMichael, 2000, *Development and Social Change: A Global Perspective*, 2nd edition, Pine Forge Press.
4. Philip McMichael, 2004, "Global Development and the Corporate Food Regime", Discussion presented for the Symposium on New Directions in the Sociology of Global Development, XI World Congress of Rural Sociology, Trondheim, July 2004. Available at https://www.iatp.org/sites/default/files/451_2_37834.pdf.

# PROLOGUE AND ACKNOWLEDGEMENTS

This book reflects my long-standing interest in the development in the region and my wish to understand the role of the state in co-shaping agrarian change in the face of contemporary agrofood globalization. I see the dramatic agrarian change in twenty-first century Latin America as the combined and complex result of world- and local-systemic rules and power relations. Both involve historically formed institutions, albeit continuously evolving and renegotiated, as well as place-based social-ecological variance. This book may have too many "facts" for most theorists, and appear too superficial for most empirical specialists studying specific aspects of this change. I remain convinced, however, that recent agrarian change in the region cannot be understood properly without a longer time frame, without consideration of relevant shifts in the international agrofood system, and without exploration of how the states, as arenas for power struggles and negotiations, have acted upon these pressures.

While the bibliography stands as a statement of my deep indebtedness to scholars from a wide range of fields, the most insightful and meaningful information in this book comes from the generous accounts patiently provided by my interviewed respondents. This book is fundamentally indebted to all of them.

I would like to thank all friends and colleagues at the *South American Institute for Resilience and Sustainability Studies* (SARAS²) for providing such a stimulating, creative and dynamic environment for thinking, talking, observing, and dreaming together—seeking not only to comprehend change in systemic and transdisciplinary ways, but also to engage in active dialogues with other arenas of society and contribute to the construction of sustainable futures. I would particularly like to express my deep gratitude to Nestor Mazzeo for support, inspiration, knowledge-sharing, and enthusiasm. From the SARAS community, I would also like to name and express my special thanks to Cristina Zurbriggen, Esteban Jobbágy, Micaela Trimble, and Manfred Steffan.

It is actually quite overwhelming to think about all the people that under this process have shared their knowledge with me, and in other ways inspired and supported me in the process of doing research and writing this book. I have had some truly fantastic talks (formal and informal) about complexity, coupled systems, change, soybeans, beef, trade-offs, timescales, historical legacies, winners and losers, and desirable pathways for Latin America. Thank you for sharing your knowledge with me. While it is impossible to name each and everyone to whom I am indebted—I think generally writers are more in debt to others than they usually know—I still want to name a special few who each in their own way has contributed to this book: Andrés Berterreche, Federico Bizzozero, Alfredo Blum, Virginia Borrat, Gastón Carro, Aurelio Enciso, Mariana Fossatti, Cecilia Gelabert, Eduardo Gudynas, Gabriella Jorge, Gervasio Piñeiro, Ruben Puentes, Mario Torres Jarrín, and Henrik Österblom.

My special thanks are extended to the staff that provided me with written information and/or interviews at MAGyP Argentina, MGAP Uruguay, MAG Paraguay, INTA Argentina, SEAM Paraguay, MVOTMA Uruguay, INIA Uruguay, SENAVE Paraguay, ANII Uruguay, University of the Republic Uruguay, University of Buenos Aires Argentina, UNDP Paraguay, IICA Paraguay, IICA Uruguay, WWF Paraguay, IICA Argentina, CLACSO Argentina, CNFR Uruguay, Marfrig Southern Cone, MTO Uruguay, El Tejar Uruguay, ALUR Uruguay.

Besides knowledge and inspiration, a book is ultimately a body of text that has to make sense in itself. It should be able to communicate insights in a correct, meaningful, coherent, and understandable way. In this respect, I am deeply indebted to many valuable comments and suggestions on drafts presented at the Department of Economic History and International Relations, and at the Institute of Latin American Studies, both at Stockholm University. I would like to express my very great appreciation to Daniel Berg, Gloria Gallardo Fernández, Maria-Therese Gustafsson, Erik Hegelund, Sandra Hellstrand, Ulf Jonsson, Magnus Lembke, Markus Lundström, María Mancilla Garcia, Daniel Silberstein and Paulina de los Reyes, for advice, support, comments, and critique, in this respect. I would also like to say *gracias* to Gonzalo Pozo-Martín for encouragement, inspiration and friendship. But special in all respects is my debt to Lisa Deutsch. She has read several drafts of the manuscript and contributed with constructive suggestions, from straightening out facts, to clarifying expressions or following the rules of grammar, to enhance theoretical consistency.

The editors of the book series, Rebecka Villanueva Ulfgard, and César Villanueva have been of great support and valuable feedback. I am likewise indebted to the thorough work of the peer reviewers of an earlier form of the manuscript. I am also grateful to Mary Fata, Editorial Assistant at Palgrave Macmillan for patience and support. This research was made possible by a postdoc grant from Anna Ahlström & Ellen Terserius Foundation.

This book would never have been written without the love and unwavering support of Andrés Rivarola, whom I have been engaged in a continuous dialogue about Latin America, history and development for so many years that I cannot fully distinguish my own ideas from his anymore. I would also like to express my gratitude to my sister Carolina Baraibar, my mother Anna Norberg and my father Julio Baraibar for all their support, love and inspiration. Lastly, thank you Alma and Astor for making everything—including this book—more meaningful.

Matilda Baraibar Norberg

# CONTENTS

1  Introduction to the Political Economy of Agrarian
   Change in Latin America                                    1

2  Changes and Continuities in Agrofood Relations,
   1870–1970s                                                57

3  Agrofood Globalization: The Global Soybean
   and Beef Commodity Chains                                 117

4  Regulative Shifts Paving the Way for Agrarian Change      165

5  Regulative Shifts and Agrarian Change
   of the Twenty-First Century                               209

6  Conclusion: State Autonomy and Capacity
   in a Comparative Light                                    327

Appendix A                                                   375

Appendix B                                                   385

Index                                                        389

# ABBREVIATIONS

| | |
|---|---|
| AAI | American Antitrust Institute (US) |
| AAPRESID | Argentine Association of Direct Seeding Producers |
| ACSOJA | Association of the Soybean Chain (Argentina) |
| ADM | Archer Daniels Midland |
| AoA | Agreement on Agriculture (WTO) |
| APROSEMP | Association of Seed Producers of Paraguay |
| APS | Paraguayan Association of Producers of Soybeans, Cereals, and Oilseeds |
| ARP | Rural Association of Paraguay |
| ARU | Rural Association of Uruguay |
| ASP | Association of Producers of Paraguay |
| BNF | National Development Bank (Paraguay) |
| BSE | Bovine Spongiform Encephalopathy |
| CAH | Agricultural Credit Agency (Paraguay) |
| CAN | National Agrarian Council (Uruguay) |
| CAP | Coordinator of Rural Producers' Organizations of Paraguay |
| CAPECO | Chamber of Grain and Oilseed Exporters and Traders (Paraguay) |
| CARI | *Argentine* Council for International Relations |
| CAS | Southern Agricultural Council (Mercosur) |
| CBoT | Chicago Board of Trade (US) |
| CEO | Chief Executive Officer |
| CEPAL | Economic Commission for Latin America and the Caribbean (UN) |
| CFK | Cristina Fernández de Kirchner (Argentina) |
| CLOC | Latin American Coordinator for Rural Organizations |

| | |
|---|---|
| CNFR | National Commission for Rural Development (Uruguay) |
| CNI | National Inter-sectorial Coordinator (Paraguay) |
| COFCO | China National Cereals, Oils and Foodstuffs Corporation |
| CONINAGRO | Inter-Cooperative Agrarian Federation (Argentina) |
| CRA | Argentine Rural Confederations |
| DCEA | Office of Agrarian Censuses and Statistics of MAG (Paraguay) |
| DGDR | General Office of Rural Development of MGAP (Uruguay) |
| DGSA | Division of Agricultural Services of MGAP (Uruguay) |
| DIEA | Division of Agrarian Statistics of MGAP (Uruguay) |
| DINAGUA | National Directorate of Waters of MVOTMA (Uruguay) |
| DINAMA | National Directorate of the Environment of MVOTMA (Uruguay) |
| DOJ | Department of Justice (U.S.) |
| EFSA | European Food Safety Authority (EU) |
| EPA | Environmental Protection Agency (U.S.) |
| EPP | Paraguayan People's Army |
| EU | European Union |
| FA | Frente Amplio (Uruguay) |
| FAA | Argentine Agrarian Federation |
| FAGRO | Faculty of Agronomy of Udelar (Uruguay) |
| FAO | Food and Agriculture Organization (UN) |
| FAUBA | Faculty of Agronomy of UBA (Argentina) |
| FECOPROD | Federation of Cooperatives of Agrarian Production (Paraguay) |
| FNC | National Peasant Federation (Paraguay and Argentina) |
| FONAF | National Forum for Family Farming (Argentina) |
| FPV | Front of Victory (Justicialist Party, Argentina) |
| FR | Rural Federation (Uruguay) |
| GATT | General Agreement on Tariffs and Trade |
| GCC | Global Commodity Chain |
| GM | Genetically Modified |
| HT | Herbicide Tolerant (GM) |
| IAPI | Argentine Institute for the Promotion of Trade |
| IBR | Rural Welfare Institute (Paraguay) |
| IDB | Inter-American Development Bank |
| IFAD | International Fund of Agricultural Development (UN) |
| IFONA | National Forest Institute (Argentina) |
| IICA | Inter-American Institute of Cooperation on Agriculture (OAS) |
| ILO | International Labour Organization (UN) |
| IMF | International Monetary Fund (UN) |
| INAC | National Institute of Meat (Uruguay) |
| INDERT | Rural Development and Land Institute (Paraguay) |
| INFONA | National Forest Institute (Paraguay) |

| | |
|---|---|
| INIA | National Agrarian Research Institute (Uruguay) |
| IO | International Organization |
| IPC | International Products Corporation |
| IPR | Intellectual Property Rights |
| ISAAA | International Service Acquisition |
| ISGA | International Soybean Growers Alliance |
| ISI | Import Substitution Industrialization |
| LAC | Christian Agrarian Leagues (Paraguay) |
| LATU | The Technological Laboratory of Uruguay |
| LCTA | London Corn Trade Association (UK) |
| LDC | Louise Dreyfus Commodities |
| LULCC | Land Use and Land Cover Change |
| M&A | Mergers and Acquisitions |
| MADES | Ministry of Environment and Sustainable Development (Paraguay) |
| MAG | Ministry of Agriculture and Livestock (Paraguay) |
| MAGyP | Ministry of Livestock, Agriculture and Fishery (Argentina) |
| MAP | Agrarian and Popular Movement (Paraguay) |
| MAyDS | Ministry of Environment and Sustainable Development (Argentina) |
| Mercosur | Common Market of the South |
| MGAP | Ministry of Livestock, Agriculture, Forestry and Fishery (Uruguay) |
| MTO | Technological Oilseeds Table (Uruguay) |
| MVOTMA | Ministry of Housing, Land Management and Environment (Uruguay) |
| NASA | National Aeronautics and Space Administration (US) |
| NGO | Non-Governmental Organization |
| OECD | Organization for Economic Co-operation and Development |
| OLT | Organization for Land Struggle (Paraguay) |
| ONCCA | Office for Control of Agricultural Business (Argentina) |
| PAN | National Autonomist Party (Argentina) |
| PCC | First Command of the Capital (Brazilian mafia in Paraguay) |
| PEA | Strategic Agrofood and Agroindustry Plan (Argentina) |
| REAF | Permanent Meeting of Family Agriculture (Mercosur) |
| REDD | Reducing Emissions from Deforestation and forest Degradation (UN) |
| RR | Roundup Ready (soybeans tolerant to the herbicide glyphosate) |
| SAGPyA | Secretariat for Agriculture, Livestock, Fisheries and Food (Argentina) |
| SDGs | Sustainable Development Goals (UN) |

| SEAM | Secretary of the Environment (Paraguay) |
| SENACSA | National Service for Animal Quality and Health (Paraguay) |
| SENASA | National Service of Sanitation and Food Quality (Argentina) |
| SENAVE | Division of Plant and Seed Health (Paraguay) |
| SINASIP | National System of Protected Areas (Paraguay) |
| SNA | National Society of Agriculture (Paraguay) |
| SNAP | National System of Protected Areas (Uruguay) |
| SRA | Argentine Rural Association |
| SSDRAF | Sub-Secretariat of Rural Development and Family Farming of SAGPyA (later MAGyP) |
| TNC | Transnational Corporations |
| TRIPS | Trade-Related Aspects of Intellectual Property Rights (WTO) |
| UBA | University of Buenos Aires (Argentina) |
| UCESI | Unit of Coordination and Assessment of Domestic Consumption Subsidies (Argentina) |
| UDELAR | University of the Republic (Uruguay) |
| UGP | Union of Producer Associations (Paraguay) |
| U.K. | United Kingdom |
| UNDP | United Nations Development Programme |
| UNFCCC | United Nations Framework Convention on Climate Change |
| UPOV | Union for the Protection of New Varieties of Plants |
| U.S. | United States |
| USDA | United States Department of Agriculture |
| VAT | Value-Added Tax |
| WB | World Bank |
| WHO | World Health Organization |
| WIPO | World Intellectual Property Organization |
| WTO | World Trade Organization |
| WWF | World Wildlife Foundation |

# METRICS

| | |
|---|---|
| B | Billion |
| Bha | Billion hectares |
| C | Capita |
| GDP | Gross Domestic Products |
| Ha | Hectare |
| Kg | Kilo |
| M | Million |
| Mha | Million hectares |
| MMT | Million Metric Tons |
| PPP | Purchasing Power Parity |
| T | Ton |
| USD | United States Dollar |
| Yr | Year |

# LIST OF FIGURES

Map 1.1  Map over the Southern Cone of South America, including
the three distinct regions; *Pampas, Gran Chaco* and the
*Atlantic forest* (*Source* Author's own illustration on map
from © OpenStreetMap contributors)                            12

Chart 3.1  Soybean exports per country, 2017                          127
Chart 3.2  Frozen bovine meat exports (including beef, steak,
red meat) per country, 2017                              128
Chart 3.3  Soybean imports per country                                130
Chart 3.4  World bovine meat imports, 2017                            133
Chart 6.1  GDP/c Argentina, Paraguay, Uruguay, and World              344
Chart B.1  Argentina 2002—total export value USD 25.3 billion         386
Chart B.2  Argentina 2017—total export value USD 53 billion           386
Chart B.3  Paraguay 2002—total export value USD 1.13 billion          387
Chart B.4  Paraguay 2017—total export value USD 6.45 billion          387
Chart B.5  Uruguay 2002—total export value USD 1.97 billion           388
Chart B.6  Uruguay 2017—total export value USD 7.32 billion           388

# CHAPTER 1

# Introduction to the Political Economy of Agrarian Change in Latin America

This is like a bulldozer, or perhaps a big combiner, you see? The change has been impressive from all points of views; technological, social, economic - everything! I always say that we will never be able to halt this machine. What we can do is to find a way to situate ourselves so that it does not run over us. We should let the machine pass, and try to make it harvest in such a way that we manage to use some of what it produces. That is basically my theory. (Agronomist and Family Farmer, Dolores, 5 March 2010)

A bulldozer is haunting Latin America—the bulldozer of distantly driven agrarian change. The most dramatic change is in the landscape. Agropastoral area has increased faster in this region than anywhere else in the world during the twenty-first century. Between 2001 and 2013, cropland expansion in Latin America was 44 million ha (Mha) and pastureland increased almost 100 Mha (Graesser et al. 2015). The main driver behind recent *Land Use and Land Cover Change* (LULCC) is the rising global demand for animal products, which requires both more pasturelands for livestock grazing and more croplands for high-protein vegetable feed grains, particularly soybeans.[1] Exports of soybeans from Latin America more than tripled between 2005 and 2015, and exports of beef more than doubled under the same period (Cepal, FAO, and IICA 2015, 14; ECLAC 2017; OECD/FAO 2016a, b; Ray and Gallagher 2016). Soybeans and beef dominate the drivers of agrarian transformations in

© The Author(s) 2020
M. Baraibar Norberg, *The Political Economy of Agrarian Change
in Latin America*, Governance, Development, and Social Inclusion in
Latin America, https://doi.org/10.1007/978-3-030-24586-3_1

1

the region during the past decades, which is why this book focuses on these two different, but highly interconnected, complexes.

The soybean areas in cultivation in Brazil, Argentina, Paraguay, Bolivia, and Uruguay have increased from 30 Mha in 2000 to more than 60 Mha in 2016 (USDA 2018). The soybean chain in the region is the most integrated to world trade of all agricultural commodities, with the lion's share of harvest destined to international markets. Soybeans have expanded over other land uses and land covers, such as agricultural crops and extensive pastures, as well as natural grasslands, savannas, and native forests (Aide et al. 2013; Caldas et al. 2015; Graesser et al. 2015, 5).[2] The soybean expansion has resulted in increased competition for land in fertile areas raising incentives for land-use intensification, which has significantly augmented the pressures on natural resources. In a parallel way, pasture areas to graze cattle have expanded over forests and other natural and seminatural areas, such as grassland/savannah ecosystems (ECLAC 2017; Graesser et al. 2015; Lambin and Meyfroidt 2011; de Waroux et al. 2017; Lipton 2009). For example, the dry tropical forests of the *Gran Chaco region* are currently disappearing faster than any other forests in the world and increased ranching is the main cause (Caldas et al. 2015; Graesser et al. 2015, 5). Considering that unexploited agricultural land areas are getting scarcer throughout the world, while there are still significant reserves in Latin America, it is probable that this region will continue to be of crucial importance as a global food provider.[3]

The dramatic LULCC in the wake of soybean-beef expansion/dislocation has come hand in hand with changes in *forms of production* and *social relations*. While there is a significant amount of variance, increased competition for land have pushed farming systems toward increased use of productivity-boosting and labor-sparing technologies, product specialization, and economies of scale. These shifts have occurred, to varying degrees, in both crop and livestock systems, but the soybean sector has been a forerunner in bringing in new game-changing technologies and organizational models into agriculture. This has in turn caused multiple social-ecological effects, ranging from loss of native forests and savannas, grassland and cropland farming intensification (causing concerns over biodiversity loss, erosion and pollution of waterways), to agribusiness advancement and displacement of family farmers (causing concerns over land concentration, urbanization and loss of food sovereignty). While the region has a long history of export-driven agropastoral transformations, in addition to the dramatic changes in land areas, the current scope and pace of change are historically unprecedented.

However, is it simply increased foreign demand for meat/animal products driving these changes? Due to the export-driven character of recent shifts in agricultural production, LULCC in Latin America is often analyzed in a context of contemporary globalization. In this literature, the Latin American state is often either entirely neglected or assumed to simply comply (by debt-driven necessity or consent) with the rules set by transnational agribusiness firms, international organizations, and/or the advanced economies of the world system. At the other end of the spectrum, innumerable studies and reports address different national regulatory frameworks and policies, without any consideration of exogenous drivers, as if the states were in complete and exclusive control over the changes in their territory. There is a need to move beyond overly simplistic assumptions of Latin American states either as powerless victims, completely failed states or in perfect control.

The point of departure of this book is that the state remains an important actor co-shaping agrarian change. Although the power of transnational agribusiness has increased, the state still has authority over legal regulatory framework in its territory, for example the design of environmental laws, the fiscal system, public investments in roads and port infrastructure, farming support systems and public spending on agricultural research, development, and education. These public regulations can facilitate, accept, renegotiate, ignore, or resist the changes in social relations and production models brought by the distantly driven agropastoral expansion, intensification, and concentration processes. The state is, nevertheless, not the only game in town. It is only one of three interdependent pillars that all need to be addressed simultaneously in order to understand recent agrarian change in Latin America: (1) Legacies of the past: the historical formation of institutions in Latin America; and (2) Contemporary agrofood globalization and in particular the global commodity chains (GCC) of soybeans and beef; as well as (3) National regulatory systems of the state, including the negotiations and power relations they express.

This book takes on the important task of a deep, historicized, contextualized, and comparative exploration of agrarian change. It is grounded in the Southern Cone countries; Argentina, Paraguay, and Uruguay, with specific emphasis on the interface of state action and mediation between different interest groups and the transformative pressures brought by agrofood globalization. These countries are important cases to explore and compare, since they are at the very heart of the present expression of distantly driven agrarian change. While giant Brazil is the largest soybean producer in the region and in the world in absolute terms,[4] the relative increases of soybean cultivation areas between 2000 and 2015 were actually the greatest in Uruguay (over ten times), followed by Argentina (almost three times) and Paraguay (two times) (Graesser et al. 2015, 4–5).

In addition, these are the countries with the highest rate of export specialization in agrofood products—particularly in soybeans and beef—which makes their economies highly dependent on the development of the international agrofood market (CAS 2018, 7). Soy-based products (including whole soybeans, soybean cakes, meal, and oil) have become the by far the most important export items for Argentina and Paraguay, and rank third, after beef (including frozen, fresh, or chilled beef and livestock cattle) and forest products (mainly cellulose from eucalyptus and pine plantations), in the Uruguayan export basket.[5] Beef exports have risen rapidly in Paraguay and now rank second, after soy-based products (CAS 2018, 48–51). At the same time, beef exports from the traditional livestock country Argentine declined between 2008 and 2016 because of policies discouraging exports of beef—indicating the importance of national policies to understand how exogenous pressures for agrarian change play out in each country. See a visualization of the export profiles of Argentina, Paraguay, and Uruguay, 2002 and 2017, including shifts in value and composition, in Appendix B, of this book. China plays a unique and extraordinary role as the main buyer of Uruguayan and Argentine exports. While an important part of the agrarian-based exports from landlocked Paraguay also ends up in China, it is typically reexported from Argentina, Brazil, and Uruguay, and is therefore not visible in the trade statistics (CAS 2018, 51–54).

While export-driven agrarian change has been dramatic in all three countries, the articulation of change involves both similarities and differences across these cases, and they have diverging capacities to cope with the pressures from contemporary agrofood globalization. It is, for example, important to bear in mind that Argentina—more than 6 times bigger than Paraguay and more than 15 times bigger than that of Uruguay—is a much more important global agrofood player, with higher bargain capacity than its neighbors. Another important difference is that Paraguay has some of the highest poverty and income inequality rates in Latin America, while Argentina and Uruguay have the lowest.[6] All three countries have a staggering inequality of access to land. They also share some similar experiences of distantly driven land-use change in the past—although the course of history took a rather different form in Paraguay from the *Rio de la Plata*.[7] By limiting the comparison to these three countries, a deep and historically informed penetration of each country is made possible, as well of the identification of region-specific patterns. This is deemed necessary to gain a fuller understanding of the contemporary processes of agrarian change. Thus, while the latest wave of agropastoral expansion and land-use intensification took off around the Southern Hemisphere's agricultural year of 2001/2002, it is analyzed in the light of a much longer history of distant drivers, policy regulations, and agrarian transformations.

A thorough understanding of the political economy of agrarian change is enabled through a *historical approach*, in which recent agrarian change is situated in a longer time frame; a *world systemic approach*, in which soybeans and beef production in Argentina, Paraguay, and Uruguay are situated in the wider global soybean and beef chains; and a *regulatory approach*, in which national regulatory frameworks are scrutinized in a comparative light. In this way, this book helps clarify what states can and cannot do in relation to exogenously driven agrarian change, engaging in the discussion about differentiated state capacity and autonomy in the present era of agrofood globalization.

*Outline of the Chapter*

This chapter serves as a short presentation to the dramatic agrarian change that has occurred in the region over the past two decades, and an introduction to the theories and methodological approaches that underpin the rest of the book. The remaining part of this chapter is organized as follows: Sect. 1.1 presents in more detail the soybean and beef driven agrarian change in Argentina, Paraguay, and Uruguay, 2002–2018. This section underscores the dramatic changes in forms of production and social relations that have emerged in the wake of agrarian change in all three countries, albeit to varying degrees. It presents the three main pathways of agrarian change in the region, associated with three distinct biomes: the *Pampas*, the *Gran Chaco*, and the *Atlantic forest*, coupled with the main emerging social-ecological concerns. This is followed by an exploration of analytical and empirical gaps in existing literature on the political economy of agrarian change (Sect. 1.2). This section further explains how this book will fill these gaps through the above-mentioned tripod of perspectives on recent agrarian change in Latin

America, namely: *a historical approach*, in which *food regimes* functions as a chronological periodization (Sect. 1.2.1); *a world systemic approach*, in which *GCC* analysis of soybeans and beef functions as a guide into the main tenants of agrofood globalization (Sect. 1.2.2); and *a regulatory approach*, in which national regulatory frameworks are examined to address how Argentina, Paraguay, and Uruguay cope with the transformative pressures brought by agrofood globalization (Sect. 1.2.3). The following section discusses how the examination of regulations cross-nationally and over time can shed new light on the discussion about constrained state autonomy and capacity (1.3). This chapter ends with an outline of the remaining chapters of the book.

## 1.1    Soybeans and Beef in Argentina, Paraguay, and Uruguay

Expansion and intensification of soybean and beef productions is at the very heart of recent dramatic agrarian change. Both soybean and beef production, and accompanying processing and commercialization activities, have brought significant changes in land-use, in forms of production and in social relations. These changes have in turn brought dramatic social-ecological consequences.

### 1.1.1    Shifts in Land-Use, Forms of Production, and Social Relations

While there is a significant amount of variance, there has been a general move toward increased use of productivity-boosting and labor-sparing technologies, product specialization and economies of scale and labor specialization. These shifts have occurred, to varying degrees, in both crop and livestock farming systems, but the soybean sector has been a forerunner in bringing in new game-changing technologies and organizational models into agriculture. One major role is played by the new technological package for soybeans centered in Monsanto's genetically modified soybeans *Roundup Ready* (RR). The genetically modified soybean is here designed to be combined with a specific weed killer—glyphosate (*Roundup*)—and no-tillage farming. The cheap glyphosate—used instead of tilling to "clear the land" of weeds—reduces the costs of labor and fuel; no-tillage farming allows for increased yields in less fertile soils.[8] This allows expansion of cultivation into areas previously considered unsuitable due to heavy weed infestations, erosion, or high risk of water deficits (Acosta Reveles 2008; Bisang et al. 2009; Garcia 2015; Prieto and Ernst 2010). Soybeans RR were authorized first in Argentina in 1996, and then rapidly spread the region (in some cases legally, illegally in others). The adoption rate was very fast and today almost all soybean production in the region is genetically modified. With this technological package and rising soybean prices, soybean cultivations have expanded over natural pastures, savanna, native forest, mixed systems of crops in rotation with pastures and cropland—driving ecological simplification—"*monoculturalization*" (Bisang et al. 2009; Garcia 2015; Prieto and Ernst 2010). In absolute terms, the soybean area in Argentina jumped from 6.6 Mha in 2000 to 29 Mha in 2017 (CAS 2018; USDA 2018).[9] In the same years, soybean areas grew from 0.1 Mha to more than 1.3 Mha in Uruguay and from 1.2 to 3.5 Mha in Paraguay (FAO, 13 May 2018, Comtrade Statistics). Besides enabling expansion, the technological package enabled production increases due to increased average yields throughout the region (MAGP and IICA 2012, 6). It also involves an increased use of fertilizers, herbicides, and pesticides (Aparicio et al. 2017; Lepori et al. 2013; Ramírez 2010), and specialized, sophisticated large-scale and more efficient machinery (seeding drills, harvesters, airplanes), irrigation and intensive use of information technology (Lapegna 2016a; Teubal 2008).

The soybean expansion brought increased competition for land and out-competed pastures for beef and dairy production in fertile areas close to ports. While pastureland retracted in these areas, it expanded into nonagricultural land, particularly forest land (ECLAC 2017; Graesser et al. 2015; Lambin and Meyfroidt 2011; de Waroux et al. 2017; Lipton 2009). Ranching and soy are the main causes to the current dramatic deforestation of the native forests throughout the region, not least the hotspot of the *Gran Chaco*, as presented in further detail in the section of different patterns of agrarian change.

Increased competition for land drives up land prices. When land is expensive, the economic incentives rise to invest more capital and incorporate more technology into the land in order to make it produce more. As soybeans expanded over the most fertile pastureland, the traditional extensive ranching systems intensified. The productivity per land area was incremented by greater incorporation of improved pastures (using more fertilizers), use of supplement feed; ranging from finishing to supplement grazing with feed formulated of vegetable protein meals to speed up their growth before slaughter, to continuous concentrated animal feed in enclosure such as feedlots (Modernel et al. 2018; Richardson 2009). Some big meat companies made strategic alliances with the "network" crop firms, in which contract farmers linked to the meat companies get access to feed below market price (CEO of El Tejar; the chief of sales of ADP; the CEO of Marfrig Beef of the Southern Cone, in Young, Dolores and Montevideo, 2008 and 2010). Feedlot farming was almost nonexistent in the region a couple of decades ago, but now expanded rapidly, freeing land for crops (Leguizamon 2014, 63; Paolino et al. 2014, 16). In this way, average yields have risen and beef production has moved toward higher input, higher output, production (García Préchac et al. 2010). Fertilizer use in the region (traditionally low) caught up with the rest of the world (Richardson 2009). In this way, the commodity boom represented some kind of second wave "green revolution" in a region where the first wave was rather limited. The use of modern breeding and disease prevention also increased.

The increased use of capital-intensive and labor-sparing technologies have brought upon huge shifts in social relations. Big capital intensive farming units, managing great amounts of land under short-term leasing contracts, have advanced at an unprecedented pace (Bisang et al. 2009; Wesz Jr. 2016). Latin America already had a concentrated land structure, inherited since colonial times and reinforced under the first globalization wave (Chapter 2 in this book), but concentration had become dramatically accentuated. In response to increasing economies of scale in agriculture, a new type of farming unit emerged and thrived throughout the region alongside the expansion of soybeans over millions and millions of hectares in all three countries, called *pools de siembra* or network firms. These firms "pool" resources from a wide range of sources, particularly financial capital through trust funds, to lease large amounts of land to plant mainly soybeans on, and split the returns among investors. They manage vast amounts of cropland through a handful of agronomists, while the actual agricultural work—to plant, fertilize, spray, harvest, and transport the crop—is subcontracted to specialized service providers, *contratistas*. Accordingly, the rural wageworker became increasingly superfluous in this new production model—intensive in technology, but labor sparing—and where the little required on-farm work is performed by the *contratistas*.

This new model of doing agriculture further spread to other farming units and other agrarian sectors. Contractual relations, specialization, professionalization (using "expert" knowledge, rather than experience), intensified use of on-time market information and information technology increased in the livestock sector (Baraibar 2014; Gras and Hernández 2013, 92; Oddone 2015). One could say that the soybean expansion pushed farming systems away from family labor and experience-based management to more technology-driven, specialized, and business-oriented units and increased concentration (Wesz Jr. 2016).

At the same time, family farmers have increasingly left agriculture. Many property-holders of small parcels of land cannot afford to invest in the inputs and services required to be able to compete successfully with the big firms, and thus often prefer to lease out their land to the soybean firms instead of working it themselves. Moreover, sharecroppers have generally lost access to land as purchase and leasing prices have skyrocketed. The social structure of agriculture in the region has thus become more polarized, with expanding, competitive, "modern" capitalized agribusiness firms at one end, and retracting family producers and peasants producing at the other (Acosta Reveles 2008). In short, the highly cost-effective agribusiness actors out-compete the family farmers and smallholders, accentuating the tendency toward land concentration (Bisang et al. 2008). Besides displacement of producers, the new forms of production displace wage laborers. Critical scholars argue that the soybean model has resulted in a transformation of the sector into "agriculture without farmers" (Teubal 2008, 6).

Ranchers still represent a much more diverse and numerous group of farmers, ranging from the poorest family farmers to the biggest transnational meat companies, and they apply more diverse production and commercialization practices. However, as mentioned, the whole agrarian sector has moved toward more industrial forms of production, which in turn results in greater specialization, and capital intensive technologies—reducing demand for labor—raising barriers to entry, excluding smallholders and pastoralists from participating in the sector (CEO of Marfrig Southern Cone Montevideo, 7 March 2010). Poor livestock farmers typically cannot afford to buy more protein-rich feed. Instead, they increasingly specialize in the less profitable early phases of cattle rearing and sell the cattle to be fattened before slaughtering. More and more cattle ranchers have in addition become contract farmers, receiving inputs from the large companies and selling the fattened animals for processing and distribution to the same slaughterhouses. Moreover, the pastureland expansion into forest area has brought tensions and conflicts with native communities that use the forests for food, shelter, and medicine.

While farming is becoming increasingly concentrated, where the biggest crop firms transcend nation-state boundaries and cultivate in various countries in the region, the rates of concentration and domination of transnational agribusiness firms in inputs (seeds, agrochemicals), processing (crushing plants, slaughterhouses), and trading (exporters) are higher still. A handful of multinational traders dominate the soybean trade, as well as the crushing industries in Argentina and Paraguay (Baraibar 2014; Pedretti 2004; Teubal 2008). The biotechnological event is patented by another handful of transnational companies who also increasingly dominate the seed and agrochemical market (Chapter 3 in this book). Moreover, these companies have increasingly adopted strategies of vertical integration. In this way, big transnational traders have gradually expanded into processing, infrastructure, and transport, as well as invested in the farming stage. At the same time, biotech, seeds, agrochemicals, and inoculant markets are progressively concentrated and integrated. The slaughterhouses are also powerful agents in the meat chains since the late nineteenth century (Chapter 2 in this book), but size and market concentration have increased significantly. The biggest beef companies in the world—JBS and Marfrig—of Brazilian origin, have expanded rapidly throughout the region.

The abovementioned general trends can be observed throughout the region, but they are articulated slightly divergently in different areas. While new technology such as seed genetic traits, no-tillage techniques, and irrigation make commercial agriculture more and more independent from specific place-based biophysical conditions, *place* still matters for agriculture. Not everything can be produced everywhere, and the consequences of the same land-use practices can vary in different places. Broadly, three pathways of agrarian change can be associated with three distinct biomes: the *Pampas*, the *Gran Chaco*, and the *Atlantic forest*.

### 1.1.2    Three Different Pathways of Agrarian Change: Pampas, Gran Chaco, and Atlantic Forest

The solid lines in Map 1.1 indicate the area of the fertile South American lowland called the *Pampas* region. This biome includes the eastern parts of Argentina, the whole of Uruguayan territory and the southern part of Brazil.[10] The dashed lines is the *Gran Chaco* region in northern Argentina, western Bolivia, southwestern Brazil, and the western region of Paraguay. It contains South America's second-largest forest, behind only the *Amazon*, and it is one of the continent's last agricultural frontiers; land with potential agricultural use, but not (yet) under agriculture. The crossed pattern area is the *Atlantic forest* region. This forest extends along the Atlantic coast of Brazil from Rio Grande do Norte in the north to Rio Grande do Sul in the south, and the Misiones province of Argentina, and inland into the smaller, but densely populated Eastern region of Paraguay. This book focuses exclusively on change in Argentina, Paraguay, and Uruguay, not in Brazil and Bolivia. Each region obviously involves a significant amount of variance in terms of soil, weather, land-use patterns and farmers' communities, but broadly speaking, they represent three distinct patterns of agrarian change, causing distinct social-ecological concerns. While these three main regions represent three main pathways of agrarian change, with important differences between them, focus here in the introductory chapter is on the broad trends of recent agrarian change across the region, and the specificities involved in the articulation of agrarian change in each biome is here in only mentioned briefly.

*Pampas*
The fertile soils of the Argentine Pampas is the country's traditional crop area, where soybeans have been cultivated since the 1970s. Cultivations have typically been integrating mixed systems with pastures. With the adoption of soybeans RR and the soybean "boom", however, there was a break with the previous model. Rotation schemes were simplified into continuous cropping with very little productive diversification; soybeans were most commonly either rotated with wheat (as a winter crop), or without any rotation crop at all—soybeans over soybeans (Garcia 2015; García Préchac et al. 2010; Pérez Bidegain et al. 2010; MAGP and IICA 2012; Modernel et al. 2016; Teubal 2008, 7). While soybeans in Argentina started to expand outside of the Pampas region during the 1990s, still today 85% of cultivated soybean area in Argentina is within the Argentine Pampas region (Craviotti 2016). After 2002, Argentine crop firms and farmers expanded into Uruguay, attracted by lower land values and absence of export taxes at the other side of the river (Arbeletche and Gutiérrez 2010; Guibert et al. 2011; Gutiérrez 2009). Soybean production in Uruguay increased from almost nothing to the number one crop in less than a decade, particularly on the fertile soils of the Litoral (Baraibar 2014).

While no-tillage farming reduces the risks of erosion and soil degradation, continuous cropping, still eventually degrades soils. Moreover, several ecosystem services are lost in this shift, as illustratively explained by a researcher at the Department of Ecology, Faculty of Agronomy, University of Buenos Aires (FAUBA):

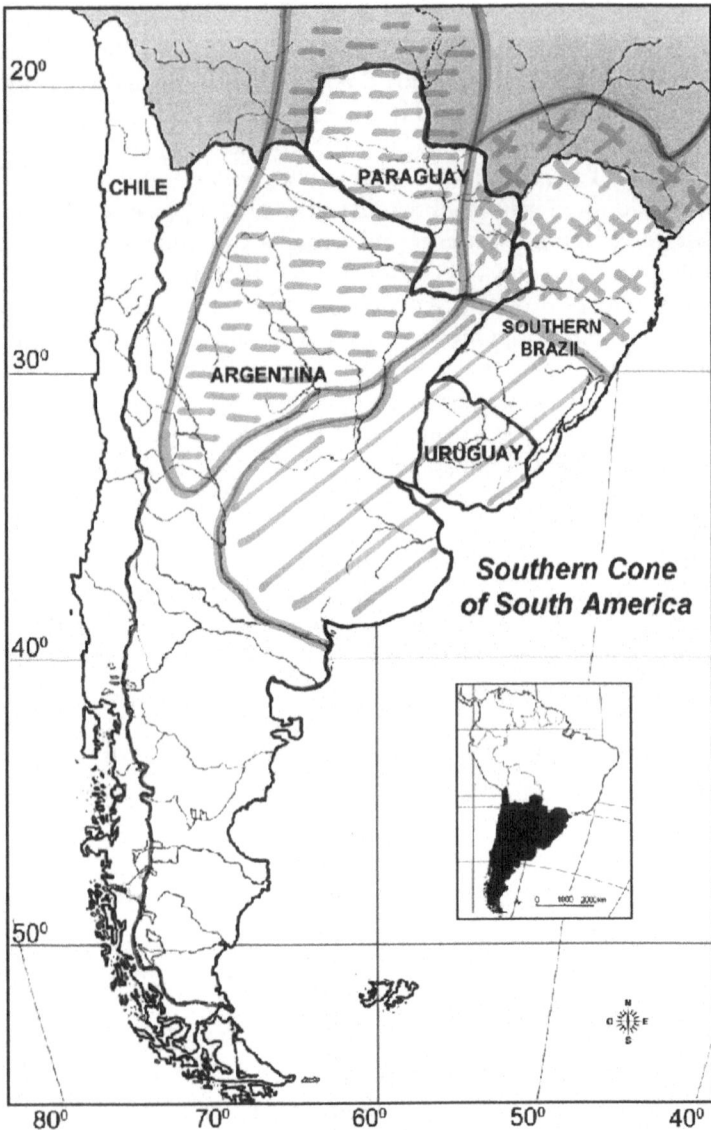

**Map 1.1**    Map over the Southern Cone of South America, including the three distinct regions; *Pampas, Gran Chaco* and the *Atlantic forest* (*Source* Author's own illustration on map from © OpenStreetMap contributors)

In the traditional crop-grazing system of the Pampas, erosion was under control because of rotation with pastures. That technology [rotations] also fulfilled the purpose of weed control. In terms of ecosystem services, the four years of pastures followed by perennial crops provided a lot of different ecosystem services. [...] What happened was a change in the price ratio and suddenly you were losing a lot of money if you put pastures instead of soybeans. The whole model switched into continuous cropping. The first years endured, but after a few years, problems emerged. Ecosystems are generally bundled, but specialized conventional agriculture unbundle and simplify the system to produce more grains. So all the other services are subordinated to maximize the soybean, ok? But, when you simplify you lose nitrogen and phosphorus, and contaminate the layers. Then the lands yield less, which makes farmers apply more and more fertilizers, herbicides and pesticides. There we are now. Either you rotate with pastures, or you have to plant a so-called service crop[11] in between the harvests. (Researcher FAUBA and Agricultural Extensionist, Buenos Aires, 24 February 2017)

As described in above quote from the FAUBA researcher, who has also been working for many years with research and extension services on both sides of the *Rio de la Plata*, the *sojization* of the Pampas brings important collateral effects for the ecosystems. The soybean "boom" also created increased competition for land, driving up land purchase and leasing prices at a snowballing rate. The cattle ranchers responded to the new scenario in different ways. Some moved their cattle rearing to more marginal lands with lower land values, such as the Argentine Chaco and into Paraguay (see the next section), others intensified production by increasing the use of external inputs and decreasing the size of land, still others—particularly small livestock farmers—could not manage to remain competitive and left the activity. Of all the new cropland that emerged between 2000 and 2015, previous pastureland accounted for 79% in Uruguay, and 40% in Argentina (Graesser et al. 2015, 4–6). Soybeans particularly displaced pastures from the most fertile land. The beef production remaining on the Pampas became increasingly intensified; producing more beef on less land, incorporating more technology and external inputs (DIEA-MGAP 2015, 8; ECLAC 2017, 108; Graesser et al. 2015, 4–6; Gras and Hernández 2013, 61). In this way, cattle production managed to compensate for the lost hectares to crops without losing production volume (Graesser et al. 2015, 4–6; Gras and Hernández 2013, 61).[12] In Uruguay, beef production actually increased significantly between 2000 and 2017, at the same time as cattle lost more than one Mha of its most fertile pastureland (DIEA-MGAP 2015, 8; ECLAC 2017, 108).[13] Livestock intensification also increased pressures on natural resources; soil and water.

The land values in the Pampas were already before the commodity "boom" among the highest in Latin America, particularly on the fertile lands of the humid Pampas in Argentina. Moreover, most Pampas-farmers were already mechanized, market-integrated, and managing relatively large amounts of land per productive unit. The soybean expansion, however, provoked a further increase in land values, at the same time as the form of production involves important amounts of economies of scale. This increased concentration rates further (Cepal, CIRAD, and IICA 2014; Manzanal and Schneider 2011; Sabourin et al. 2014; Alianza Internacional de las organizaciones de productores familiares de soya 2008). Small farmers have by large lost access to land. Between 1988 and 2000, the Argentine Pampas region lost 60,000 farmers, meanwhile the average size of the productive units increased from 391 to 531 ha (Acosta Reveles 2008, 9). Meanwhile, the *pools de siembra* expanded fast, transcending nation-state boundaries, and achieved high economic returns. The traditional landowners of the Pampas—often ranchers—were also able to make money on the soybean "boom", by leasing out land for increasingly high rents to the new firms without having to take any risk or do any work for it (Baraibar 2014; Caligaris 2017; Teubal 2008; Varrotti and Frederico 2018). As land rents rose rapidly and as landowners often preferred to lease the land to the new agribusiness firms, however, many traditional sharecroppers could not afford to pay the rents and lost access to land (Sandoval 2016). Around half of all the sharecroppers in the Uruguayan Litoral abandoned the activity between 2000 and 2009 (Arbeletche Favat and Cividanes Hernández 2012; Arbeletche and Gutiérrez 2010). Many of these; former crop producers, instead became *contratistas*—using their machinery to offer drilling, spreading, and harvesting services to the new firms (Baraibar 2014, 181; Manzanal and Schneider 2011, 11; Pedretti 2004; Rossi 2010, Teubal 2008).

## Gran Chaco

The *Gran Chaco* eco-region has been classified as of global and regional importance due to the existing biological diversity and endemism (Imbach 2016, 6; Walcott et al. 2014, 15). It includes forests forming part of the world's largest contiguous tract of dry Neotropical forest, as well as wetlands and savannas. This region hosts one the last large global reserves of uncultivated fertile soils (Jobbágy et al. 2015). However, agropastoral expansion, mainly soy and beef, has turned the region into a global deforestation hotspot (ECLAC 2017, 95; Goldfarb and van der Haar 2016; Kuemmerle et al. 2017; Vallejos et al. 2015). While this area still represents one of the world's largest forest reserves and "unexploited" fertile land, deforestation is leading to loss of natural habitats and fragmentation of the landscape (Vallejos et al. 2015, 7–8, Walcott et al. 2014; Imbach 2016; Hansen et al. 2013).

In the Argentine side of the Gran Chaco, agricultural expansion has resulted in the loss of almost 6 Mha of forest between 1977 and 2010 (Torrella et al. 2018). The pace of loss accelerated after the turn of the millennium, when both RR soybeans and "displaced" ranching activities (from the Pampas region) started to enter massively in the poor northern provinces of Santiago del Estero, Salta, Chaco, and Formosa. With the increased valorization of the land, many previously absentee landowners started to claim the land with the goal of clearing the forest and renting or selling it for agricultural production, in spite of the fact that there were people living on it (Leguizamon 2014, 73). The rapid deforestation and the role of soybeans and beef as the central drivers behind it, led to a strong societal debate, which eventually resulted in 2007 in a native Forest Protection Law (26.331), thoroughly explored in Chapter 5 of this book.

The Paraguayan side of the Gran Chaco, often referred to as the Western region, includes the vast and sparsely populated departments of Boquerón, Alto Paraguay, and of Presidente Hayes (Caldas et al. 2015; Graesser et al. 2015, 5). The forest area had remained rather intact until the turn of the millennium, but is currently exhibiting one of the most rapid deforestation rates in the world (Graesser et al. 2015; Hansen et al. 2013; Vallejos et al. 2015, 4).[14] Between 2000 and 2015, around 5.5 Mha of this low-lying, thorn-forested dry tropical forest was lost—almost as much as the total loss on the Argentine side during the past 40 years (Py/SEAM/INFONA/FAPI 2016, 9).[15] The motor behind deforestation forest is big and extensive systems of farming and fattening of livestock (Py/SEAM/INFONA/FAPI 2016).[16] Forest to pastureland conversion accounted for 62% of the 0.82 Mha of new pastureland in Paraguay from 2001 to 2013 (Baumann et al. 2016).[17] Besides forests, the *Gran Chaco* also includes non-forest habitats such as grasslands, scrublands, and wetlands that favor the persistence of a wider range of species in higher densities, that are also becoming increasingly lost (Grau et al. 2015).[18]

The low land prices of the Paraguayan Chaco, in combination with relative high beef prices, made livestock production increasingly profitable, in spite of the poorly developed infrastructure in the region and its remoteness from any international harbor. While soybean cultivation has not been a direct driver of LULCC, it is still indirectly linked to the deforestation. Many Uruguayan and Argentine ranchers sold their land on the Pampas to soybean producers and for the same price bought five to nine times more land in the Paraguayan Chaco (Paraguay's Deputy Minister of Agriculture; Innovation Specialist IICA Paraguay; Director of Social Studies INTA Argentina; Uruguay's former Minister of MGAP; Asunción, Buenos Aires and Montevideo, February–March, 2017). Over 2 Mha of land in the Paraguayan Chaco is estimated to be owned by Uruguayan investors alone (USDA 2017). There are also an important number of Brazilian ranchers expanding in the region. Moreover, the pressures on deforestation of the Paraguayan Chaco forest increased after Paraguay passed a still running moratorium of deforestation in the Eastern region of the country in 2004, and Argentina passed a new forest law in 2007, suggesting a leakage effect. In addition, rapid soybean expansion in the eastern region of Paraguay (the Atlantic forest area) dislocated ranching to the western region (the Chaco):

> In the 1980s, Alto Paraná and Itapúa were mainly livestock land. Now a lot of livestock activity has left the area and moved to the Western region, as the Eastern region is increasingly dominated by immense soybean fields. (Director of DCEA-MAG, Asunción, 21 February 2017)

The Paraguayan Arid Chaco forests represents a third of national territory, but less than 3% of the country's population lives there (Walcott et al. 2014, 31).[19] The Argentine side of the Gran Chaco is more densely populated than the Paraguayan side, and has many non-capitalized small farmers, peasants, *campesinos*, and indigenous communities living in the area. Many of these lost access to land as soybeans and large-scale beef farming expanded into the area (Colla 2017; Lapegna 2016b; Leguizamon 2014, 129–135; Muñoz 2016; Teubal 2008). However, in both Paraguay and Argentina, several indigenous groups live in the Gran Chaco forests.[20] The native communities use the forest for food, medicine, and spiritual motives, but often lack formal property rights to the lands they occupy (Human Rights Council 2015; Walcott et al. 2014, 2; Volante and Seghezzo 2018). Given their reliance on their lands, these peoples are extremely vulnerable to LULCC (Specialists UNDP Paraguay and Public Policy Officer WWF Paraguay, Asunción, 20 and 23 February 2017).

*Atlantic Forest*
The other pattern of agrarian change in Paraguay is the soybean expansion and general land-use intensification in the densely populated (former) Atlantic forest area—the eastern region of Paraguay.[21] Once this area was dominated by intact forest cover, but after decades of massive deforestation and fragmentation, today only an estimated 11.7% of forest's original area remains (Aide et al. 2013; Caldas et al. 2015; Graesser et al. 2015, 5). Besides agropastoral activities—mainly soybeans, cotton, and beef—degradation of the forest can be attributed to demand for fuelwood and land squatting. More than 97% of Paraguay's over six million inhabitants live in this area, while it makes up just 39% of the total land area of the country (Mansourian et al. 2014).

Dramatic agrarian change in this area started in the 1960s and 1970s, driven both by Stroessner's agrarian programs for frontier expansion and "modernization", and by Brazilian soybean and beef farmers crossing the border and rapidly expanding in the region (Chapter 2 in this book). The pace of soybean expansion accelerated with the (illegal) adoption of RR soybeans and rising international prices since the late 1990s. The great majority of the agrochemicals used in the Paraguay are utilized for RR soybean production (Ezquerro-Cañete 2016). The public institutions and international organizations write manuals for "good agricultural practices" and work with extension services to farmers try to get out the message that in the long-run it is cheaper to do proper rotations than do plant only soybeans:

> We try to show the producers how the soil loses productivity if they don't do good rotations. They already know that many times, but they still find it hard to plant anything else than the crop with the highest margins and many have debts to pay. They pray so that things will work out anyway and they chose to plant whatever provides the highest returns in the short-term. (Senior technical advisor UNDP Paraguay, Asunción, 20 February 2017)

The soybean "boom" has not only created specialized, monocultural, farming systems with high reliance on agrochemicals, fertilizers, and continuous cropping, but also generated higher land values, pushing toward land-use intensification in all farming systems. In a similar way as in the

Pampas region, land prices have been skyrocketing in agricultural areas due to the soybean "boom"; from USD 200/ha in 2000 to USD 1400/ha in 2014.[22] Accordingly, the use of artificial pastures, fertilizers and grains for feed is incrementing in livestock systems (USDA 2017).[23]

Soybeans have not only expanded over extensive pastures and forestland, but also over cotton production and peasant food farming (Aide et al. 2013; Caldas et al. 2015; Graesser et al. 2015, 5). The social consequences of this agrarian change are dramatic due to the high population density and many peasant farmers. The land structure, already very unequal, has become increasingly concentrated during the past decades. Paraguay's Deputy Minister of Agriculture vividly illustrated this process, in the following way:

> The substitution process was fast and easy. You imagine, I, as a big soybean farmer offer you money in the hand in quantities you have never seen before. I buy your hectares and you leave agriculture for the city or to land that is more marginal. So, I buy five ha here, ten ha there and fifteen over here. So, the extensions become bigger. The small farmers leave. Hunger joined forces with the will to eat. (Paraguay's Deputy Minister of Agriculture, Asunción, 27 February 2017)

Undoubtedly, the rapid expansion of soy is linked to the displacement of *campesino* agriculture in the area (Ezquerro-Cañete 2016). Thousands of small farmers left agriculture and a growing number of landless have emerged. There are several material constraints facing small producers trying to engage in soybean production, since it has developed into an activity with important economies of scale, and because of poorer soils and inadequate technologies. Besides displacement of small farmers and rural communities, the soybean expansion in the region has largely displaced the much more labor-intensive cotton production. Thus, the labor-sparing agribusiness model has contracted employment in the agricultural sector (Ezquerro-Cañete 2016).[24] According to *the National Peasant Federation of Paraguay* (FNC), around 900,000 Paraguayans left the countryside, mainly displaced by mechanized soybean agriculture, and settled in precarious living conditions in the outskirts of Asunción.[25]

There are, however, still many non-capitalized small peasants, campesinos, and indigenous communities living in the Atlantic forest region. There has also been a tendency of sub-fractioning among small farmers, and thus as families have expanded, small plots of 4 ha, have been divided into two. In this way, average unit size actually decreased in some area (Masterson 2007; Nuñez 2009). Some of them are also involved in soybean production. Estimations show that around 20% of soybean area (0.68 Mha) is still managed by 38,000 small producers, who produce on plots smaller than 20 ha each (Nuñez 2009). Besides considering the number of hectares managed, however, it is important to consider the quality of the land, where smallholders are leaving fertile land and moving to poorer land areas (Director of DCEA-MAG, Asunción, 21 February 2017).[26] *Latifundio*[27] and agribusiness—with a clear domination of soy and beef production—represent a small percent of the producers, but manage 85% of agricultural land. A particular feature of the extreme concentration of soybean production in the region is that a specific group—*Brasiguayos*[28]—produce the great majority of all soybeans (Lambert 2016). There have been many occupations of soybean land in the hands of *Brasiguayos*—often resulting in violent evictions by the police, protecting the soy producers.[29] There have also been several violent attacks and even assassinations of soybean producing *Brasiguayos* during the past years.[30] The tensions and conflicts get fuel from the extreme land concentration. This is not only considered unfair, but often also completely illegitimate because of the well-known occurrence of false titles, double titling, corruption, and contradictory land legislation.

### 1.1.3 Social-Ecological Concerns in the Wake of Agrarian Change

Changes in the forms of production in the wake of soybean/beef expansion and intensification have contributed to several serious social-ecological effects everywhere, but to varying degrees. The combined effects of LULCC increased social-ecological pressures and created a heated debate about social-ecological hazards caused by snowballing agrochemical use.[31] This causes adverse effects on pollinators, soil nutrients, food safety (chemical residues in food exceeding safety limits), soil and water resources, and biological diversity (Avila-Vazquez et al. 2018; Díaz-Zorita et al. 2002; Modernel et al. 2016; Novelli et al. 2017; Primost et al. 2017; Ronco et al. 2016). More specific concerns address the wiping out of natural enemies and other biological control agents by insecticides leading to more resistant and secondary plagues showing up on the fields and more toxic agrochemicals have to be applied to the crops (Bruno 2007). There are also public health concerns over the agrochemical exposure among farmers and communities nearby the sprayed fields causing health problems (Lapegna 2016b; Teubal 2008). *Sojization* also leads to rising groundwater levels, limiting soil water storage (Kroes et al. 2019), excessive nutrient extraction, and increase vulnerability and exposure to erosion (Bouza et al. 2015; Ernst and Siri-Prieto 2009; García-Préchac et al. 2004; Pérez Bidegain et al. 2010; Prieto and Ernst 2010). The concentrated rates of animal wastes and pollution from effluents also cause water quality impairment and contamination of drinking water (Demetrio 2012; Nardo 2011). Overgrazing may also lead to soil degradation (Leguizamón 2016; Oliveira and Hecht 2016).

An important amount of concerns is particularly linked to the rapid deforestation of the Gran Chaco region. The landscape fragmentation and decreased connectivity have caused the weakening or loss of several ecosystem services; for example, climate regulation, water and soil regulation services, wildlife, and livelihoods for people living in the forest (Aide et al. 2013; Allan et al. 2015; Baumann et al. 2016; Caldas et al. 2015; Cartes et al. 2015; ECLAC 2017, 136–142; Graesser et al. 2015, 5; Jobbágy et al. 2015; Kuemmerle et al. 2017; Malkamäki et al. 2016; ONU-REDD+/SEAM/INFONA/FAPI 2016, 13–18; Py/SEAM/INFONA/FAPI 2016, 17–19; Ronco et al. 2016; Torrella et al. 2018; Vallejos et al. 2015, 6).[32] In this way, the forests contribute, among other things, to maintain the fertility of soils and to prevent water-induced soil erosion as well as wind erosion (ONU-REDD+/SEAM/INFONA/FAPI 2016, 7, 13). When deep-rooted native forest has been replaced by crops the water cycle has been disrupted, the soil physical properties that regulate water circulation and storage change, and deeply stored soil salts come to the surface by raising water tables (Amdan et al. 2013; Jobbágy et al. 2015; Mereles and Rodas 2014). Moreover, problems of flooding emerged in some provinces with a sudden appearance of a network of new rivers, as subterranean water flow increased in speed after deforestation (Magliano et al. 2017). Lastly, deforestation also provokes a significant amount of greenhouse gas emissions (Baumann et al. 2016, 11).[33]

The overall changes in social relations have also rendered concerns over loss of national sovereignty, transnational agribusiness dominance and "reprimirization" of the export structure (Aide et al. 2013; ECLAC 2017, 136; Vergara-Camus and Kay 2017, 231; Wesz Jr. 2016). Family producers, peasants and rural workers have decreased everywhere, but producers were already rather big and capitalized in the Pampas region (almost no units managing less than 100 ha left in activity), while social effects of displacement in Paraguay and in the poor northern provinces of Argentina (where the majority manages less than 20 ha) are severe. Land-related conflicts have become more frequent. Land occupations have turned violent as both private and paramilitary-like forces bulldoze the homes of the peasants and threaten their leaders (Dominguez and Sabatino 2008, 38–39; Leguizamon 2014, 130). Peasant and indigenous organizations claim rights to the land on customary principles (Giarracca 2008, 20). There has also emerged a discussion about reduced food security as soybeans have replaced some domestic food crops (Teubal 2008). In general, there has been an increase of social-ecological tensions throughout the region (de Castro et al. 2016; de Waroux et al. 2017). Besides the aggravated unequal land structure in an already unequal region, the advancement of agribusiness firms from Brazil, Uruguay, and Argentina in Paraguay, as well as Argentine firms in Uruguay, have provoked concerns over foreignization of land (Coordinadora de Derechos Humanos del Paraguay 2007, 19). Many NGOs and social movements claim that the interests of peasant societies and agribusiness are fundamentally irreconcilable, as are sustainability and monoculture (García-López and Arizpe 2010).

However, the agricultural boom has also brought rising income and export revenues (and land values), improvements in infrastructure (ports and roads), foreign direct investment, technology transfer, professionalization of agriculture agricultural productivity growth and economic growth (de la Torre et al. 2015, 11–16; Turzi 2016). The remaining chapters of this book explore the political economy of aforementioned dramatic changes through an integrated analysis of the intersecting roles of historical formed institutions, international pressures, and national regulations in co-shaping agrarian change.

## 1.2 Perspectives on Agrarian Change: Exploring the Gaps

Soybean and beef-led agrarian change in Argentina, Paraguay, and Uruguay brought impressive change across various domains—indeed a bulldozer or a harvester, as referred by the agronomist and family farmer from Dolores in Uruguay, at the very beginning of this book. How did this happen? This section presents how previous research has conceptualized this change—typically either focusing on distant drivers and world systemic explanations, or myopically centering the analysis on formal state regulations as if recent agrarian change could be exclusively governed by the nation-states. This is followed by subsections presenting how this book conceptualizes the bulldozer; by exploring the roles of: (1) history (national and regional agrarian history situated in world history), (2) contemporary agrofood globalization (with focus on the global soybean and beef chains), and (3) national regulations—including the power struggles they represent.

A growing body of literature is situating agrarian change in Latin America in a context of globalization and/or land teleconnections (Baumann et al. 2016; Caldas et al. 2015; Graesser et al. 2015; Leguizamón 2016; Malkamäki et al. 2016; McMichael 2013; Meyfroidt et al. 2013; Oliveira and Hecht 2016; Ronco et al. 2016; Seto et al. 2012; Van der Ploeg 2014; Wesz Jr. 2016).[34] Many studies argue the need to acknowledge multiple governing scales across international borders. In practice, however, the role of nation-states in agrarian transformations is largely ignored. This is particularly so within the influential traditions of the "Food Regime" (FR) approach and the "GCC" literature, both originating in world systems theory.[35]

The main narrative of these approaches is that an increasingly institutionalized form of a globally integrated agrofood market (sometimes referred to as the corporate food regime), has gradually replaced the postwar "national developmentalist" regime (McMichael 2013). The current globalization process is characterized by increasingly spatially integrated transnational production systems, coordinated through a variety of market and nonmarket forms (Bair 2005, 172; Gereffi and Memedovic 2003, 4; Raikes et al. 2000, 393). This process is seen as the result of liberalization and accompanying deregulation, in which the state has surrendered much of its regulatory role to agribusiness and transformed itself into a facilitating agent to accommodate transnational capital (McMichael 2013).[36] Businesses, it is argued, have taken over large tracts of policy space previously occupied by nonmarket arrangements, including those held with the state, and become the main agents of change of the international agrofood system (Clapp 2016, 102–103; FAO 2012, 22).

Moreover, the approaches argue that agribusiness firms have further enhanced their power to restructure, coordinate, and govern the agrofood system by adopting strategies of both vertical integration, from biotechnology to processing, as well as widespread outsourcing of productive functions (Buttel 1987, 173; Clapp 2016, 97–99; Lang and Heasman 2015; McKay and Colque 2016; Neilson et al. 2014). The financialization of agriculture is found to have further increased distance and abstraction of production/social-ecological resources from sites of consumption, allowing investors to make profits while their risks and costs are transferred to geographically disparate locations. Accordingly, the dirtiest, most extractive, and damaging agricultural activities are seen to be outsourced to poorer countries with the greatest natural resource endowments. "Free" trade thus involves an unequal socio-ecological exchange (Eisenmenger and Giljum 2007, 289; Frank 2007, 305; Vega 2009, 52). Thus, not only is the state in general considered to be increasingly powerless, but the state in the periphery of the world system is seen as particularly devoid of policy space, with no other option but to largely accept the rules set by agribusiness, international organizations and the core states of the world system (Clapp 2016; Lang and Heasman 2015; McMichael 2013; Sassen 2013, 26–27).

Furthermore, there is a prevalent notion of Latin American states and public institutions as particularly weak or even failed, although to differing degrees (Díez and Franceschet 2012, 12–13). Latin American countries, with their Iberian colonial heritage, are often described as lacking *state autonomy*—the ability to control all their territory and to act independently from outside pressures. For example, foreign debt loan conditionality, multilateral agreements that institutionalize protection for foreign investors vis-à-vis the states and other types of external dependence put constraints on state autonomy. Moreover, Latin American countries are described as lacking *state capacity*—the ability to effectively perform, implement, and enforce decisions, such as tax collection and environmental laws (Díez and Franceschet 2012, 12–13). Widespread corruption, political instability, rent-seeking behavior among elites, co-option of the state by strong pressure groups (for example, the agribusiness sector), budget constraints, inefficiency, and lack of enforcement mechanisms, are often proposed institutional weaknesses and explanations for the lack of state capacity (Acemoglu 2008; Blyde and Fernández-Arias 2006). Against this backdrop, the state is often assumed to comply with agribusiness interests, regardless of whether it is due to inadequate state autonomy or capacity.

In accordance with a view of the state in Latin America as weak or with no agency to influence the dramatic changes within their territory, some studies have instead focused entirely on local level policy making and stakeholder participation. Within some critical traditions, social movements, peasants, and indigenous communities are given significant scholarly attention and often depicted as the strongest sources of resistance to neoliberal agrofood globalization (Dawson 2010; Desmarais 2007; Evans 2012; Sandbrook 2011; Vergara-Camus 2013; Worth 2013). Food sovereignty and *agroecology* have been suggested as possible alternative responses to capitalist agrofood globalization (Altieri and Nicholls 2017; Altieri and Toledo 2011; McMichael 2012; Shiva 2009, 22; Stahler-Sholk and Vanden 2011). These radical counterpoints can express an interesting critique of the mainstream system, but the initiatives are typically too small and fragmented to have any impact on the larger system. Accordingly, this book does not primarily focus on them.

The a priori negligence of the meso level—with the state at the center of the analysis—is problematic. Although the power of transnational agribusiness has certainly increased, the state still has authority over legal, tax, and other regulatory frameworks within its territory. The state remains a primary site for power struggles and negotiation, as well as an important actor co-shaping agrarian change. While it is clear that agrofood globalization is increasingly driving land-use change, a wide range of public policies have clearly also been intimately involved in articulating business activity. National regulatory systems thus play decisive roles in how the international impetus of agrofood globalization ends up articulated in each country, ranging from facilitating, negotiating, modifying, ignoring, constraining, or resisting the social relations and forms of production brought by transnational agribusiness.

There exists, in contrast to the traditions mentioned, contrasting studies and reports that instead exclusively consider state activity regarding agrarian change. They typically provide empirically detailed accounts of specific regulatory responses to recent agrarian transformation, such as the emergent forestry conservation initiatives, soil management policies, and actions in support of family farming (Granato et al. 2016; de Waroux et al. 2017; Min-Venditti et al. 2017; Minaverry 2016; Sabourin et al. 2014; Silvetti et al. 2013; Zurbriggen and Sierra 2017). There are also an innumerable amount of reports and whitepapers on behalf of international organizations and the public institutions themselves that deal at length with different public regulatory frameworks and policies (Bértola 2015; Cepal, FAO, and IICA 2015; ECLAC 2017, 136–137; Garcia 2015; Landini and Riet 2015; Paolino et al. 2014; Paolino 2015; Piera Valdés 2016).

These empirically rich studies and reports have brought new knowledge to the table, but they are mostly written within the framework of methodological nationalism and focus on public regulations and policy without addressing the exogenous pressures from agrofood globalization.[37] Policies are thus analyzed in isolation, failing to see the role of distant drivers and governing bodies, as well as possible regionally interacting policy effects.[38] In this way, these studies tend to conceptualize agrarian change through the lens of a purely Westphalian state system, with its presumption of the nation-state as completely sovereign and in control of its territory. Consequently, the constraints on state action imposed by the wider processes of agrofood globalization are largely ignored. The limitations of such a myopic, state-centered view are rather striking considering that recent agrarian change in Latin America is, incontestably, intimately intertwined with growing international demand, the advancement of transnational agribusiness actors and technological packages (developed and patented in the "North"), which even follow similar patterns across the countries in the region (Chapters 3 and 4 in this book). Thus, domestic policies hardly determine agrarian change on their own. In fact, the forces transcending national boundaries are greater today than ever before, thus putting new constraints on both state capacity and state autonomy.

The two approaches: (1) an overemphasis on global forces with a mischaracterization of the role of the Latin American state and public regulations as well as (2) a state-centrist approach lacking connection to the globe and region are incomplete approaches to understand the political economy of current agrarian change in Latin America. This book aims to move beyond existing clichés of Latin American states and instead provide a deep and comparative exploration of how agrarian change in Argentina, Paraguay and Uruguay has been shaped by a combination of forces from agrofood globalization (in particular the specific articulations of the global soybean and beef chains) and their own historically formed national institutions, including national regulations.

The main contention of this book is the continuing relevance of the state in the current phase of agrofood globalization and in developing countries. Public policies matter, but their abilities to regulate and balance exogenous forces are constrained by several factors, including changes in power relations in the international agrofood system. In which ways the policy space related to agrarian change is constrained, and how the limitations in public regulative capacity vary between countries, parts of the territory and policy areas, are issues that need to be addressed empirically. By bringing the state back in to the analysis, through the various aspects of the national regulatory frameworks, and by connecting them to both macro- (international) and micro- (specific areas of territories) levels, this book aims to strike a balance between two extremes: on the one hand, the state-centric approach that assumes the state has the capacity and ability to regulate various aspects of agrarian transformation without acknowledging pressures from agrofood globalization, and on the other hand, the world systems approach that assumes the agribusiness-led world market largely determines agrarian change and that government policy has very limited autonomous capacity and just adopts standard neoliberal rules.[39]

Moreover, this book further contends that to proper balance consideration of globalization and the states' capacities to regulate various aspects of agrarian transformation is, in fact, not enough. To understand the political economy of current agrarian change in Latin America one must also understand the strong legacies of the past, because, as is explored in the next section, history matters.

### 1.2.1    A Historical Approach: History Matters

Although contemporary global demand for natural resources is higher than ever before, and the pace of agrarian change is historically unprecedented, Argentina, Paraguay, and Uruguay already have a long history of government mediation between different interests in the midst of a changing international agrofood system. The tensions between export-oriented interests and other interests, as well as between large landholders and peasant farmers already existed during colonial rule and were reinforced under "the first international food regime" which emerged in the late nineteenth century. Argentina and Uruguay became rapidly and thoroughly embedded in the international food system as suppliers of meat, hides, wool, and grain to the industrial workers in Europe, particularly Great Britain, as well as to the slaves on the sugar plantations of the Caribbean. They also managed to develop relatively strong independent state apparatus, while their room for maneuver was constrained to varying degrees by foreign powers and strong domestic pressure groups. Landlocked Paraguay, on the other hand, had its economy severely damaged by the extremely bloody and devastating invasion by its neighbors (Argentina, Brazil, and Uruguay) during the War of the Triple Alliance, 1864–1870. After the war, the indebted state entered into a weak and dependent position vis-à-vis its big neighbors and other foreign interests, a pattern that became a continuous thread in Paraguayan history (Chapter 2 in this book).

As the export-oriented model that had dominated the region started to erode and finally cracked with the Great Depression, a strong impetus for industrialization and inward-oriented development strategies emerged. Meanwhile, the advanced economies adopted an industrial agricultural model and domestic farm subsidies that soon generated food surpluses, creating downward pressure on food prices (Clapp 2016, 27–34). After WWII, the share of Latin American agrofood products declined, while US food exports skyrocketed. With the debt crisis and the neoliberal turn, however, export-led growth in line with "comparative advantage" became the development recipe for resource-rich developing countries (Clapp 2016, 64–67; Wolford et al. 2013, 2). Argentina, Paraguay, and Uruguay, to varying extents and at slightly different times, liberalized their agrofood markets, strengthened immaterial private property rights, opened up for foreign investment, and cut price supports to farmers (Baraibar 2014; Bárcena Ibarra and Prado 2015, 19; Katz 2015, 243). Without these regulatory shifts, the current post-2000 commodity boom would not have been possible (Chapter 4 in this book).

As illustrated by the above short historical snapshot, a thorough understanding of the past is necessary to understand the present. For example, the rather differentiated consequences in Argentina, Paraguay, and Uruguay of contemporary agrarian change cannot be explained without knowing the development trajectories of the past that have resulted in the different existing social institutions and capacities to deal with change.

Moreover, national regulations reflect the concerns and power relations of the particular historical time and context in which they emerged. Existing regulatory frameworks include and build on regulations that emerged in the past. Old legislation is still valid (even though it may not always be used in practice) until it is explicitly modified or rescinded. Since regulatory practices and traditions are historically formed and slowly evolving, concerns and power relations of the past are partially present in current regulations. There is, accordingly, a significant amount of path dependency and time lag involved simultaneously with pressures for change, which is one of the reasons why a historical backdrop is necessary for an analysis of regulations. However, the present is not fully determined by its past. While some historical structures seem strong and full of inertia, a deeper look at history shows that Latin American states have at times showed the capacity to change development directions, although the palette of real options has always been constrained.

In this book, agrarian change and continuity in Latin America since 1870, with emphasis on Argentina, Paraguay, and Uruguay, is situated in the broader world history of capitalist agrofood relations, as outlined in the *Food Regimes* approach. A food regime is defined as a period in which there is a temporary constellation shaped by a particular global division of labor, and a relatively stable set of rules and power relations between states, enterprises, farmers, and populations (McMichael 2009, 140). The chronological periodization provides a way to structure and order the complex agrofood history, and by relating the national and regional trajectories to the international food regimes, the macro–micro dynamics become apparent. The norms and rules of the wider agrofood system have always had a huge impact on the region, although in some periods interests seemed even more aligned than during others.[40]

In the original publication by Harriet Friedmann and Philip McMichael (1989), the history of global food and agricultural systems was categorized into two distinct food regimes; 1870–1914 under British hegemony, and 1950s–1973 under US hegemony, with a period of reorganization in between. Later, McMichael proposed that a third regime had emerged with the neoliberalization impetus of the 1980s and is still ongoing under increasingly corporate control (Burch and Lawrence 2009; McMichael 2009, 2013). While Friedmann argues that there is too much contestedness involved in today's agrofood system to qualify as a proper regime (Friedmann 2005, 2016), the proposed periodization by McMichael is used herein as a framework within which the regional and domestic agrarian history and regulative frameworks (for the three countries) are situated (Chapters 2, 4 and 5 in this book).

### 1.2.2    A World Systemic Approach: Agrofood Globalization Expressed in Two Global Commodity Chains—Soybean and Beef

Recent agrarian change in Argentina, Paraguay, and Uruguay has been mainly driven by increased global demand for land to support animal source foods—cropland for the production of soybeans as a high-protein vegetable feed grain and pastureland for livestock grazing—ultimately driven by increased global meat and dairy consumption. Since domestic policies do not solely determine agrarian change on their own and forces transcending national boundaries are greater today than ever before, it is important to thoroughly understand the main tenets of contemporary agrofood globalization, and this book addresses, in particular, how they are materialized in the global soybean and beef chains.

An important part of this book thus pays attention to the main shifts in the international agrofood system and discusses the implications of these alterations for recent agrarian change in Latin America. Of relevance here are, for example, shifts in: international regulations (trade liberalization, financial deregulation and strengthened property rights), demography (population growth and urbanization), technology (cheaper container-shipping, new biotechnological traits, improved storage techniques and intensification of animal farming), and economic geography (Asia's increased purchasing power and relevance as global food buyer) (Clapp 2016, 61; ECLAC 2017; Morgan et al. 2006, 10; OECD/FAO 2016b; Oosterveer and Sonnenfeld 2012, 61, 73–77).

Since the lion's share of recent agrarian change in Argentina, Paraguay, and Uruguay is linked to soybeans and beef, it becomes relevant to take a deep look into the wider GCC in which these products are embedded. The structure of these global chains co-shapes (together with national regulations and historical shaped institutional structures) the ways agrarian change in the region is articulated. A historically informed *GCC* approach is used as a tool to map the full set of interlinked places, actors, assets, and activities and dynamic interactions in different stages of the commodity chain (Hopkins and Wallerstein 1977; Gereffi and Korzeniewicz 1994). This is a way to investigate the interconnected processes of cross-border production, trade, and consumption, and to avoid the pitfalls of traditional state-centric approaches, failing to grasp the wider systems dynamics. This includes an overview of the flow of capital, services, and goods, as well as the complex webs of power relations among different actors with particular emphasis on those firms that are identified as chain drivers—those playing a lead role in the construction and management of the configuration of the chain (Bair 2009, 7–9, 11). Through commodity chain analysis, it becomes possible to grasp the international division of labor and its unequal distribution of costs and benefits, encapsulated in specific commodities (Bair 2009).

Thus, not only are the shifting geographies of production, trade, and consumption in soybeans and beef chains outlined and explored, but GCC also puts attention on the relations between different social actors involved along the long-distant supply chains at different places and scales; farmers, trading companies, seed companies, slaughterhouses, and financial intermediaries. More concretely, the overall organization of the production and distribution systems of soybeans and beef is explored in the light of their main actors, rules, technologies, organizational models, business strategies, and power relations involved both upstream (agro-chemical inputs, biotechnology, and seeds) and downstream (trading and processing) in the commodity chains.

The soybean and beef chains interact, and a historically informed commodity chain analysis shows how the soybean was intentionally con-verted into a main ingredient in the meat production systems in the US almost a century ago, and how this model spread throughout the world.[41] There are nevertheless also important differences between the chains. The soybean chain encapsulates the main tenants and trends of what food regime scholars call the *third food regime*, which include extreme market concentration, transnationalization, vertical integration, strengthened intellectual property rights, and financialization (Burch and Lawrence 2009; McMichael 2009, 2013). Since the trade of meat products is far more regulated, restricted, and taxed than the soybean trade, meat products are still mostly consumed locally (while the ani-mals in Asia and Europe are increasingly fed on imported soybean meal). While meat is traded less "freely" than soybeans, it includes far more value added and can get high price premiums in higher quality segments. The business model of soybeans, on the other hand, is centered in keep-ing costs down and no quality differentiation. Soren Schroder, the chief executive of the global grain merchant Bunge, illustratively expressed this sentiment to Reuters in May 2018: "You can arbitrage the soybean sourcing from any origin" (Chapter 3 in this book). In short, the struc-tures of the GCC of soybeans and beef naturally have important implica-tions for how soybeans and beef production are articulated in Argentina, Paraguay, and Uruguay.

### 1.2.3   Exploring National Regulations (As Complex Compromises) Through History and Comparison

While the general process of liberalization of international agricultural trade and the specific features of the global soy and beef chains shape the ways the soybean and beef complexes are articulated in the region, they are not the only relevant factors. As previously mentioned, national regulatory frameworks can modify and rebalance the social relations and forms of production brought by the distantly driven agropastoral expan-sion and intensification processes in different ways. However, national regulatory frameworks do not represent one stable and coherent interest, but rather the negotiated results emanating from an arena full of dispute.

Regulations bring together actors and institutions across different sites and scales, from national to regional and international and represent accordingly complex compromises between different actors and interests.[42] It is, therefore, important to pay attention to how social groups and organizations (including producers' organizations, environmental organizations, social movements, international organizations as well as agribusiness actors) strive to put their interests and concerns on the agenda by influencing the regulatory framework. Moreover, since regulations are, as previously mentioned, historical products, existing regulatory frameworks can include and build on regulations that emerged in the past. The regulations have been modified, reframed and sometimes rescinded to match new concerns, ideals, and power relations (or to address perceived shifts in the reality the regulations are supposed to deal with). Thus, public regulations are full of inconsistencies, overlaps, and even contradictions, reflecting both past and present concerns and shifting power relations between different parts of the state apparatus. In this way, regulations constitute a unique lens into macro–micro interactions since they bring together actors and institutions across different spatial and temporal scales.

In this book, the connections from the national to the regional and international have been tracked by following the trajectory, or genealogy, of specific policies, that is how they emerged on the agenda, were negotiated, institutionalized, translated into practice, and interpreted divergently across actors. Regulations provide in this way a powerful tool to connect structure and agency, as well as to link global processes at the macro-level to the micro-level changes in forms of production and social relations at specific places. Thus, besides discussing the actual content of regulations in a comparative light, some of the negotiations that both preceded and followed their implementation are also addressed, as well as the enforcement of regulations (Chapters 4 and 5 in this book). An important and unique source into the "backstage" of the regulatory trajectories in this book, is in-depth personal interviews made with central stakeholders representing different positions in policy formulation and implementation processes.[43] See Appendix A for a detailed presentation of all respondents interviewed and a note on the main written sources used in this book.

Moreover, in order for the regulatory frameworks to say something beyond their own explicit aims and objectives, they need to be related to alternative ways that the same issue could have been addressed and tackled. Thus, as the major overall public regulations in Argentina, Paraguay, and Uruguay relevant for recent agrarian change have been identified from the plethora of rules and public policies, they have been compared to each other. Inspired by Marc Bloch's idea of "comparative history", national regulations in one country are critically contrasted and analyzed through the lens of those in the others. In this way, the temptation to regard specific regulations as logical or natural is avoided (Bloch 1953). Comparison also enables us to avoid the trap of reading too much into the explanatory value of specific national regulations on agrarian change that follow similar patterns in a country that has not adopted the same regulation. In this way, comparison can help us to better understand the effects of national regulation and be able to separate them from other factors and stressors. At the same time, the comparison enables us to widen our horizon of expectations by showing a broader variance of possible regulations than what can be grasped if only considering regulations in one country. Argentina, Paraguay, and Uruguay are both similar and different enough for a comparison to nuance each country's recent attempts to regulate agrarian change. In addition to comparing the ways these countries have regulated and handled similar pressures from agrofood globalization in the past and their more recent history, the national regulatory frameworks are also further related to regional and international agrofood relations and regulations (Chapters 2, 4, 5 and 6 in this book).

## 1.3    State Autonomy and Capacity to Regulate Agrarian Change in a Comparative Light

Argentina, Paraguay, and Uruguay have both similarities and some important differences in their regulatory approaches toward agrarian activities, in general, and how they have coped with social-ecological concerns in the wake of the soybean and beef expansion and intensification (Chapters 4 and 5 in this book). These variances are partly the result of differences in policy goals and visions, but also due to differentiated state capacity and autonomy to effectively implement chosen paths (Chapter 6 in this book).

There are always some limitations to both state autonomy and capacity, but as an ideal type, state autonomy can be defined as the ability of the state to control its territory and act freely and independently from outside pressures and interference. This involves the capacity to control its borders, have legitimate monopoly of the means of violence, and the ability to handle different external actors, such as transnational agribusinesses, foreign states, and international organizations, and prevent them from having overly considerable influence. Foreign debts, poverty, international trade agreements, co-option, and corruption can severely limit state autonomy. State capacity can be defined as the ability to formulate, perform, implement, and enforce effective, legitimate, and coherent rules and decisions in line with established national development targets. This is often discussed in terms of government effectiveness, rule of law, and absence of corruption (Lange 2015). It includes the capacity to generate on-the-ground impacts, including everything from effective tax collection to compliance with national environmental laws. It also involves the capacity to: align targets, manage field enforcement, make an independent analysis and agenda, encourage participation of affected actors, and systematically monitor and evaluate the effects, including indirect effects.

Western modern and consolidated states are often portrayed as the norm in the literature on states, and developing countries in the *global south* are consequently characterized with multiple deficits and diagnoses (Risse 2015, 3). While this type of western-centrism needs to be avoided, it is clear that countries in the global south often have more severe limitations to exercise full autonomy due to a more dependent position in the world economy, where both richer countries and powerful transnational corporations may have an advantage in negotiating agreements and/or trade conditions because of their greater resources. States may also be dependent in relation to foreign donors or loan providers, such as the powerful international organizations, private banks, other countries' governments, and nongovernmental organizations (NGOs) both in direct ways via loan conditionality or through more subtle influence. In addition, poor states have more constrained resources to adequately prepare and staff the state, and, despite that they in fact oftentimes need to play a more active role in economic management and production in their pursuit of national development and industrialization as domestic capital is limited (Lange 2015, 683). Moreover, when states are the best source of valuable resources in a resource-scarce environment, the incentives for their capture are higher. Poor states also lack resources to buy support and restrain anti-state resentment by providing their citizens with valuable goods and services (Lange 2015). Widespread corruption, political instability, co-option by strong pressure groups (for example, by the landed elites and/or by agribusiness), budget constraints, inefficiency, and lack of enforcement mechanisms also pose potential limitations in statehood (Acemoglu 2008; Blyde and Fernández-Arias 2006; Díez and Franceschet 2012, 12–13).

While limited statehood is the rule, there is a wide variation of how well (or poorly) specific countries fulfill the ideals of state autonomy and capacity. Moreover, limitations may vary in different policy areas and specific parts of national territory; there can be "areas of limited statehood" where the state has less capacity to enforce its own decisions than in the country as a whole (Lange 2015). Both state autonomy and state capacity are nevertheless relative concepts, which is why this book discusses state autonomy and capacity in Argentina, Paraguay, and Uruguay in both a cross-national and historically comparative light (Chapter 6 in this book).

In this book, state autonomy and capacity is discussed based on a comparative analysis of the historically situated regulations that emerged in Argentina, Paraguay, and Uruguay in the past decades in relation to the strong, distantly driven transformative pressures. While the historical approach makes it possible to grasp relative changes in state autonomy and capacity over time, the cross-national comparison of how the states have dealt with agrarian change provides a powerful lens to explore how different limitations to state autonomy and capacity are in fact articulated across policy areas and places. In this way, the book provides depth and nuance to the discussion of limited state autonomy and capacity in contemporary agrofood globalization. This allows for a broader discussion of the possibilities and constraints of governmental policy space in these countries under contemporary agrofood globalization, and also feeds the theoretical discussion of constrained state autonomy and capacity in the Global South, beyond Latin America.

## 1.4    OUTLINE OF THE REMAINING CHAPTERS OF THIS BOOK

In conclusion, this introductory chapter, has in broad brush strokes presented the dramatic agrarian change that has taken place in the region, led by the expansion and intensification of soybean and beef production. The chapter has also presented how these processes in Argentina, Paraguay, and Uruguay have transformed both social relations and forms of production in a relatively similar way in all three countries, although there are also significant differences. In addition, this chapter has served as an introduction to the various perspectives, theories, and methodological approaches that have been applied in this book to explore the political economy of recent agrarian change in Latin America in new and fruitful ways, particularly, the integrated tripod of approaches: history, world systems, and regulations that form the backbone of this book. The remaining chapters of this book are organized as follows:

Chapter 2 provides a historical backdrop to the recent wave of distantly driven agrarian changes. The political economy of agrarian change and continuity in Latin America since 1870, with emphasis on Argentina, Paraguay, and Uruguay, is situated in the broader world history of capitalist agrofood relations, as outlined in the *Food Regimes* approach. The chapter shows how the wider international agrofood system always has had a huge impact on the region, providing shifting opportunities and constraints, at the same time as shifting local norms and power-figurations have also influenced the development in different ways. The chapter also shows how some institutions, such as concentrated land structure, strong pressure groups linked to export-oriented agriculture and an underdeveloped countryside emerged, became consolidated and remained strong over time.

Chapter 3 presents the intertwined global soybean and beef commodity chains, using the lens of GCC analysis. The chapter explains how the soybean was converted into the main ingredient in the meat production systems in the US and how this model spread throughout the world. It also outlines the shifting geographies of production, trade and consumption, as well as the structures of upstream (agrochemical inputs, biotechnology, and seeds) and downstream stages (trading and processing) of the long, distant supply chains. In short, the overall organization of the production and distribution systems of soybeans and beef are explored in the light of their main actors, regulations, technologies, and power relations, as well as the wider context of contemporary agrofood globalization in which they are embedded.

Chapter 4 delves into the national regulative shifts in the region of the 1980s and 1990s, with specific focus on Argentina, Paraguay, and Uruguay. The chapter includes a presentation of relevant *Washington Consensus* inspired regulative shifts and their role in setting the conditions and shaping the posterior intensified wave of agrarian change after the new millennium. The chapter discusses the relation between the new policy orientation and the accelerated pace of increments in production, extension, and land-use intensification, including their dramatic effects on *forms of production* and *social relations*.

Chapter 5 presents the main public regulative shifts that emerged after the turn of the millennium, when the increasing discontent with neoliberalism provoked the rise of Latin American *neostructuralism*, including the so-called pink tide. The chapter includes an exploration of how Argentina, Paraguay, and Uruguay tackled the new social-ecological concerns that emerged in the wake of shifts in *forms of production* and *social relations*, and how public policies have shifted and evolved until today. By following the trajectory of specific regulative frameworks, a deeper and fuller understanding of their actual role for agrarian change, including their acceptance and enforcement, is here duly addressed and critically compared between the countries.

Chapter 6 concludes the book with a comparative reflection over the ways the three nations differently have coped with the strong distantly driven transformative pressures and analyzes what they may say about constrained state autonomy and capacity. More concretely, it focuses on the main ways in which the three have acted in order to make agrarian change contribute to broader national development aims. This involves making agrarian change: (1) more "developmental", through value-added incorporation and upgrading of the productive structure; (2) more socially inclusive, thorough redistribution, such as targeted support programs to family producers and other vulnerable groups; and (3) more sustainable, through long-term care of natural resources. The chapter also shows how a historical and comparative exploration and analysis of agrarian regulations in Argentina, Paraguay, and Uruguay can yield new fruitful insights into the discussion of constrained state autonomy and capacity—ultimately policy space—in Latin America within the realm of contemporary agrofood globalization.

Appendix A is a detailed presentation of all respondents interviewed and a note on the main written sources used in this book.

Appendix B is a visualization of the export profiles of Argentina, Paraguay, and Uruguay, 2002 and 2017, including shifts in export value in dollars and composition per commodity, from the Observatory of Economic Complexity, based on data from Comtrade FAO statistics.

## NOTES

1. While soybeans stand out as the by far most important crop in the region, there has been a general expansion of crops of multiple and flexible uses ranging from food to feed and energy—flex crops—such as soybeans,

maize, and sunflower (Borras Jr. et al. 2013, 162; Borras et al. 2014; Margulis 2014, 8).

2. Considering the changes since 2000, the soybean area expanded mainly over native forests in in the Argentine Chaco, the Paraguayan and Brazilian Atlantic forest, the Bolivian eastern lowlands forest and in the Brazilian Amazon. It has also expanded over Savannah and other grassland in Argentina and Uruguay, and to some degree over other crops (Aide et al. 2013; Caldas et al. 2015; Graesser et al. 2015, 4–5).

3. Some scholars argue that the last international commodity boom was between 2003 and 2012 and is therefore over (Vivares 2017, 5). However, others claim that while prices have moved downwards, productivity improvement has been able to compensate for much of the downfall. Moreover, while volatile, food prices seem to be on the rise again. The FAO food price index has risen almost 7% between December 2018 and June 2019. See *Financial Times* "Cheese and pork drive food prices higher in 2019—FAO", published 6 June 2019, https://www.ft.com/content/6b022e9c-884a-11e9-a028-86cea8523dc2. Accessed 27 June 2019.

4. Brazil has rightfully received a lot of worldwide attention, not least for the rapid agropastoral expansion into the Brazilian Amazon forest and *Cerrado* tropical savanna. Expansion is currently however increasing in pace and scope under the new pro-agribusiness government of President *Jair Bolsonaro* (2019–2023).

5. Argentina exported soybean products (soybeans, soybean oil and soybean oil cake) for a value of USD 5.57 billion (B) in 2002 (22% of total export value) and for USD 13.78B in 2017 (26%). It exported beef for 0.29B in 2002, and for 1.25B in 2017. Paraguay exported soy products for USD 0.5B in 2002, and for 2.99B in 2017—more than 46% of total export value. Paraguayan beef exports rose from 0.07B in 2002, to 1.03B in 2017. Uruguay exported soy products for 0.034B in 2002, and for 0.98B in 2017. Uruguayan beef exports increased from 0.27B in 2002, to 1.52B in 2017. Live cattle added up with an extra 0.27B. Source: Observatory of Economic Complexity, data from Comtrade FAO statistics. Interactive versions are available at atlas.media.mit.edu.

6. Earning less than USD 5.50 (2011 PPP) per day in 2016: In Argentina—7.8%; in Paraguay—20.6%; in Uruguay—3.7%. Data, Poverty, the World Bank Group, https://data.worldbank.org/topic/poverty?locations=UY-AR-PY. Accessed 22 December 2018.

7. This includes the colonial history under the Spanish empire, and the multiple territorial conflicts between different competing foreign powers (posing constant threats of invasion) and in different types of alliances with rival "domestic" groups even long after independence.

8. Since RR soybeans (GTS-40-3-2) started to spread the region from 1996 and onward, several new biotechnological traits and seed varieties, designed to be combined with specific herbicides and pesticides (instead of, or in combination with, glyphosate), have emerged (ISAAA 2017).

9. The soybean area in Argentina in 2014 represented 60% of total land cultivated (Comtrade FAO statistics, http://www.fao.org/faostat/en/#-data/QC. Accessed 13 May 2018).

10. The Pampas region spans over an important part of the following Argentine provinces: Buenos Aires, Santa Fe, Córdoba, and La Pampas (Acosta Reveles 2008). The whole Uruguayan territory is under the Pampas biome, but the most fertile land is in western Uruguay (Littoral area) with the most soybeans spans over the departments of Soriano, Río Negro, Colonia, and southwest of Paysandú (DIEA-MGAP 2015, 14).

11. Service crops are crops grown during fallow periods with the aim of providing nonmarketed ecosystem services, such as to fix Nitrogen from atmosphere, to control weeds, to cover the soil against erosion, to reduce soil compaction, and to increase soil organic matter.

12. In Argentina, between 1995 and 2006, more than 5 Mha of pastureland were lost to crops, without falling beef production volumes (Gras and Hernández 2013, 61). After 2006, however, beef production stagnated and even declined due to governmental policies to keep domestic beef prices down (Chapter 5 in this book). Pastureland is the by far the largest land-use type in the region and a significant part of it is still *campo natural*. Out of Argentina's 277 Mha of land, 142 Mha are under pastures. Uruguay has the highest proportion of pastureland, with 13.3 Mha out of 17.6 Mha of land. In Paraguay, 20 Mha, out of 41 Mha of land, is pastureland (CAS 2018).

13. The annual bovine meat production augmented from 0.453 mmt in 2000, to 0.593 mmt in 2017 (ECLAC 2017, 108).

14. The loss during the past one and a half decade represents a reduction of the forest of more than 20%. Between 2006 and 2007, deforestation tripled and has since remained high. According to the national forest Institute, INFONA, of the Paraguayan government, the dry Chaco forest in Paraguay covered 11.561 Mha of land in Paraguay, representing 60% of all Paraguayan forest land, and 28.4% of total Paraguayan land area. Infona, http://www.infona.gov.py/application/files/2114/3093/5539/BNB2011_6ESTRATOS.jpg. Accessed 11 October 2018.

15. In only one year, in 2014, the Paraguayan Chaco lost 287.000 ha of forest (Imbach 2016, 6). This contrasts to the post-2007 slowdown of agricultural expansion into the Argentine Chaco and the Brazilian Amazon (Graesser et al. 2015).

16. Mennonite communities have been engaged in dairy production in the areas for almost one hundred years, but it is not until after the new millennium that large cattle agribusiness have become a major driver to deforestation.

17. However, it is not exclusively beef producers behind deforestation, but to some degree also poor peasants using the forests as firewood (Santagada 2013). Satellite imagery show, however, that during recent years, patterns of deforestation have changed from small parcels cleared with chain saws or chaining over long time periods, to rapidly cleared rather big areas of square pattern, suggesting a change in clearing technology to bulldozers and other heavy machinery to remove the forest (Caldas et al. 2015).

18. While there are many initiatives at international, national, and private governance levels aiming to protect the forests, loss of natural grasslands to crops have received less attention (Grau et al. 2015).

19. While population is sparse, it increased between 1962 and 2002, from 74,129 to 135,186 people (Caldas et al. 2015).

20. The indigenous population in the Paraguayan Chaco is of approximately 40,000, belonging to the communities of Ayoreo, Chamacoco, Enxet, Nivakle, Manjuy, Maka'a, Toba Qom, Nandeva, and the Guarayo. Some communities of the native groups living in the Paraguayan Chaco have chosen voluntary isolation from the outside world, and live in regions containing the ample supply of water and the richest biodiversity in the region, although it has remained poorly known (Yanosky 2013, 115–118). In Argentina, there are: Toba, Wichi, Cuom, and Mocovi (Walcott et al. 2014, 2; Volante and Seghezzo 2018).

21. This includes the departments of Asunción, Concepción, San Pedro, Cordillera, Guairá, Caaguazú, Caazapá, Itapúa, Misiones, Paraguarí, Alto Paraná, Central, Ñeembucú, Amambay, and Canindeyú.

22. See *El País*, "La codicia por la tierra en Paraguay", published 2 March 2017, https://elpais.com/elpais/2017/02/07/planeta_futuro/1486488199_675583.html. Accessed 10 March 2018.

23. An expansion in grain feeding in the livestock sector in Paraguay is expected over the next few years, since the country has excess production of animal feed. The local industry estimates that there are some 300,000–500,000 head of cattle finished with grains, and the majority in the eastern region. Alongside these shifts, animals grow faster into slaughter weight. While 10–12 years ago most steers were slaughtered with about 30–40 months of age, they are now slaughtered with roughly 20–24 months of age (USDA 2017). The difference in technology and management, and thus productivity and weaning ratio, is nevertheless very big between small and big producers, with small farmers scoring very low (USDA 2017).

24. The livestock sector is also increasingly concentrated. Between the census of 1991 and 2008, 16.5% of livestock farmers left activity, at the same time as the average size of livestock farmers increased with 37.6% (Nuñez 2009).
25. See El País, "La codicia por la tierra en Paraguay", published 2 March 2017, https://elpais.com/elpais/2017/02/07/planeta_futuro/1486488199_675583.html. Accessed 10 March 2018.
26. "In areas where grain is produced, as in Alto Paraná and Itapúa, the trend is that the big are getting bigger and the small migrate. A wave of migration of smallholders is moving into land that is more marginal. So the small producers that sold their hectares in the areas of grain production go and migrate to Asunción or to poorer land areas" (Director of DCEA-MAG, in Asunción, 21 February 2017).
27. Large commercial estates of land established under colonial times, that have dominated the land tenure structure of Latin America, see Chapter 2.
28. The development of commercial farming in Paraguay, particularly of soybeans, is closely tied to the important influx of farmers from Brazil, so-called *Brasiguayos*, attracted by Paraguay's low land prices and fertile land.
29. See various reports from NGOs and social movements; La Coordinadora Latinoamericana de Organizaciones del Campo (CLOC)—Vía Campesina, https://clocvcparaguay.wordpress.com/2014/01/22/el-reino-del-latifundio/; Base IS, http://www.baseis.org.py/wp-content/uploads/2017/02/guahory-vale.compressed.pdf; Biodiversidad, Paraguay, http://www.biodiversidadla.org/Paises/Paraguay/(offset)/27 and the Environmental Justice Atlas https://www.ejatlas.org/print/the-guahory-crisis, and Farm land grab https://www.farmlandgrab.org/post/view/20455. All accessed 21 October 2018.
30. See several news articles in national press: Ultima hora 2018, https://www.ultimahora.com/disputa-parcela-tierra-deja-un-muerto-y-dos-heridos-n2704854.html; Ñandutí 2016, http://www.nanduti.com.py/2016/12/27/nuevo-desalojo-en-guajhory-un-muerto-y-otro-en-grave-estado1/; Vanguardia 2016, http://www.vanguardia.com.py/2016/10/31/matan-a-productor-cuando-realizaba-trabajos-de-fumigacion/; ABC 2004, http://www.abc.com.py/edicion-impresa/policiales/campesinos-asesinan-a-brasiguayo-y-a-su-guardia-a-escopetazos-en-itakyry-767248.html; ABC 2011, http://www.abc.com.py/edicion-impresa/policiales/degellan-a-colono-brasiguayo-206200.html; Noticde 2016, http://www.noticde.com/2016/10/itakyry-asesinan-al-hijo-de-un.html. All accessed 22 October 2018.
31. See statistics for Argentina from the Ministry of agroindustry, the division for Agri-Food Health and Quality (SENASA), http://www.senasa.

gob.ar/informacion/productos-veterinarios-fitosanitarios-y-fertilizantes/ productos-fitosanitarios-y; for Uruguay from the division of agricultural services (DGSA-MGAP), http://www.mgap.gub.uy/profit/pantalla.aspx and for Paraguay, authorized agrochemical products from the division of Plant and seed health (SENAVE-MAG), http://www.senave.gov.py/registros.html.

32. Several exclusive plant and animal species in the Chaco are considered to be threatened of global extinction and changes in forest area and configuration have strong negative impacts on wildlife habitat, resulting in a greater vulnerability of various endangered species as well as a wide range of fungi, birds, mammals, and plants (Mereles and Rodas 2014; Torrella et al. 2018; Walcott et al. 2014, 18).

33. Forest loss emitted on average of $99.4tCO_2e/yr$, 1990–2014, i.e., 77% of total GHG. Global Forest Watch. "Forest reforestation in Paraguay", https://www.globalforestwatch.org/dashboards/country/PRY. Accessed 23 June 2018.

34. Within the transdisciplinary field of land-use science, the concepts of teleconnections and telecouplings are increasingly used to address the increasing importance of distal connections, flows, and feedbacks characterizing change in land systems (Friis et al. 2016).

35. The food regime approach was first developed by Harriet Friedmann and Philip McMichael, identifying relatively stable periods of international agrofood rules and relations, as a lens to the investigation of economic transformations (1989). The GCC approach was coined by Terence Hopkins and Immanuel Wallerstein (1977, 1994) as a way to understand where, how, and by whom wealth is accumulated in transborder commodity chains.

36. Both multilateral agreements, such as the 1995 *Agreement on Agriculture* of the *World Trade Organization*, and hundreds of bilateral and regional trade and investment agreements, have opened up markets, reduced risks and costs for firms engaging in international trade and decreased farm support (Aksoy 2005, 5; McMichael 2013, 45).

37. Methodological nationalism takes (explicitly or implicitly) the state as central unit of analysis in both theoretical and empirical works.

38. There can for example be negative spillovers involved, since stricter environmental regulation in one area might generate increased exploitation in another area through different types of leakage effects. The poster child and schoolbook example is the 2006 *Amazon Soy Moratorium* and Brazil's *Forest Code* that markedly reduced soybean expansion into the Amazon Brazil, but probably also increased the pressures for deforestation in places with less strict regulations.

39. While Food Regime and GCC approaches have not provided the developing state with sufficient attention, they have still provided important insights about the restructuring of the international food system. In this book, the food regime approach serves as a tool for chronological periodization of agrarian history (Chapters 2 and 4), while the GCC assists in mapping the global soybean and meat chains (Chapter 3).

40. During the second food regime, many Latin American states explicitly intended to reduce foreign influence, with at least some partial success (Chapter 2 in this book).

41. As thoroughly argued when exploring the gaps in the literature about agrarian change in Latin America, approaches from world systems theory have over emphasized global forces and missed the roles played by the Latin American states in recent agrarian change. Clearly this does not take away the fact that recent agrarian change is to a high degree exogenously driven and forms part of a wider process of agrofood globalization.

42. In addition, different bodies of the state apparatus, from the congress enacting legislation, to government departments, secretariats, divisions, and offices, and public governing bodies at local level, are involved in different ways in formulating and implementing regulations and policies. Power relations and traditions may also vary between these different parts of the public arena.

43. The following questions have provided guidance to grasp content, relations, and negotiations, as well as perceived efficacy among different stakeholders of regulations: How have states acted through public policies and regulations in relation to the transformative pressures of agrofood globalization since re-democratization to present day? What have been the roles of different actors in the policy formulation process? Which actors have been able to put their particular concerns on the agenda and how are these incorporated into the regulatory framework? What have been the main tools and strategies used by the state to balance new social relations and negotiate new forms of production and business in order to align them with the wider development objectives of the country? Which regulations show the largest discrepancies between targeted aims and actual results, and why?

## References

Acemoglu, Daron. 2008. Oligarchic Versus Democratic Societies. *Journal of the European Economic Association* 6 (1): 1–44.

Acosta Reveles, Irma L. 2008. "Capitalismo agrario y sojización en la pampa Argentina. Las razones del desalojo laboral." *Lavboratorio, Cambio*

*Estructural y Desigualdad Social (CEyDS)* / *Facultad de Ciencias Soicales, UBA* 10 (22): 8–12.

Aide, T. M., D. Clark, H. R. Grau, M. Levy, D. Carr, D. Redo, and M. Andrade. 2013. Deforestation and Forest Expansion in Latin America: 2001–2010. *Biotropica* 45 (262): e271.

Aksoy, M. Ataman. 2005. "Global Agricultural Trade Policies." In *Global Agricultural Trade and Developing Countries*, edited by M. Ataman Aksoy and John C. Beghin. Washington, DC: The World Bank.

Alianza Internacional de las organizaciones de productores familiares de soya. 2008. *Declaración de Asunción*. IV Encuentro Internacional de Pequeños Productores de Soya y la Sociedad Civil, 6, 7, y 8 de Febrero del 2008. Asunción: Tekokatu, Ser, Probioma and Solidaridad.

Allan, Eric, Pete Manning, Fabian Alt, Julia Binkenstein, Stefan Blaser, Nico Blüthgen, Stefan Böhm, Fabrice Grassein, Norbert Hölzel, and Valentin H. Klaus. 2015. Land Use Intensification Alters Ecosystem Multifunctionality Via Loss of Biodiversity and Changes to Functional Composition. *Ecology Letters* 18 (8): 834–843.

Altieri, Miguel A., and Clara I. Nicholls. 2017. Agroecology: A Brief Account of Its Origins and Currents of Thought in Latin America. *Agroecology and Sustainable Food Systems* 41 (3–4): 231–237.

Altieri, Miguel A., and Victor Manuel Toledo. 2011. The Agroecological Revolution in Latin America: Rescuing Nature, Ensuring Food Sovereignty and Empowering Peasants. *Journal of Peasant Studies* 38 (3): 587–612.

Amdan, M. L., R. Aragón, E. G. Jobbágy, J. N. Volante, and J. M. Paruelo. 2013. Onset of Deep Drainage and Salt Mobilization Following Forest Clearing and Cultivation in the Chaco Plains (Argentina). *Water Resources Research* 49 (10): 6601–6612.

Aparicio, Virginia Carolina, Jose Luis Costa, Gonzalo Mayoral, and Eliana Soledad. 2017. *Plaguicidas en el ambiente*. Buenos Aires: INTA Ediciones.

Arbeletche, Pedro, and Gonzalo Gutiérrez. 2010. "Crecimiento de la agricultura en Uruguay: exclusión social o integración económica en redes." *PAMPA, Revista Interuniversitaria de Estudios Territoriales* 6: 113–138.

Arbeletche Favat, Pedro, and José Luis Cividanes Hernández. 2012, Thursday 31 May. "Análisis del agro-negocio desde la perspectiva de gestión empresarial: el caso de Uruguay." *XIV Reunión Economía Mundial - Internacionalización en tiempos de crisis*. Jaén: Universidad de Jaén.

Avila-Vazquez, Medardo, Flavia S. Difilippo, Bryan Mac Lean, Eduardo Maturano, and Agustina Etchegoyen. 2018. Environmental Exposure to Glyphosate and Reproductive Health Impacts in Agricultural Population of Argentina. *Journal of Environmental Protection* 9 (3): 241.

Bair, Jennifer. 2005. Global Capitalism and Commodity Chains: Looking Back, Going Forward. *Competition and Change* 9 (2): 153–180.

Bair, Jennifer. 2009. Global Commodity Chains: Genealogy and Review. In *Frontiers of Commodity Chain Research*, edited by Jennifer Bair, 1–281. Stanford: Stanford University Press.

Baraibar, Matilda. 2014. "Green Deserts or New Opportunities? Competing and Complementary Views on the Soybean Expansion in Uruguay, 2002–2013." Doctoral thesis Monograph, Economic History, Stockholm University. Stockholm Studies in Economic History 64.

Bárcena Ibarra, Alicia, and Antonio Prado. 2015. "Introducción." In *Neoestructuralismo y corrientes heterodoxas en América Latina y el Caribe a inicios del siglo XXI*, edited by Alicia Bárcena Ibarra and Antonio Prado. Santiago de Chile: CEPAL.

Baumann, Matthias, Ignacio Gasparri, María Piquer-Rodríguez, Gregorio Gavier Pizarro, Patrick Griffiths, Patrick Hostert, and Tobias Kuemmerle. 2016. "Carbon Emissions from Agricultural Expansion and Intensification in the Chaco." *Global Change Biology*, 23 (5): 1902–1916.

Bértola, Luis. 2015. "Patrones de desarrollo y Estado de Bienestar en América Latina." In *Neoestructuralismo y corrientes heterodoxas en América Latina y el Caribe a inicios del siglo XXI*, 261–295. Santiago de Chile: CEPAL.

Bisang, R., G. Anlló, and M. Campi. 2008. Una revolución (no tan) silenciosa. Claves para repensar el agro en Argentina. *Desarrollo Economico* 48 (190): 165–207.

Bisang, R., M. Campi, and V. Cesa. 2009. Biotecnología y desarrollo. In *Documento de proyecto*. Santiago de Chile: Comisión Económica para América Latina y el Caribe (CEPAL).

Bloch, Marc. 1953. "Toward a Comparative History of European Societies." In *Enterprise and Secular Change: Readings in Economic History*, edited by Frederick C. Lane and Jelle C. Riermersma, 494–521. Homewood: Irwin.

Blyde, Juan S., and Eduardo Fernández-Arias. 2006. "Why Does Latin America Grow More Slowly?" In *Sources of Growth in Latin America: What Is Missing?* edited by E. Fernández-Arias, R. Manuelli, and J. S. Blyde. Washington, DC: Inter-American Development Bank, IADB.

Borras Jr., Saturnino M., Jennifer C. Franco, and Chunyu Wang. 2013. "The Challenge of Global Governance of Land Grabbing: Changing International Agricultural Context and Competing Political Views and Strategies." *Globalizations* 10 (1): 161–179.

Borras, Saturnino M., Jennifer C. Franco, Ryan Isakson, Les Levidow, and Pietje Vervest. 2014. Towards Understanding the Politics of Flex Crops and Commodities: Implications for Research and Policy Advocacy. In *Think Piece Series on Flex Crops & Commodities*. Amsterdam: TNI.

Bouza, Mariana, Adriana Aranda-Rickert, Maria Brizuela, Marcelo Wilson, Maria Sasal, Silvana Sione, Stella Beghetto, Emmanuel Gabioud, José Oszust, Donaldo Bran, Virginia Velazco, Juan Gaitan, Juan C. Silenzi, Nora

Echeverría, Martín P. De Lucia, Daniel E. Iurman, Juan Ignacio Vanzolini, Federico J. Castoldi, Joaquin Etorena Hormaeche, and Ephraim Nkonya. 2015. "Economics of Land Degradation in Argentina." In *Evaluating Global Land Degradation Using Ground-Based Measurements and Remote Sensing*, edited by E. Nkonya, A. Mirzabaev, and J. von Braun, 291–326. Cham: Springer.

Bruno, Alfredo. 2007. *Plaguicidas usados en el cultivo de soja. Evolución de su uso y estimación de su impacto ambiental.* Río Negro: Coyuntura agropecuaria.

Burch, David, and Geoffrey Lawrence. 2009. Towards a Third Food Regime: Behind the Transformation. *Agriculture and Human Values* 26 (4): 267.

Buttel, Frederick H. 1987. New Directions in Environmental Sociology. *Annual Review of Sociology* 13 (1): 465–488.

Caldas, Marcellus M., Douglas Goodin, Steven Sherwood, Juan M. Campos Krauer, and Samantha M. Wisely. 2015. Land-Cover Change in the Paraguayan Chaco: 2000–2011. *Journal of Land Use Science* 10 (1): 1–18.

Caligaris, Gastón. 2017. Las grandes empresas agropecuarias en Argentina: los casos de Cresud y El Tejar. *Cuadernos de Economía* 36 (71): 469–488.

Cartes, José L., J. J. Thompson, and A. Yanosky. 2015. "El Chaco paraguayo como uno de los últimos refugios para los mamíferos amenazados del Cono Sur." *Paraquaria Natural* 3 (2): 37–47.

CAS. 2018. *Anuario de Comercio Exterior de base agraria de los países CAS 2013–2017.* Edited by M. Ackermann and L. Gorga. Montevideo: Consejo Agropecuario del Sur.

Cepal, CIRAD, and IICA. 2014. *Políticas públicas y agriculturas familiares en América Latina y el Caribe: Balance, desafíos y perspectivas.* Edited by Eric Sabourin, Mario Samper, and Octavio Sotomayor Echenique. Santiago: CEPAL-ECLAC.

Cepal, FAO, and IICA. 2015. Perspectivas de la agricultura y del desarrollo rural en las Américas: Una mirada hacia América Latina y el Caribe 2015–2016. In *Perspectivas de la agricultura y del desarrollo rural en las Américas.* San José: Comisión Económica para América Latina y el Caribe (CEPAL), Organización de las Naciones Unidas para la Agricultura y la Alimentación (FAO), Instituto Interamericano de Cooperación para la Agricultura (IICA).

Clapp, Jennifer. 2016. *Food,* 2nd ed. Cambridge: Polity Press.

Colla, Julia Lucía. 2017. "La territorialidad campesina indígena y la disputa por el territorio en el Chaco (Argentina)." *Geografícando* 13 (2): 2–16.

Coordinadora de Derechos Humanos del Paraguay. 2007. *Informe Chokokue: Informe al Relator Especialsobre las ejecuciones extrajudiciales, sumarias o arbitrarias del Consejo de Derechos Humanos de Naciones Unidas sobre las violaciones al derecho a la vida en contra de miembors y dirigentes de las organizaciones campesinas en el contexto de reforma agaria en Paraguay (1989–2005).* Asunción: Codehupy.

Craviotti, C. 2016. Which Territorial Embeddedness? Territorial Relationships of Recently Internationalized Firms of the Soybean Chain. *The Journal of Peasant Studies* 43 (2): 331–347.

Dawson, Ashley. 2010. Climate Justice: The Emerging Movement Against Green Capitalism. *South Atlantic Quarterly* 109 (2): 313–338.

de Castro, Fabio, Barbara Hogenboom, and Michiel Baud. 2016. *Environmental Governance in Latin America*. Hampshire: Palgrave Macmillian.

de la Torre, Augusto, Tatiana Didier, Alain Ize, Daniel Lederman, and Sergio L. Schmukler. 2015. *Latin America and the Rising South: Changing World, Changing Priorities*. Edited by World Bank Group, *Latin America and Caribbean Studies*. Washington, DC: © World Bank.

Demetrio, Pablo Martín. 2012. "Estudio de efectos biológicos de plaguicidas utilizados en cultivos de soja RR y evaluación de impactos adversos en ambientes acuáticos de agroecosistemas de la región pampeana." Doctoral dissertation, Facultad de Ciencias Exactas, Universidad Nacional de la Plata, Argentina.

Desmarais, Annette Aurelie. 2007. *La Vía Campesina: Globalisation and the Power of Peasants*. Ann Arbor, MI: Pluto Press.

de Waroux, Yann le Polain, Rachael D. Garrett, Jordan Graesser, Christoph Nolte, Christopher White, and Eric F. Lambin. 2017. "The Restructuring of South American Soy and Beef Production and Trade Under Changing Environmental Regulations." *World Development* 121: 188–202. https://doi.org/10.1016/j.worlddev.2017.05.034.

Díaz-Zorita, Martín, Gustavo A. Duarte, and John H. Grove. 2002. "A Review of No-Till Systems and Soil Management for Sustainable Crop Production in the Subhumid and Semiarid Pampas of Argentina." *Soil and Tillage Research* 65 (1): 1–18. https://doi.org/10.1016/S0167-1987(01)00274-4.

DIEA-MGAP. 2015. *Regiones Agropecuaroas del Uruguay*. Edited by DIEA Estadísticas Agropeciarias. Montevideo: Ministerio de Ganaderia, Agricultura y Pesca.

Díez, Jordi, and Susan Franceschet. 2012. *Comparative Public Policy in Latin America*. Vol. 38. Toronto: University of Toronto Press.

Dominguez, Diego, and Pablo Sabatino. 2008. "La conflicitividad en los espacios rurales de Argentina." *Lavboratorio, Cambio Estructural y Desigualdad Social (CEyDS) / Facultad de Ciencias Soicales, UBA* 1m0 (22): 38–52.

ECLAC, FAO, IICA. 2017. *The Outlook for Agriculture and Rural Development in the Americas: A Perspective on Latin America and the Caribbean 2017–2018*. San José, Costa Rica: Economic Commission for Latin America and the Caribbean (ECLAC), Food and Agriculture Organization of the United Nations (FAO), Inter-American Institute for Cooperation on Agriculture (IICA).

Eisenmenger, Nina, and Stefan Giljum. 2007. "Evidence from Societal Metabolism Studies for Ecological Unequal Trade." In *The World System and the Earth System: Global Socioenvironmental Change and Sustainability Since the Neolithic*, edited by Alf Hornborg and Carole L. Crumley, 288–302. Walnut Creek: Left Coast Press.

Ernst, Oswaldo, and Guillermo Siri-Prieto. 2009. Impact of Perennial Pasture and Tillage Systems on Carbon Input and Soil Quality Indicators. *Soil and Tillage Research* 105 (2): 260–268.

Evans, Peter. 2012. *Counter-Hegemonic Globalization*. The Wiley-Blackwell Encyclopedia of Globalization. New York: Blackwell Publishing Ltd.

Ezquerro-Cañete, Arturo. 2016. Poisoned, Dispossessed and Excluded: A Critique of the Neoliberal Soy Regime in Paraguay. *Journal of Agrarian Change* 16 (4): 702–710.

FAO. 2012. "The State of Food and Agriculture 2012: Investing in Agriculture for a Better Future." In *The State of Food and Agriculture*. Rome: Food and Agriculture Organization of the United Nations.

Frank, Andre Gunder. 2007. "Entropy Generation and Displacement: The Nineteenth-Century Multilateral Network of World Trade." In *The World System and the Earth System: Global Socioenvironmental Change and Sustainability Since the Neolithic*, edited by Alf Hornborg and Carole L. Crumley, 303–316. Walnut Creek, CA: Left Coast Press.

Friedmann, Harriet. 2005. "From Colonialism to Green Capitalism: Social Movements and Emergence of Food Regimes." In New *Directions in the Sociology of Global Development*, Volume 11 (Research in Rural Sociology and Development), edited by Frederick H. Buttel and Philip McMichael, 227–264. Bingley: Emerald Group Publishing Limited.

Friedmann, Harriet. 2016. Commentary: Food Regime Analysis and Agrarian Questions: Widening the Conversation. *The Journal of Peasant Studies* 43 (3): 671–692.

Friedmann, Harriet, and Philip McMichael. 1989. Agriculture and the State System: The Rise and Fall of National Agricultures, 1870 to the Present. *Sociologia Ruralis* 29 (2): 93–117.

Friis, Cecilie, Jonas Østergaard Nielsen, Iago Otero, Helmut Haberl, Jörg Niewöhner, and Patrick Hostert. 2016. From Teleconnection to Telecoupling: Taking Stock of an Emerging Framework in Land System Science. *Journal of Land Use Science* 11 (2): 131–153. https://doi.org/10.10 80/1747423X.2015.1096423.

Garcia, Fernando. 2015. Agricultura en el Cono Sur ¿Qué se conoce, qué falta por conocer? *Siembra, Repositorio Digital UCE* 2: 103–115.

García-López, Gustavo A., and Nancy Arizpe. 2010. Participatory Processes in the Soy Conflicts in Paraguay and Argentina. *Ecological Economics* 70 (2): 196–206. https://doi.org/10.1016/j.ecolecon.2010.06.013.

García-Préchac, F., O. Ernst, G. Siri-Prieto, and J. A. Terra. 2004. Integrating No-Till into Crop-Pasture Rotations in Uruguay. *Soil and Tillage Research* 77 (1): 1–13.

García Préchac, Fernando, Oswaldo Ernst, Pedro Arbeletche, Mario Pérez Bidegain, Clara Pritsch, Alejandra Ferenczi, and Mercedes Rivas. 2010. *Intensificación agrícola: oportunidades y amenazas para un país productivo y natural.* Colección Art 2. Montevideo: Udelar.

Gereffi, Gary, and Miguel Korzeniewicz. (Eds.). 1994. *Commodity Chains and Global Capitalism.* Westport: Praeger.

Gereffi, Gary, and Olga Memedovic. 2003. *The Global Apparel Value Chain: What Prospects for Upgrading by Developing Countries?* Vienna: United Nations Industrial Development Organization (UNIDO).

Giarracca, Norma. 2008. "La Argentina y la democratización de la tierra." *Lavboratorio, Cambio Estructural y Desigualdad Social (CEyDS) / Facultad de Ciencias Soicales, UBA* 10 (22): 18–21.

Goldfarb, Lucía, and Gemma van der Haar. 2016. The Moving Frontiers of Genetically Modified Soy Production: Shifts in Land Control in the Argentinian Chaco. *The Journal of Peasant Studies* 43 (2): 562–582. https://doi.org/10.1080/03066150.2015.1041107.

Graesser, Jordan, T. Mitchell Aide, H. Ricardo Grau, and Navin Ramankutty. 2015. Cropland/Pastureland Dynamics and the Slowdown of Deforestation in Latin America. *Environmental Research Letters* 10 (3): 034017.

Granato, Leonardo, Carlos Nahuel Oddone, and Matías Panelo Simón. 2016. "Política, economía y energía frente al siglo veintiuno: la integración regional y el fortalecimiento de la democracia en Paraguay." *Población y Desarrollo* 18 (33): 64–79.

Gras, Carla, and Valera Hernández. 2013. *El agro como negocio: producción, sociedad y territorios en la globalización.* Buenos Aires: Editorial Biblos.

Grau, H. Ricardo, N. Ricardo Torres, Ignacio Gasparri, Pedro G. Blendinger, Sofía Marinaro, and Leandro Macchi. 2015. Natural Grasslands in the Chaco: A Neglected Ecosystem Under Threat by Agriculture Expansion and Forest-Oriented Conservation Policies. *Journal of Arid Environments* 123: 40–46.

Guibert, Martine, Susana Grosso, María Eva Bellini, and Pedro Arbeletche. 2011. "De Argentina a Uruguay: espacios y actores en una nueva lógica de producción agrícola." *PAMPA, Revista Interuniversitaria de Estudios Territoriales* 7 (Special Issue: Impactos territoriales asociados a la reconfiguración del sistema productivo primario): 13–38.

Gutiérrez, Gonzalo. 2009. "Crecimiento y nuevas formas de gestión en la agricultura." In *Anuario 2009*, edited by OPYPA. Montevideo: MGAP.

Hansen, Matthew C., Peter V. Potapov, Rebecca Moore, Matt Hancher, S. A. Turubanova, Alexandra Tyukavina, David Thau, S. V. Stehman, S. J. Goetz, and T. R. Loveland. 2013. "High-Resolution Global Maps of 21st-Century Forest Cover Change." *Science* 342 (6160): 850–853.

Hopkins, Terence K., and Immanuel Wallerstein. 1977. Patterns of Development of the Modern World-System. *Review (Fernand Braudel Center)* 1 (2): 111–145.

Hopkins, Terence K., and Immanuel Wallerstein. 1994. Commodity Chains in the Capitalist World-Economy Prior 1800. In *Commodity Chains and Global Capitalism*, edited by Gary Gereffi and Miguel Korzeniewicz, 17–50. Westport: Praeger.

The Special Rapporteur on the rights of indigenous peoples, Victoria Tauli-Corpuz. 2015. "Addendum the Situation of Indigenous Peoples in Paraguay." A/HRC/30/41/Add.1 Human Rights Council, Thirtieth Session, Agenda Item 3. New York: United Nations—General Assembly.

Imbach, P., J. Robalino, J. Zamora, C. Brenes, C. Sandoval, E. Pacay, M. Cifuentes-Jara, and G. Labbate. 2016. *Escenarios de deforestación futura en Paraguay*. Edited by FAO/PNUD/PNUMA. Asunción: PNC ONU-REDD+ Py/SEAM/INFONA/FAPI.

ISAAA. 2017. *Global Status of Commercialized Biotech/GM Crops: 2016*. Ithaca, NY: ISAAA.

Jobbágy, E. G., H. R. Grau, J. M. Paruelo, and E. F. Viglizzo. 2015. Farming the Chaco: Tales from Both Sides of the Fence. *Journal of Arid Environments* 123: 1–2. https://doi.org/10.1016/j.jaridenv.2015.07.011.

Katz, Jorge. 2015. "La macro- y la microeconomía del crecimiento basado en los recursos naturales." In *Neoestructuralismo y corrientes heterodoxas en América Latina y el Caribe a inicios del siglo XXI*, edited by Alicia Bárcena Ibarra and Antonio Prado, 243–259. Santiago: CEPAL.

Kroes, Joop, Jos van Dam, Iwan Supit, Diego de Abelleyra, Santiago Verón, Allard de Wit, Hendrik Boogaard, Marcos Angelini, Francisco Damiano, Piet Groenendijk, Jan Wesseling, and Ab Veldhuizen. 2019. Agrohydrological Analysis of Groundwater Recharge and Land Use Changes in the Pampas of Argentina. *Agricultural Water Management* 213: 843–857. https://doi.org/10.1016/j.agwat.2018.12.008.

Kuemmerle, Tobias, Mariana Altrichter, Germán Baldi, Marcel Cabido, Micaela Camino, Erika Cuellar, Rosa Leny Cuellar, Julieta Decarre, Sandra Díaz, and Ignacio Gasparri. 2017. Forest Conservation: Remember Gran Chaco. *Science* 355 (6324): 465.

Lambert, Peter. 2016. The Myth of the Good Neighbour: Paraguay's Uneasy Relationship with Brazil. *Bulletin of Latin American Research* 35 (1): 34–48. https://doi.org/10.1111/blar.12410.

Lambin, Eric F., and Patrick Meyfroidt. 2011. Global Land Use Change, Economic Globalization, and the Looming Land Scarcity. *Proceedings of the National Academy of Sciences* 108 (9): 3465–3472.

Landini, Fernando, and Leonardo Riet. 2015. "Extensión rural en Uruguay: problemas y enfoques vistos por sus extensionistas." *Mundo Agrario* 16 (32): 3–20.

Lang, Tim, and Michael Heasman. 2015. *Food Wars: The Global Battle for Mouths, Minds and Markets*, 2nd ed. London: Routledge.

Lange, Matthew. 2015. "States in the Global South: Transformations, Trends, and Diversity." In *The Oxford Handbook of Transformations of the State*, edited by Stephan Leibfried, Evelyne Huber, Matthew Lange, Jonah D. Levy, Frank Nullmeier, and John D. Stephens, 673–690. New York: Oxford University Press.

Lapegna, Pablo. 2016a. "Genetically Modified Soybeans, Agrochemical Exposure, and Everyday Forms of Peasant Collaboration in Argentina." *The Journal of Peasant Studies* 43 (2): 517–536. https://doi.org/10.1080/0306 6150.2015.1041519.

Lapegna, Pablo. 2016b. *Soybeans and Power: Genetically Modified Crops, Environmental Politics, and Social Movements in Argentina.* Oxford: Oxford University Press.

Leguizamon, Amalia. 2014. "Roundup Ready Nation: The Political Ecology of Genetically Modified Soy in Argentina." PhD dissertation, Sociology, City University of New York—CUNY Academic Works.

Leguizamón, Amalia. 2016. Disappearing Nature? Agribusiness, Biotechnology and Distance in Argentine Soybean Production. *The Journal of Peasant Studies* 43 (2): 313–330. https://doi.org/10.1080/03066150.2016.1140647.

Lepori, Edda C. Villaamil, Graciela Bovi Mitre, and Mirtha Nassetta. 2013. "Situación actual de la contaminación por plaguicidas en Argentina." *Revista Internacional de Contaminación Ambiental* 29: 25–43.

Lipton, M. 2009. From Policy Aims and Small-Farm Characteristics to Farm Science Needs. *World Development* 38 (10): 1399–1412.

Magliano, Patricio N., Roberto J. Fernández, Eva L. Florio, Francisco Murray, and Esteban G. Jobbágy. 2017. Soil Physical Changes After Conversion of Woodlands to Pastures in Dry Chaco Rangelands (Argentina). *Rangeland Ecology & Management* 70 (2): 225–229.

MAGP, and IICA. 2012. *Estudio comparativo entre el cultivo de soja genéticamente modificada y el convencional en Argentina, Brasil, Paraguay y Uruguay,* edited by Pedro Jesús Rocha and Victor M. Villalobos. San José, CR: Ministerio de Agricultura, Ganadería y Pesca, Argentina, Instituto Interamericano de Cooperacion para la Agricultura.

Malkamäki, Arttu, Anne Toppinen, and Markku Kanninen. 2016. Impacts of Land Use and Land Use Changes on the Resilience of Beekeeping in Uruguay. *Forest Policy and Economics* 70: 113–123. https://doi.org/10.1016/j.forpol.2016.06.002.

Mansourian, Stephanie, Lucy Aquino, Thomas Erdmann, and Francisco Pereira. 2014. A Comparison of Governance Challenges in Forest Restoration in Paraguay's Privately-Owned Forests and Madagascar's Co-Managed State Forests. *Forests* 5 (4): 763–783.

Manzanal, Mabel, and Sergio Schneider. 2011. "Agricultura Familiar y Políticas de Desarrollo Rural en Argentina y Brasil (análisis comparativo, 1990–2010)." *Revista Interdisciplinaria de Estudios Agrarios* 34 (CIEA, FCE, UBA): 35–71.

Margulis, Matias E. 2014. Trading Out of the Global Food Crisis? The World Trade Organization and the Geopolitics of Food Security. *Geopolitics* 19 (2): 322–350.

Masterson, Thomas. 2007. "Land Rental and Sales Markets in Paraguay." Economics Working Paper Archive No. 491, Levy Economics Institute.

McKay, Ben, and Gonzalo Colque. 2016. Bolivia's Soy Complex: The Development of 'Productive Exclusion'. *The Journal of Peasant Studies* 43 (2): 583–610. https://doi.org/10.1080/03066150.2015.1053875.

McMichael, Philip. 2009. A Food Regime Genealogy. *The Journal of Peasant Studies* 36 (1): 139–169. https://doi.org/10.1080/03066150902820354.

McMichael, Philip. 2012. The Land Grab and Corporate Food Regime Restructuring. *The Journal of Peasant Studies* 39 (3–4): 681–701.

McMichael, Philip. 2013. *Food Regimes and Agrarian Questions*. Edited by Kate Kennedy. Agrarian Change & Peasant Studies. Winnipeg, MB: Fernwood Publishing.

Mereles, María Fátima, and Oscar Rodas. 2014. Assessment of Rates of Deforestation Classes in the Paraguayan Chaco (Great South American Chaco) with Comments on the Vulnerability of Forests Fragments to Climate Change. *Climatic Change* 127 (1): 55–71. https://doi.org/10.1007/s10584-014-1256-3.

Meyfroidt, Patrick, Eric F. Lambin, Karl-Heinz Erb, and Thomas W. Hertel. 2013. Globalization of Land Use: Distant Drivers of Land Change and Geographic Displacement of Land Use. *Current Opinion in Environmental Sustainability* 5 (5): 438–444. https://doi.org/10.1016/j.cosust.2013.04.003.

Minaverry, Clara María. 2016. "Consideraciones sobre la regulación jurídica ambiental de los servicios ecosistémicos en Argentina." *Estudios sociales (Hermosillo, Son.)* 26 (48): 43–66.

Min-Venditti, Amelia A., Georgianne W. Moore, and Forrest Fleischman. 2017. "What Policies Improve Forest Cover? A Systematic Review of Research from Mesoamerica." *Global Environmental Change* 47 (Supplement C): 21–27. https://doi.org/10.1016/j.gloenvcha.2017.08.010.

Modernel, Pablo, Santiago Dogliotti, Stéphanie Alvarez, Marc Corbeels, Valentin Picasso, Pablo Tittonell, and Walter A. H. Rossing. 2018. Identification of Beef Production Farms in the Pampas and Campos Area That Stand Out in Economic and Environmental Performance. *Ecological Indicators* 89: 755–770.

Modernel, Pablo, Walter A. H. Rossing, Marc Corbeels, Santiago Dogliotti, Valentin Picasso, and Pablo Tittonell. 2016. Land Use Change and Ecosystem Service Provision in Pampas and Campos Grasslands of Southern South America. *Environmental Research Letters* 11 (11): 113002.

Morgan, Kevin, Terry Marsden, and Jonathan Murdoch. 2006. *Worlds of Food: Place, Power and Provenance in the Food Chain, Oxford Geographical and Environmental Studies Series*. Oxford, UK: Oxford University Press.

Muñoz, Roberto. 2016. "Organizaciones campesinas en la provincia de Chaco, Argentina. Una aproximación a su composición social a partir de sus acciones de protesta: el caso de la Unión Campesina de Chaco (UCC), 2002–2011." *e-l@tina. Revista electrónica de estudios latinoamericanos* 14 (55): 23–43.

Nardo, Daniela. 2011. "Estudio del impacto de plaguicidas utilizados en el cultivo de soja y en otras actividades agrícolas sobre las especies acuáticas de consumo humano en el Área Protegida Laguna de Rocha." Master's thesis, department of Nutrition, Universidad católica del Uruguay, UCUDAL, Montevideo.

Neilson, Jeffrey, Bill Pritchard, and Henry Wai-chung Yeung. 2014. Golbal Value Chains and Global Production Networks in the Changing International Political Economy: An Introduction. *Review of International Political Economy* 21 (1): 1–8.

Novelli, Leonardo E., Octavio P. Caviglia, and Gervasio Piñeiro. 2017. Increased Cropping Intensity Improves Crop Residue Inputs to the Soil and Aggregate-Associated Soil Organic Carbon Stocks. *Soil and Tillage Research* 165: 128–136.

Nuñez, Edgardo. 2009. Realización y Resultados del Censo Agropecuario del 2008 en Paraguay. Dirección de Censo y Estadísticas Agropecuarias (DCEA), Ministerio de Agricultura y Ganadería (MAG).

Oddone, Gabriel. 2015. "Sustento del fuerte dinamismo del sector agropecuario." In *El desarrollo agropecuario y agroindustrial de Uruguay: Reflexiones en el 50 aniversario de la Oficina de Programación y Política Agropecuaria (OPYPA-MGAP)*, edited by Unidad de Comunicación Organizacional y Difusión and Diego Campoy, 32–36. Montevideo: MGAP.

OECD/FAO. 2016a. *OECD-FAO Agricultural Outlook 2016–2025.* Paris: OECD Publishing.

OECD/FAO. 2016b. "Oilseeds and Oilseed Products." In *OECD-FAO Agricultural Outlook 2016–2025*, 13. Paris: OECD Publishing.

Oliveira, Gustavo, and Susanna Hecht. 2016. Sacred Groves, Sacrifice Zones and Soy Production: Globalization, Intensification and Neo-Nature in South America. *The Journal of Peasant Studies* 43 (2): 251–285. https://doi.org/10.1080/03066150.2016.1146705.

ONU-REDD+/SEAM/INFONA/FAPI, PNC. 2016. Mapeo de los benificios múltiples de REDD+ en Paraguay: analisís adicionales para orientar la toma de decisiones sobre políticas y medidas REDD+. Edited by Programa de las Naciones Unidas para el Medio Ambiente. Asunción: FAO/PNUD/PNUMA.

Oosterveer, Peter, and David A. Sonnenfeld. 2012. *Food, Globalization and Sustainability.* London: Routledge.

Paolino, Carlos. 2015. "La política públca y el apoyo al sector agropecuario." In *El desarrollo agropecuario y agroindustrial de Uruguay: Reflexiones en el 50 aniversario de la Oficina de Programación y Política Agropecuaria (OPYPA-MGAP)*, edited by Unidad de Comunicación Organizacional y Difusión and Diego Campoy, 37–45. Montevideo: MGAP.

Paolino, Carlos, Lucía Pittulaga, and Mario Mondelli. 2014. Cambios en la dinámica agropecuaria y agroindustrial del Uruguay y políticas públicas. In *Estudios y Perspectivas*. Santiago: CEPAL.

Pedretti, Ricardo. 2004. Inversión en programas de diferenciación y diversificación de productos oleaginosos en Paraguay. In *Proyecto FAO Mercosur ampliado (TCP/RLA/2910), Apoyo a la integracion del sector agropecuario del conosur para contribuir a las politicas de seguridad alimentaria / FAO*. Asunción: Oficina Regional para America Latina y el Caribe, FAO. Original edition, Project Report, PROYECTO FAO TCP/RLA/2910.

Pérez Bidegain, M., F. García Préchac, M. Hill, and C. Clérci. 2010. "La erosión de suelos en sistemas agrícolas." In *Intensificación agrícola: oportunidades y amenazas para un país productivo y natural*, 67–88. Montevideo: Udelar.

Piera Valdés, A. 2016. Consultoría Nacional: Análisis del Marco Legal e Institucional Vigente para la Implementación de REDD+ en Paraguay. In *Program ONU-REDD+ Paraguay*. Asunción: FAO, PNUD, PUMA, INFONA, SEAM, FAPI.

Prieto, G. S., and O. Ernst. 2010. "Manejo del suelo y rotación con pasturas: Efecto sobre la calidad del suelo, el rendimiento de los cultivos y el uso de insumos." *Informaciones Agronómicas del Cono Sur* (45): 22–26.

Primost, Jezabel E., Damián J. G. Marino, Virginia C. Aparicio, José Luis Costa, and Pedro Carriquiriborde. 2017. Glyphosate and AMPA, "Pseudo-Persistent" Pollutants Under Real-World Agricultural Management Practices in the Mesopotamic Pampas Agroecosystem, Argentina. *Environmental Pollution* 229: 771–779.

Py/SEAM/INFONA/FAPI, PNC ONU-REDD+. 2016. *Paraguay: cambio de uso e suelo y costos de oportunidad. Sinergias entre REDD+ y la Ley de Valoración y Retribución de Servicios Ambientales*. Ciudad de Panama: FAO/ PNUD/PNUMA.

Raikes, P., M. F. Jensen, and S. Ponte. 2000. Global Commodity Chain Analysis and the French Filiere Approach: Comparison and Critique. *Economy and Society* 29 (3): 390–417.

Ramírez, Omar Javier. 2010. Percepción del riesgo del sector agroindustrial frente al uso agrícola de plaguicidas: la soja transgénica en la Pampa Argentina. *Ambiente y desarrollo* 14 (26): 35–62.

Ray, Rebecca, and Kevin P. Gallagher. 2016. "China in Latin America: Environment and Development Dimensions." *Revista Tempo do Mundo (RTM)* 2 (2): 131–154.

Richardson, Neal P. 2009. Export-Oriented Populism: Commodities and Coalitions in Argentina. *Studies in Comparative International Development* 44 (3): 228.

Risse, Thomas. 2015. "Limited Statehood: A Critical Perspective." In *The Oxford Handbook of Transformations of the State*, edited by T. Leibfried, E. Huber, M. Lange, J. D. Levy, and J. D. Stephens. Oxford: Oxford University Press.

Ronco, A. E., et al. 2016. Water Quality of the Main Tributaries of the Paraná Basin: Glyphosate and AMPA in Surface Water and Bottom Sediments. *Environmental Monitoring and Assessment* 188 (8): 1–13.

Rossi, Virginia. 2010. La producción familiar en la cuestión agraria uruguaya. *Revista Nera* 13 (16): 63–80.

Sabourin, Eric, Mario Samper, Jean François Le Coq, Gilles Massardier, and Octavio Sotomayor. 2014. El surgimiento de políticas públicas para la agricultura familiar en América Latina: trayectorias, tendencias y perspectivas. *Cadernos de Ciência & Tecnologia* 31 (2): 189–226.

Sandbrook, Richard. 2011. Polanyi and Post-neoliberalism in the Global South: Dilemmas of Re-embedding the Economy. *New Political Economy* 16 (4): 415–443. https://doi.org/10.1080/13563467.2010.504300.

Sandoval, Lazaro. 2016. "Oilseeds and Products Annual: Uruguay." In *GAIN Report*, edited by Global Agricultural Information Network. USDA.

Santagada, Ezequiel. 2013. *Reporte sobre la problemática de la tenencia de la tierra en el Paraguay de cara a la implementación del Programa REDD+.* Edited by ONU-REDD PNC Paraguay. Asunción: Programa de Naciones Unidas para el Medio Ambiente (PNUMA).

Sassen, Saskia. 2013. Land Grabs Today: Feeding the Disassembling of National Territory. *Globalizations* 10 (1): 25–46.

Seto, Karen C., Anette Reenberg, Christopher G. Boone, Michail Fragkias, Dagmar Haase, Tobias Langanke, Peter Marcotullio, Darla K. Munroe, Branislav Olah, and David Simon. 2012. Urban Land Teleconnections and Sustainability. *Proceedings of the National Academy of Sciences* 109 (20): 7687–7692.

Shiva, Vandana. 2009. Soil Not Oil. *Alternatives Journal* 35 (3): 19.

Silvetti, Felicitas, Gustavo Soto, Daniel M. Cáceres, and Diego Cabrol. 2013. "¿Por qué la legislación no protege los bosques nativos de Argentina?: Conflictos socioambientales y políticas públicas." *Mundo agrario* 13 (26): 1–21.

Stahler-Sholk, Richard, and Harry Vanden. 2011. A Second Look at Latin American Social Movements Globalizing Resistance to the Neoliberal Paradigm Introduction. *Latin American Perspectives* 38 (1): 5–13. https://doi.org/10.1177/0094582x10384204.

Teubal, Miguel. 2008. "Soja y agronegocios en la Argentina: la crisis del modelo." *Lavboratorio, Cambio Estructural y Desigualdad Social (CEyDS) / Facultad de Ciencias Sociales, UBA.* 10 (22): 5–8.

Torrella, Sebastián A., María Piquer-Rodríguez, Christian Levers, Rubén Ginzburg, Gregorio Gavier-Pizarro, and Tobias Kuemmerle. 2018. "Multiscale Spatial Planning to Maintain Forest Connectivity in the Argentine Chaco in the Face of Deforestation." *Ecology and Society* 23 (4). https://doi.org/10.5751/es-10546-230437.

Turzi, Mariano. 2016. *The Political Economy of Agricultural Booms: Managing Soybean Production in Argentina, Brazil, and Paraguay*. Cham, Switzerland: Springer.

USDA. 2017. "Paraguay: Livestock and Products Annual." In *USDA Foreign Agricultural Services*. Global Agricultural Information Network—GAIN.

USDA. 2018. "Oilseeds: World Market and Trade". In *World Agricultural Reports*. United States Department of Agriculture, Foreign Agricultural Services.

Vallejos, María, José N. Volante, María J. Mosciaro, Laura M. Vale, M. Laura Bustamante, and José M. Paruelo. 2015. Transformation Dynamics of the Natural Cover in the Dry Chaco Ecoregion: A Plot Level Geo-Database from 1976 to 2012. *Journal of Arid Environments* 123: 3–11. https://doi.org/10.1016/j.jaridenv.2014.11.009.

Van der Ploeg, Jan Douwe. 2014. "Peasant-Driven Agricultural Growth and Food Sovereignty." *Journal of Peasant Studies* 41 (6): 999–1030.

Varrotti, Andrea Patricia Sosa, and Samuel Frederico. 2018. "Las estrategias empresariales del agronegocio en la era de la financiarización. El caso de El Tejar." *Mundo Agrario* 19 (41): e086–e086.

Vega, Gerardo Cerdas 2009. "Monocultures and Agrofuels: Key Elements for Debate." In *Red Sugar, Green Deserts*, edited by Maria Silva Emanuelli, Jennie Jonsén, and Sofia Monsalve Suárez. Halmstad, Sweden: FIAN International, FIAN Sweden, HIC-AL and SAL.

Vergara-Camus, Leandro. 2013. Rural Social Movements in Latin America: In the Eye of the Storm. *Journal of Agrarian Change* 13 (4): 590–606. https://doi.org/10.1111/Joac.12030.

Vergara-Camus, Leandro, and Cristobal Kay. 2017. "New Agrarian Democracies: The Pink Tide's Lost Opportunity." *Socialist Register* 54 (54): 211–230.

Vivares, Ernesto. (Ed.). 2017. *Regionalism, Development and the Post-commodities Boom in South America*. Cham: Springer.

Volante, José Norberto, and Lucas Seghezzo. 2018. Can't See the Forest for the Trees: Can Declining Deforestation Trends in the Argentinian Chaco Region Be Ascribed to Efficient Law Enforcement? *Ecological Economics* 146: 408–413. https://doi.org/10.1016/j.ecolecon.2017.12.007.

Walcott, J., J. Thorley, G. Casco, L. Coronel, V. Kapos, L. Miles, R. Blaney, and S. Woroniecki. 2014. *Mapeo de los beneficios múltiples de REDD+ en Paraguay: el uso de la información espacial para apoyar la planificación del uso de la tierra*. Cambridge: Programa de las Naciones Unidas para el Medio Ambiente.

Wesz Jr., Valdemar João. 2016. "Strategies and Hybrid Dynamics of Soy Transnational Companies in the Southern Cone." *The Journal of Peasant Studies* 43 (2): 286–312. https://doi.org/10.1080/03066150.2015.1129496.

Wolford, W., S. M. Borras Jr., R. Hall, and I. Scoones. 2013. "Governing Global Land Deals: The Role of the State in the Rush for Land." In *Governing*

*Global Land Deals: The Role of the State in the Rush for Land*, edited by W. Wolford, S. M. Borras Jr., R. Hall, and I. Scoones, 1–22. Malaysia: Wiley-Blackwell.

Worth, Owen. 2013. Polanyi's Magnum Opus? Assessing the Application of the Counter-Movement in International Political Economy. *The International History Review* 35 (4): 905–920. https://doi.org/10.1080/07075332.2013.817464.

Yanosky, Alberto. 2013. "Paraguay's, Challenge of Conserving Natural Habitats and Biodiversity with Global Markets Demanding for Products." In *Conservation Biology: Voices from the Tropics*, edited by Luke Gibson Navjot S. Sodhi and Peter H. Raven, 113–119. Asunción: Wiley.

Zurbriggen, C., and M. Sierra. 2017. Innovación colaborativa: el caso del Sistema Nacional de Información Ganadera. *Agrociencia Uruguay* 21: 140–153.

# CHAPTER 2

# Changes and Continuities in Agrofood Relations, 1870–1970s

Global demand for natural resources is at levels never seen before. Meanwhile, Latin America, with its extensive plains, fertile soils, freshwater reserves, and deep forests, maintains its historic role of world supplier of agrofood products and other natural resource-based commodities. This history started already under European colonialism, which included a widespread transfer of animals and plants between the American and Afro-Eurasian hemispheres. With the increase of overseas commerce, the new territories became increasingly specialized as agrofood providers, which allowed for the formation of the modern capitalist integrated world system (Wallerstein 1974).

This chapter paints, in broad brushstrokes, one hundred years of agrarian change and continuity in Latin America (1870 to the 1970s), with particular emphasis on the roles that states have played in agrarian development and agrofood relations. Latin America has a long history of government mediation between different national and international interests in the midst of changing international agrofood systems. This chapter's historical review is situated in the broader world history of capitalist agrofood relations, as conceptualized by the *Food Regimes* approach mentioned in the introductory chapter. This influential world system's inspired framework was first developed by Friedmann and McMichael (1989). It categorizes the complex capitalist agrofood history into food regimes, which are defined as temporary constellations of relatively stable sets of rules and power relations between firms, states,

© The Author(s) 2020                                                      57
M. Baraibar Norberg, *The Political Economy of Agrarian Change
in Latin America*, Governance, Development, and Social Inclusion in
Latin America, https://doi.org/10.1007/978-3-030-24586-3_2

and farmers (McMichael 2009, 140). When contradictions and contestation grow, the regime starts to erode and eventually it falls apart in a crisis. This is followed by a period of reorganization and eventually the emergence of a new regime (McMichael 2013).

In the seminal work of Friedmann and McMichael from 1989, two main food regimes were discerned in international agrofood history (1989). The *first food regime* emerged out of the industrialization and urbanization processes in Europe of the late nineteenth century. It was by large dominated by Great Britain. It started to erode with the outbreak of the First World War, and finally fell apart with the Great Depression. After a reorganization period, including the Second World War, a new state-regulated *second food regime*, emerged. This regime was under hegemony of the US and lasted until the fall of the Bretton Woods system in 1973. According to later research by McMichael, and many other food regime scholars, a *third food regime* emerged in the 1980s, characterized by increased corporate control (Burch and Lawrence 2009; Holt Giménez and Shattuck 2011; McMichael 2009, 2013; Otero 2012).[1]

The focus of this chapter is on Latin American economic and agrarian history. This forms a necessary backdrop for the understanding of the legacies involved in recent agrarian change in Argentina, Paraguay, and Uruguay. It is also vital for the understanding of the differentiated capacities of these three countries to regulate agrarian change. The two major food regimes proposed in the original work of Friedmann and McMichael are used as an overall framework in which the Latin American economic and agrarian history is situated within. The remainder of this chapter is divided into two main sections; one about *the first food regime* (Sect. 2.1), and the other about *the second food regime* (Sect. 2.2). Both sections involve a presentation of the regimes at three scales; international, regional, and national (for Argentina, Paraguay, and Uruguay) and both present the articulation of each regime from emergence to fall, and to the post-regime period of reorganization.

## 2.1   The First International Food Regime

According to the *Food Regime* approach, the first proper international food regime in the world emerged around 1870, started to decay with the First World War, and finally fell apart with the Great Depression. This section is organized the following way. First comes a brief presentation of the regime's main tenants, trends, and rules (Sect. 2.1.1). This is followed by a short introduction to the way Latin America was situated within the regime (Sect. 2.1.2). The next subsection chronologically presents agrarian change and national regulations in Argentina, Paraguay, and Uruguay during this generally export-oriented regime (Sect. 2.1.3), and the section ends with a presentation of how the erosion and fall of the food regime, with the 1930 economic crisis, was articulated in the region and what happened during the period of reorganization during the 1930s with increased state intervention (Sect. 2.1.4).

### 2.1.1   Main Tenants, Trends and Rules

The world market in primary products between 1850 and the outbreak of the First World War was subject to fewer restrictions than any other time in history and it expanded significantly (Bulmer-Thomas 2003, 154). Europe, notably Britain, was characterized by relatively liberal trade policies (with tariffs and nontariff barriers progressively lowered after the repeal of the Corn Laws in 1846). World trade was facilitated by the gold standard, and largely financed by short-term credits from the London money market. The emergence of new mechanisms for standardization further facilitated trade and reduced transaction costs (Velkar 2010, 24–34).[2] At the same time, new improved storage techniques, agronomic advances, and impressive transport price reduction improved the competitiveness of natural resource based production in remote places and boosted overseas agrofood trade (Bértola et al. 2008). In this way, the first food regime was characterized by an extensive form of capitalist development, which was based upon an expansion in the resources and regional specialization.[3] Out of the new geographic specialized pattern of production and consumption, a new international division of labor emerged. Recently industrialized Europe (again mainly Britain), could meet its intensified demand for foodstuffs by imported basic grains and livestock, as well as tropical products from both colonies and newly independent republics. By providing the emerging European industrial classes with cheaper and more diverse food, real wages were allowed to be kept down (Friedmann 2005). The inflow of cheap food weakened the power of the landed aristocracy. At the same time, Europe searched for new markets to supply with industrialized products. Thus, European manufactures went in the reverse direction. The main consequence for food-exporting countries was thus the insertion into the world system as commodity providers—often based of monocultural agricultural systems—and importers of manufactures (McMichael 2009, 141). This formed the fundamental conditions for the emergence of the *first food regime*—the first worldwide constellation of relatively stable sets of rules and power relations between firms, states, and farmers. The newly independent Latin American states both influenced and were heavily influenced by this new international food regime.[4]

### 2.1.2    Latin America—The Export-Oriented Growth Model

Since colonization, Latin America played a role as world commodity supplier, cocreating the conditions for the emergence of the more integrated modern world system (Gordon and Morales 2017). Under the age of merchant capitalism, sugar, hardwoods, coffee, guano, hides and jerked beef (salted meat), and precious metals were widely traded commodities from the region (Jones 2000, 5).[5] An accumulative strategy based on very low levels of risk and costs (low use of technology, capital, and low use of labor) and unequal land tenure with huge estates of land—*el latifundio*—were established in the region under colonial rule. These colonial institutions still casted long shadows over historical trajectory after independence, when the first food regime with massive agrofood trade emerged in the era of industrial capitalism, reinforcing the role of Latin America as world food supplier.

A new consensus emerged in the new republics of Latin America in favor of export-led growth and they gradually abandoned the colonial heritage of mercantilist trade policies (Bulmer-Thomas 2003, 43). The newly independent Latin American republics tried to boost export-driven growth as much as they could. The main governmental preoccupation throughout the region, in addition to peace and the consolidation of national territory, was to increase exports of primary products. This export-oriented model adopted in Latin America thus responded to a combination of European (mainly British) openness to overseas trade, falling transport costs, more available foreign capital imports and investments in export-facilitating infrastructure, and the availability of "cheap" natural resource demanded by Europe.

The degree of export specialization of each country was extremely relying heavily on a small number of primary products.[6] This made the countries vulnerable to price fluctuations in single-commodity markets. Radically changing market conditions (supply shocks, the emergence of substitution products, or new trade policies) and new technological innovations resulted in the waning of former export stars while new-comers emerged in subsequent "booms and busts", but the concentration ratio remained high throughout the period (Bulmer-Thomas 2003, 60). In addition, specialization meant dramatic ecological simplification (monocultures and biodiversity loss), at the same time as social complexity (rising amount of middlemen and specialized activities) increased due to the growing division of labor (Friedmann 2000, 485; 2017; Martinez-Alier et al. 2016).[7]

While most countries in Latin America had consolidated independence already by the mid-nineteenth century, they were nevertheless still plagued by conflicts, civil wars (often with some degree of foreign intervention) and social unrest, until at least the early twentieth century. The economic historian Victor Bulmer-Thomas noted that "the ever-present threat of territorial dispute in the nineteenth century forced governments to keep armed forces, which not only acted as a drain on scarce fiscal resources but also impeded the efforts to establish strong civilian-controlled political institutions" (2003, 48). Due to capital constraints, all countries collected some revenue from trade duties in spite of the free trade discourse. The reason for the duties was thus not to discourage trade, but to raise revenue from the tariff system. An alternative revenue could have been to tax land, but with the exception of Uruguay under *Batlle y Ordoñez* (see below), this proved impossible due to the powerful *latifundistas* (Bulmer-Thomas 2003, 140). To overcome the difficulties encountered in mobilizing domestic resources for capital accumulation, the republics also adopted policies favorable for foreign capital sources. European—again particularly British—capital export flooded into the Latin American republics, particularly in the *Río de la Plata* (Jones 2000, 20–42). Some British trading companies became multinational business groups and diversified into a range of activities, including sourcing of natural resources, manufacturing plants, financial services, and not least railway construction (Jones 2000, 43–47). In general, the export booms in Latin America throughout the period had a high component of foreign engagement, which led to a rapid advancement of the export capacity, but it also rendered a loss of national control of the export sector (Bulmer-Thomas 2003, 170).

In addition to high external dependence, the export-oriented model reinforced inequality in an already unequal region, strengthening the landed elites (Bértola and Williamson 2003; Bértola et al. 2008). Latin America's exports of primary products competed in the world market both with each other and with production from other parts of the world. To increase competitiveness, labor costs—and thus the purchase capacity of the domestic working population—were kept down. In some places even coercion, in the form of debt-bondage or slavery, were used to keep production costs down (Bulmer-Thomas 2003, 145). Besides hindering wealth distribution, this created little incentives to invest in machinery and labor productivity was relatively low (Bulmer-Thomas 2003, 88–93).

Nevertheless, the period between 1870 and 1929 is often considered the *Belle Époque* of many Latin American countries and of the Southern Cone in particular, due to the economic growth brought by trade. In this period of export-oriented growth strategies in Latin America and phenomenal growth in the world economy, most countries in the region experienced, at least during short periods, high export growth rates. However, the Latin American exports were virtually exclusively based on primary products, whereas the main imports were manufactured goods, as well as labor and capital (Pinilla and Rayes 2017). Moreover, growth in many countries did not manage to sustain over a long period, or to transform into wider development. Only Argentina, Chile, Cuba, and Uruguay reached high annual growth rates of real GDP/c throughout the period (Bértola et al. 2008; Bulmer-Thomas 2003, 60–68, 110–111). The following section presents a more detailed picture of the focus countries of this study—Argentina, Paraguay, and Uruguay.

### 2.1.3    A Closer Look at Argentina, Uruguay, and Paraguay

The settler economies Argentina and Uruguay were among the few that managed to accumulate wealth based on expanding domestic productive capacity and agrofood export growth (Bonino-Gayoso et al. 2015). They also managed to establish a more or less stable representative political system. By 1910, Argentina and Uruguay were among the nation-states with the highest GDP/c rates in the world (Bertino and Tajam 1999; Bolt et al. 2014).[8] Their economic success can, to a certain degree, be explained by mere luck in the "commodity lottery", where the main export sectors—meat and wool, and grain (for Argentina)—benefited from being the same commodities as those produced by high-income farmers in the advanced economies, which created a high marginal price on wool, hides, meat and its extracts for a relatively sustained period of time (Bértola et al. 2008). Moreover, they had significant forward and backward linkages into other sectors, such as milling, processing, packaging, seed improvement, breeding, and so forth, allowing for more value-added incorporated into the primary products and a move toward industrialization (Bulmer-Thomas 2003, 78–79).[9] The meat packers *frigoríficos* turned meat into the first processed primary product for export from Latin America (Bulmer-Thomas 2003, 185).[10] Moreover, technological innovations allowed for a diversification of the cattle-based trade (from hides, jerky and corned beef, to chilled and frozen meat). The reduction of real freight prices and the development of harbors of Montevideo and Buenos Aires, reduced transport costs of the cattle-based exports to the rising European meat demand. At the same time, the British consumers demanded higher quality breeds, which required fencing to prevent the mixing and degeneration of stock, the livestock sector modernized significantly in both countries between 1860 and 1914 (Mederos Porto 2014, 55–56; Scobie 1964, 44–45). Moreover, the export commodities were also staples in the national diet, and thus the technological changes that brought productivity gains for the export sector did the same for the domestic sector (Bulmer-Thomas 2003, 123; Flichman 1990).

In addition, Argentina and Uruguay benefited relatively more than any other countries in the region from an important inflow of European migrants (increasing supply of labor and expertise) and foreign capital (Bértola 2000; Bértola and Williamson 2003; Williamson 2002). In both countries, British companies engaged in everything from the issuing of long-term mortgage bonds, the construction and operation of railroads, the investments in and acquisitions of meatpacking plants and refrigerated shipping fleets and other public utility enterprises (Bulmer-Thomas 2003, 74–75; Flichman 1990, 3; Jones 2000, 257).[11]

While long-term economic growth was high, both Argentina and Uruguay also underwent a series of political and economic crises, and while populations of both countries were able to earn high incomes, rates of income inequality were also high and rising between 1870 and 1913 (Álvarez and Bértola 2010; Bértola and Williamson 2003; Bértola et al. 2008; Pinilla and Rayes 2017).[12] The traditional pattern of land ownership since colonial times was largely consolidated and reinforced both *in jure* by the legal strengthening of private property rights to land, and then de facto by the fencing which facilitated the enclosure (Alonso 1981; de la Torre et al. 1971; Zum-Felde 1987). In the enclosure process, important amounts of public land (sometimes populated with native peoples) ended up in the hands of the big landlords; either sold by the state to secure public funds or simply appropriated by the ranchers. Vast herds of cattle, sheep, a few horses and even fewer men occupied pastoralist activities (Bulmer-Thomas 2003, 94; Scobie 1964, 114–115). As the meat sector was dominated by large-scale estates, and since the wealth in Argentina and Uruguay was largely agrarian-based the unequal land structure had important spillovers to the whole economy.

In Argentina, governmental policies between the 1870s and the 1930s were firmly based on liberal economic traditions, particularly considering private property rights and hostility toward state interventionism (Scobie 1964, 114). The interests of the ranching elite was, since 1866, well-organized; represented in the powerful *Argentine Rural Association* (SRA) as well as by many political and military leaders (Manzetti 1992; Scobie 1964, 115).[13] The SRA, the Catholic Church, the British investors, and the military formed a conservative power-bloc that ruled the country (Manzetti 1992, 592–594). Free trade liberalism reigned, but the necessity of the state to collect fiscal revenue was also considered, and trade was among the easiest things to tax.[14] While pastoralist interest dominated, Argentina still achieved a more diversified agricultural system than the rest of the region. Grain production came to complement the traditional livestock production as Europe started to demand more wheat (and more corn), and pay well for it.

The construction of railways from 1860s and onwards, inserted new territories in the commercial and trade activities and created conditions for grain trade. While the state started the construction of railroads, British capital eventually took over most of them (Bulmer-Thomas 2003, 143; Flichman 1990, 8).[15] In parallel with the expanding railway system, the federal government consolidated control over national territory inhabited up until then by indigenous populations and other local leaders and their loyal followers (Flichman 1990). The military confrontations with the *Pampean* Indians, under the so-called *conquest of the desert*, under the 1870s, defeated, killed, and displaced the indigenous peoples and communities, thus "freeing" new lands for agricultural exploitation (Bulmer-Thomas 2003, 92–95; Scobie 1964, 117–121). In 1878, wheat exports represented 0, 3% of Argentine export value, while in 1910, 25% of the export value came from wheat; now the country's major export product (Scobie 1964, 48–51). With the "wheat boom", the Argentine peso appreciated and costs of labor and land increased.

In contrast to the rest of the region, the introduction of new products did not eclipse the old, but led to export diversification—where wheat, maize, and other grains added to the export structure without replacing the traditional meat and wool sectors and all their derivatives (Bulmer-Thomas 2003, 60–61).[16] Moreover, while Argentine policies were not intentionally oriented toward industrialization, from the grain production emerged a rather advanced milling industry for grains, with both local and foreign millers (Scobie 1964, 102–111). Besides processing of meat and grain, beverages, wool and leather, tobacco, and glass, dominated the initial industrial boom. Later, in the 1920s, national manufacturing expanded into other sectors such as textiles, electrical appliances, and metallurgy (Flichman 1990, 4).

However, as Argentina became the breadbasket of the world, the vast majority of land never became the property of the immigrant labor force who worked it (Scobie 1964, 80–88). The land concentration was extreme, and the majority of the farm units were bigger than 1000 ha, by the turn of the century (Balsa 2004, 72–73). Instead, those actually tilling the land were smallholders (*chacareros*), or (more often) poor sharecroppers who worked the land (owned by big ranchers) on 3-year contracts for a percentage of the future harvest (Balsa 2004, 58–60, 85). As remarked by the Argentine history scholar James Scobie, no houses were built, no schools and no churches, no roads and no villages—only pastoralist interest reigned (1964, 50–53). The agricultural immigrant and the domestic small farmer found little support to improve their living conditions in Argentine land policies and legislation (Scobie 1964, 121).[17] However, compared to the rest of this highly unequal region, Argentina had some European settlement of family farms in the Pampas region, and public policies promoted settlement (Teubal 2008, 192). The situation of the tenant farmers and sharecroppers also improved slightly after staging a nationwide strike in 1912 and getting organized, forming the *Argentine Agrarian Federation* (FAA), which also marked a break with the hegemonic position of SRA in the rural sector (Manzetti 1992, 594). However, the majority of the crop farmers remained being very small property holders or poor sharecroppers, and Argentina did not anyway near develop settlement patterns with strong family farmers as did the US under the same period (Balsa 2004, 59–62). By contrast, the business structure of the grain trade was virtually monopolistic, controlled by four big firms—*Bunge y Born, Dreyfus, Weil Brothers* and *Huni y Wormser*—closely linked to the world market (Scobie 1964, 92–93).[18]

Counter intuitively, the agrarian-based society spurred urban, not rural, development. For every newcomer who tilled the soil, ten others made their living at urban pursuits stimulated by the income from the rural products (Scobie 1964, 54).[19] The "new" land became instead property of investors or speculators. The landowners, both new and traditional, made a lot of wealth on the rising land values, which on average increased with 218% throughout the cereal provinces from 1881 to 1911 (Scobie 1964, 50–52). In this way, capital accumulation originated more from land rents—rentier capitalism—than from the surplus extracted from labor (Flichman 1990).

These patterns of capital accumulation, immigration, and urbanization were even more extreme in Uruguay (Moraes 2008). Nearly all of the vast amounts of newly arriving European immigrants settled in Montevideo. Almost no land was available for settlers as it was already mostly in the hands of big ranchers (Álvarez et al. 2007; Álvarez and Bértola 2010). In addition, the pastoral countryside, characterized by extensive lands of natural pastures under *latifundio*, demanded very little labor. After fencing and the strengthening of private property rights (the establishment of rural police offices), the livestock sector demanded even less labor, as the role of safeguarding the animals became obsolete. In this way, the countryside was further depopulated and almost emptied on people (Barrán and Nahum 1981, 103; Bertino and Tajam 2000; Finch 1981, 88–90). In Montevideo, on the other hand, a substantial urban middle class and bureaucracy emerged.

In contrast to Argentina, a dynamic cultivation sector never emerged as a complement to the livestock system. Uruguayan crop farming was instead almost exclusively characterized by intensive and continuous tilling under *minifundio*, small plots of land close to the cities. Crop yields were poor, highly irregular and never competitive outside the small domestic market (Barrán and Nahum 1981; Bertino and Bucheli 2000). Continuous tilling damaged and degraded the soils and caused severe erosion (Bertino and Bucheli 2000, 8). The influential historians José-Pedro Barrán and Benjamin Nahum argued that the Uruguayan society constructed a "cattle-based civilization, which never betrayed itself and which remained faithful to the land and livestock and hostile to dirt farming" (Barrán and Nahum 1984, 656).

However, after the modernization impetus at the turn of the century (fencing, the introduction of new imported pedigreed bulls and rams, the introduction of sheep to graze with cattle and so forth) the livestock sector largely stagnated. The volume of exports was virtually unchanged between 1913 and 1926 (Astori 1984; Barrán and Nahum 1981; Finch

1981; Jacob et al. 1984). High international prices and falling transport costs, however, managed partly to hide this fact, making the sector appear more dynamic than it was (Bértola et al. 2008).

Part of the explanation for the low productivity rates per hectare in both livestock and cultivation systems was the land structure of *latifundio-minifundio*. It was rational for the big ranchers to adopt a model of low risks and costs, without much use of labor or technology, since the vast amount of land they possessed could compensate for their poor productivity performance per ha (Barrán and Nahum 1981, 103; Bertino and Tajam 2000; Bertino and Bucheli 2000; Finch 1981). The *minifundio* crop farmers, on the other hand, lacked capital to invest in technology that could improve their productivity as well as medium- and long-term returns (Astori 1984; Jacob et al. 1984). Other factors that may have contributed to the low productivity are high climate variability, erosive soils combined with inadequate soil and water conservation practices and strong livestock identity and risk aversion among the landowning elite (Barrán and Nahum 1981, 103; 1984, 663–664; Bertino and Bucheli 2000, 8; Ernst and Siri-Prieto 2011, 153; Finch 1981, 88–90). Another deterrent to a productivity increase in the livestock sector was the cartelization of the meat industry by the foreign-dominated meat-packing houses, which allowed them to set prices (Barrán and Nahum 1981; Bulmer-Thomas 2003, 173–174).

While foreigners controlled manufacturing and trading stages, the land was still in the hands of national ranchers. Since 1871, the powerful *Rural Association of Uruguay* (ARU) voiced ranchers' interests (Moraes 2008). This interest group was also well represented among late nineteenth century Uruguayan political leaders (within the two traditional parties; the conservatives—*blancos*—and the liberals—*colorados*). Yet, since 1903, it was an urban, rationalist, reformist, nationalist, and socially liberal flank of the Colorado Party—*Batllismo*—that came to dominate the Uruguayan political scene for decades. This stream of thought, named after the president José Batlle y Ordoñez—antagonized against the big ranchers, the British meatpacking plants, and railway operators and dependence on "London capital" (Barrán and Nahum 1984, 656; 1977).[20] The *Batllismo* ideal was to nationalize strategic companies, to subdivide the land and to make the country prosperous through "rational" scientific planning as opposed to policies based on particular interests (Garcé 2002, 26). The economic historian Henry Finch describes how Uruguay for the most part was governed by politicians and bureaucrats that did not necessarily represent the interests of the agrarian (livestock) activities, of which most of the economy depended, but articulated a proper urban rationalist public policy discourse (Finch 1981, 88–90).

While *Batllismo* talked about agrarian reform, adopted protectionist industrial policy (for example, the *Raw Materials Act*, 1912) and established public corporations to defend national interest against foreign capital,[21] resistance to these reformist initiatives to change was also strong (Barrán and Nahum 1981; Maubrigades 2009; Panizza 1990). Besides ARU, a new powerful producers' organization, *Federación Rural* (FR) was formed in 1915, in explicit protective response against *Batllismo*. ARU and FR, the Catholic Church and the British investors formed a conservative alliance, which resisted many of the politically proposed changes (Barrán and Nahum 1977).[22] Moreover, the state constantly lacked economic resources for reform implementation. It could for example not accumulate enough domestic capital to increase national independence from the foreign investors (Barrán and Nahum 1981, 1984). It did likewise not succeed in changing the concentrated agrarian structure. Instead, it entered a tacit contract with the ranchers, allowing the land structure to remain intact, while taxing "agrarian wealth" in order to finance the socially advanced welfare state (Finch 1981).[23]

*Batllismo* led, nevertheless, Uruguay to be among the first Latin American republics to leave the export-led growth model and strive for industrialization (Bulmer-Thomas 2003, 174–185; Finch 1981).[24] The "developmentalist" agenda of *Batllismo* managed to revert the trend of rising inequality since 1870 (with the landowning class acquiring the highest relative rise in income), and after 1910 inequality rates dropped until the 1960s (Álvarez and Bértola 2010). In the late 1920s, the state set up a meatpacking plant, the *Frigorífico Nacional* to break foreign dominance in the meatpacking industry (Panizza 1990, 153). In spite of its small domestic market, Uruguay managed by the end of the 1920s, to become the second most industrialized country in the region, after Argentina. The manufacturing sector accounted in 1929 for 20% of GDP in Argentina, and for 16% of GDP in Uruguay (Bulmer-Thomas 2003, 191). While Argentina and Uruguay became regional front-runners in the process of industrialization, their industrial sectors still suffered from the same constraints as in the rest of the region; influential landed elites and/or foreign investors constraining the public policy space, political instability, lack of capital, poor transport systems, labor shortage, shifts in international demand, and weak administrative systems (Bulmer-Thomas 2003, 153–189). Thus, the industrial sector never managed to become internationally competitive. Imports supplied most of the domestic demand for manufactured products. The units cost of production was high and domestic manufacturing firms relied heavily on imported intermediate and capital inputs throughout the period.

Paraguay's history is quite different from that of Argentina and Uruguay. In 1870, after the defeat in the *War of the Triple Alliance*, 1865–1870, against the joined forces of Brazil, Argentina, and Uruguay, the country stood at the verge of bankruptcy and faced the real threat of not being able to survive as an independent nation. Before the war (but after independence), a blend of authoritarianism, secularism, anti-elitism, nationalism, as well as self-sufficient (semi-autarchic and isolationist) economic policies in favor of infant industries had reigned (Bulmer-Thomas 2003, 44).[25] Distinct from most other Latin American countries, the state had expropriated land from the elite, the church and foreigners and leased it to the poor (Sacks 1988, 17–21). The state was the nation's largest landowner with 60% of the country's land by the mid-eighteenth hundred and the country had the largest herd in the region with around 3 million heads (Seyler 1988, 113–115). Hides, cotton, tobacco, and *yerba mate* were also important export products (Sacks 1988).

After the war, Paraguay stood with its physical infrastructure destroyed, 25% of its territory lost and its population decimated—almost half of the male population died in the war (Abente 1989; Bulmer-Thomas 2003, 48; Sacks 1988, 25–27). The Paraguayan economy was devastated, and the head of cattle reduced to only 100,000 heads (Seyler 1988, 125–127).[26] Both Brazil and Argentina increased their control and influence in the country. Brazil considered Paraguay as a buffer state (Abente 1989, 63). In order to pay substantial war debts incurred in London, the liberal governments that ruled Paraguay after the war sold large tracts of the state's vast holdings (fiscal land) to anyone that could pay for it within 12 months. The great majority of buyers were foreigners, mostly Argentinians (Abente 1989; Coordinadora de Derechos Humanos del Paraguay 2007). This "defeat" still lives in the memory of many Paraguayans:

> There is still a derogatory name for Argentineans in Guaraní—*kurepi*—which means pig skin. This is from the war, when the Argentine soldiers had pink skin boots. There is a lot of resentment with the Argentineans

since the 1878 auction in Buenos Aires, where Paraguayan fiscal land was sold. Paraguay, in order to repay the war, had to sell its land on auctions. (Senior Technical Advisor UNDP Paraguay, Asunción February 20, 2017)

Thus, in contrast to Argentina and Uruguay, foreigners typically owned Paraguay's large landholdings. Argentine, Brazilian, and British companies purchased some of Paraguay's best land and started the first large-scale production of agricultural goods for export.[27] Landlocked Paraguay also depended heavily on Argentina for access to the Atlantic Ocean through the Parana-Plata River. Besides the economic condition that forced Paraguay to sell out land, the liberal government in the country completely adhered to *laissez-faire* ideology and believed that unleashed market forces would perform miracles. Because of this belief, a profound and lasting process of privatization was launched in the 1880s (Abente 1989, 83). In this process of denationalization, the state virtually ceased to play any role in the process of capital accumulation.

The outward-oriented policies and foreign engagement in agrarian production and trade led to a gradual increase and diversification of export products. By the turn of the century, exports of tobacco, *yerba maté* (tea), hides, timber, lumber, and the bark of the *quebracho* (breakaxe) hardwood, from which tannin was extracted, and meat, had become important export products (Bulmer-Thomas 2003, 66–75; Sacks 1988). The war had nearly wiped out cotton production, but the state encouraged its renewed growth and in the 1920s it reemerged as export crop, mainly cultivated by small farmers (Seyler 1988, 116–118). However, a sizable proportion of cultivations was controlled by Argentine, British or Brazilian interests. Moreover, Argentina successfully restricted Paraguayan exports largely to "raw" products. For instance, the processing of *yerba maté* was in Argentine mills, and the processing of products from the forest sector was also made in Argentina, with for example Paraguayan *quebracho* extract shipped to Argentina to be reexported as Argentine tannin (Abente 1989, 67–69). Thus, Argentina succeeded in maintain Paraguay as commodity provider for processing in Argentina. Thus, not only was the actual production controlled by foreign actors, but Paraguay was hindered to develop into anything else than the rawest commodity provider.

While many landowners were foreigners, there was also a domestic landed elite of ranchers, of which many belonged to the ruling Colorado Party. Many national landowners were also members of the powerful producer organization from 1885, later called the *Rural Association of Paraguay* (ARP).[28] In a similar way as the earlier formation of the 1866 SRA in Argentina, and the 1871 ARU in Uruguay, the ARP represented the ranchers' interests with a lot of say in national policies. The ARP was also important in pushing for modernization of the livestock sector and export-orientation. It contributed to the rapid advancement of fencing, of the introduction of new European races, and developed systems for systematic breeding, and fencing. The stock of cattle recovered quickly after the war, partly due to imports from Argentina and Brazil). By 1900, the stock of cattle was two million heads—which was still one million less than fifty years earlier (Nickson 2015, 383). The same year, seventy-nine persons owned half of the land in the country (Sacks 1988, 31–32). The *latifundio-minifundio* structure was extreme, and the gap between small peasants, *campesinos* and large landholders widened throughout the period (Abente 1989, 63–64). The 22 largest cattle ranchers in 1911 owned together 500,000 head of cattle, representing around 20% of the total stock of cattle (Nickson 2015, 383–384). As in the case of Argentina and Uruguay, the landed elite benefitted from the export bonanza. However, gradually commerce and trading—dominated by Argentine and British capital groups—displaced ranching as the leading source of capital accumulation (Abente 1989, 75–76). The domestic elite was always weak, only as junior partners to the powerful foreign firms that controlled the exporting sectors of the economy, from the actual production, to the infrastructure and trade (Abente 1989, 65; Sacks 1988, 32–33).

The rural populace of *campesinos*, however, was little affected by shifts among the elites. They continued to subsist as it had done for centuries, eking out a meager existence under difficult conditions. Landlessness was mitigated, nevertheless, by the unexploited forest lands of both the eastern region (the Atlantic forest) and the western region (the Chaco). Estate owners typically used only a portion of their holdings, and peasants could squat on the properties without retribution. In addition, there were vast tracts of untitled land that sometimes were simply taken into use. In this way, there was no push from the countryside to the cities (as in Argentina and Uruguay) and the Paraguayan population remained mainly rural. While Paraguayan exports were typically raw and unprocessed, meatpacking industries started to be installed in the 1910s. The development of the meat processing plants was nevertheless weak and shaky, both due to restrictions from Argentina, put up in order to curb competition, and due to the deteriorating prices and reduced demand that emerged after the outbreak of First World War (Abente 1989, 70).

As mentioned, the Paraguayan governments around the turn of the century embraced *laissez-faire*. But as market forces alone did not create the growth expected, the state started to become more interventionist. However, the strong agro-exporting interest groups did not accept regulations that they found harmful, such as high export taxes (Abente 1989, 69).[29] The indebted, weakened, and politically instable postwar state had no capacity to mediate and balance between different interests.[30] Moreover, it lacked resources to control and manage its territory. Accordingly, the Paraguayan countryside was essentially outside state-control, which permitted a handful of landowners to exercise almost feudal control over the territory. Smuggling and contraband activities flourished in all border regions.

### 2.1.4   The Export-Led Model Erodes and Collapses—Reorientation Phase

After the First World War in 1914, the first *international food regime* started to show increasingly severe limitations. The demand for cheap food for the urban workers in Europe had caused downward pressure on international food prices that provoked a crisis in European agriculture. This in turn provoked a protectionist reaction. First, Continental Europe raised protection for domestic agriculture considerably and then, Britain adopted policies of imperial preference, although the meat trade remained exempted for still two more decades (Bulmer-Thomas 2003, 154–173). The suspension of the gold standard, which had acted as the foundation of the world wheat price, in 1914, resulted in falling international trade, while it also created long-time currency instability in Latin America.[31] Moreover, Latin America, and particularly the Southern Cone, suffered the end of inflows of foreign direct investment from Europe, and the general retraction of trade with Europe.

While Europe's role in the world economy was declining, the US emerged as more important for Latin American exports. However, the export performance of Argentina, Uruguay, and Paraguay was severely constrained by the US protective tariffs on their most important products: wool, hides, meat, and grains (Bulmer-Thomas 2003, 160).[32] World market prices for temperate climate commodities started to fall (Gerchunoff and Machinea 2015, 105). Paraguay, however, was still in the path of postwar recovery and export figures were rising throughout the 1920s. In addition, two companies expanded quickly in meat-packaging industry, the British company the *Liebig's Extract Meat company* (as in Fray Bentos in Uruguay), and the *International Products Corporation* (IPC), from the US (Nickson 2015, 384). The meat exporting business experienced a boost for a decade, going from an almost nonexistent export item, to account for 37% of total Paraguayan exports by 1932. For the first time, Paraguay had a modern processing industry (Nickson 2015, 384). However, soon two great pests broke out: the Great Depression and the even worse Chaco War, which made the Paraguayan export bonanza short-lived.

The collapse of US stock market prices on October 29, 1929, and the following Great Depression, shocked the world. The balance of interests that had sustained the first international food regime had now completely collapsed. Prices were free falling and international capital flows were almost nil. The regulatory responses within the advanced economies to the crisis were immediate. The US raised tariffs, and soon other countries retaliated. At the British Empire Conference (Ottawa Conference), in 1932, the British decided to give preference to wool, meat, and grains from the Commonwealth countries (Gerchunoff and Machinea 2015, 115–117). A wave of protectionism, currency devaluations, and a growing trend toward bilateralism swept the world. Costs were not divided equally among nations as the resulting tariff hikes more heavily affected foodstuffs than manufactured goods. Moreover, prices on temperate climate export products fell disproportionately, at least for the countries outside of the Commonwealth (Flichman 1990; Gerchunoff and Machinea 2015, 105; Pérez Arrarte 1982). The fall of trade also led to a dramatic drop in fiscal revenue in Latin America, since the most important tax income came from trade tariffs (Bulmer-Thomas 2003, 199). While trade and access to new foreign capital fell dramatically, the interest rate on public and private debts did not.

In macroeconomic terms, Argentine leaders during the conservative reorientation under the so-called "infamous decade" (1930–1943) prioritized fiscal discipline and full debt repayments over domestic economic protection and support (Gerchunoff and Machinea 2015, 108). While most countries in the region defaulted on external debt, Argentina, serviced internal and external (mostly British) debt in full, hoping it would in return make the British end imperial preference and ensure less discriminatory treatment of Argentine meat.[33] In 1932, Argentina was still the world's leading food exporter, supplying 36% of world agricultural exports, and was the richest country in the region, but the value of exports was significantly reduced. Between 1929 and 1932, the prices of its exports fell by 60%, the terms of trade by 35%, and GDP by 25% (Gerchunoff and Machinea 2015, 105). In 1933, Argentine and British authorities signed a three-year bilateral trade pact, the *Roca-Runciman Agreement*, later extended three more years, which yielded lower tariffs on Argentine meat, and later also cereals (Gerchunoff and Machinea 2015, 115). SRA formed an active part in the negotiations with the British authorities. While the meat interest was prioritized, the overall price Argentina had to pay for this market access was very high (Basualdo 2016).[34] The British were assured several long-going commercial benefits, but the Argentine leaders hoped that continued good relations would eventually pay off.

The state also took a series of initiatives in defense of the national grain and livestock sectors. In 1933, the *Junta Nacional de Granos* and the *Junta Nacional de Carnes* were created to set prices, quality standards and trade conditions for meat and grains (León and Rossi 2003).[35] A national board had for long been advocated by the FAA in order to defend domestic grain producers from the speculative operations of foreign buyers. SRA was given a prioritized role in the meat board, making the government finance the gap between domestic and international prices when the latter were lower than the former through minimum prices (Manzetti 1992, 596). The government also launched the *Comisión Nacional de Granos y Elevadores*, which expropriated private silos and elevators and built new ones in order to minimize storage and shipment costs (Manzetti 1992).[36] Thus, after the crisis, relations between the state and the landed interests started to change. The state started to increasingly respond to the wider interests of society compared to preceding periods (Flichman 1990, 5). The government initiated a number of countercyclical measures to mitigate the effects of worldwide recessionary periods. Several of these measures strengthened subsidies to rural production. In 1940, the Argentine government launched the *National Agrarian Council* (CAN) to allow the state to take a more proactive role in balancing the relations of force between different groups involved in the agrarian sectors and protect national interest. These steps indicate a new policy orientation that would bloom under the second food regime.

In Uruguay, a relatively conservative branch of the Colorado Party came to presidential power in the political vacuum created by the death in 1929 of José Batlle y Ordoñez, only a couple of weeks before the crash in New York (Benvenuto 1969). This conservative branch represented mainly the interest of the ranchers and the investors by postponing social legislation, lowering export tariffs on meat, and halting previous government attempts to nationalize British enterprises.[37] However, while the discourse was in favor of free-market principles, the state continued to play a direct and intervening role in the economy, not least by expanding the number of public employees and encouraging industrial development (Finch 1981; Panizza 1990, 153).[38]

The Depression hit Latin America hard, but Paraguay had even bigger problems to deal with. Since the 1920s when oil had been discovered in the Chaco area by the US company Standard Oil, Paraguay saw the snowballing threat of invasion from Bolivia (Cote 2013; Hughes 2005). Accordingly, Paraguay had been preparing for war by using all its money (and the money it did not have; debts) buying arms from Europe and the US (Hughes 2005).[39] The *Chaco war* (1932–1935) was extremely costly in both economic terms and human lives on both sides.[40] Bolivia was expected to win the war, due to bigger troops and arms that were more advanced. It had the initial successes, but Paraguay stopped Bolivia's progress and began pushing the troops back in late 1934 (Cote 2013). Paraguay won the war and for the last time settled the disputed limits of the Chaco region within the boundaries of Paraguay (Hughes 2005). However, the nation had yet again lost an important part of its male population, of which many had thirsted to death in the arid Chaco. Further, huge government war expenditures were largely financed by printing money, which created inflation (Bulmer-Thomas 2003, 249). A brief move toward social and agrarian reform emerged after the war, but was effectively curbed and replaced by a succession of authoritarian, albeit unstable, military regimes, all representing the Colorado Party (Sacks 1988, 35–38).

In short, the Great Depression resulted in a pronounced disbelief in the self-regulating market in Latin America and in the world. The main tenants of *the first food regime*—Britain at the center of the world economy, the gold standard, and liberal trade policies—were no longer prevailing. Hurt by the crises and protectionist responses from the US and Europe, Latin America concluded that dependence on external markets should be reduced and that true development and independence required industrialization. Most Latin American countries during the 1930s, gradually increased protective tariffs, which combined with public investments, became an important recovery mechanism between 1931 and 1939 for the bigger countries in the region. After this period of reorganization, a second international food regime emerged after the Second World War, which in most countries of Latin America meant a full-blown model of inward-looking development strategies.

## 2.2    THE SECOND INTERNATIONAL FOOD REGIME

The second international food regime emerged with the end of the Second World War and fell apart with the crack of the Bretton-Woods system. This section is organized in the same way as the past section. First comes a brief presentation of the regime's main tenants, trends, and rules (Sect. 2.2.1), followed by a short introduction to the way Latin America was situated within the regime (Sect. 2.2.2). Then, a chronological presentation of agrarian change and national regulations in Argentina, Paraguay, and Uruguay (Sect. 2.2.3). The section ends with a presentation of how the erosion and fall of the second food regime was articulated in the region—as import substitution industrialization (ISI) was abandoned in Argentina and Uruguay, while Paraguay was under "stable" authoritarian Stroessner rule—and what happened during the period of reorganization with debt crisis and first steps taken toward neoliberalization (Sect. 2.2.4).

### 2.2.1    *Main Tenants, Trends, and Rules*

The belief in free markets as capable of fostering long-term growth and development on their own without proper guidance from the state was dead and buried after the Great Depression. Agriculture and food in particular were seen too vital for human well-being to be exclusively in the hands of whimsical market forces. After World War II, the advanced countries decided to take back control and renationalize the food systems in order to secure domestic food supplies for food security and in order to protect domestic farmers.[41] Agriculture was accordingly exempted in the terms of reference of the *General Agreement on Tariffs and Trade* (GATT) rounds and did not form part in the general trade liberalization agenda of the advanced economies (Friedmann 1982, 2009; Friedmann and McMichael 1989).[42] From 1947 until the fall of the Bretton Woods system in 1973, a rather stable system centered in strong state intervention and industrialization of agriculture for increased food supply is referred to as *the second food regime*, or mercantile industrial food regime (Friedmann and McMichael 1989).

The US, now the largest economic power, had immediately after the *Depression* adopted a policy orientation built on public promotion and diffusion of new agrarian technologies for "modernization" of agriculture, price stabilization schemes, and higher restrictions on agricultural imports for protection of domestic farmers (Chang 2006, 2009, 504; Flichman 1990, 6).[43] In Europe, with the memory of war food scarcity fresh in mind, boosting national food supplies for food security was high priority.[44] Both the farm support programs of the US and Europe mainly benefitted highly specialized and mechanized, large-scale producers (Morgan et al. 2006). Thus, the protectionist and "productivist" agricultural policies in the advanced emphasized increased supply over all other features, such as biodiversity, small-scale models, enhanced micro-nutritional qualities, variety of food or taste (Lang and Heasman 2015). In this way, supply rose rapidly in the Global North, starting to produce important amounts of surplus food, which created downward pressure on international food prices (Friedmann and McMichael 1989; Margulis 2014, 2; McMichael 2013). The surplus was rerouted to developing states through a combination of subsidies and food aid (Chang 2009; Clapp 2016, 27–34; Friedmann 1982).[45] The greatest beneficiaries of the food aid programs were the increasingly powerful grain traders (Howard 2016, 76–78; McKeon 2015, 15).

The model of industrial agriculture was diffused to the developing world through the so-called *green revolution*. The introduction and development of new high-yielding seed varieties to increase production in developing countries was at the center of this "revolution" (Thiesenhusen 1972, 8).[46] For the new seed varieties to de facto produce high yields, they had to be combined with agrochemicals (specific fertilizers, herbicides, pesticides, and insecticides), agrarian machines, and irrigation.[47] This technological package rendered significant productivity growth per ha, but this capital-intensive and labor-saving model of industrial farming, also caused a wide range of negative social-ecological externalities, not lest the expulsion of poor farmers that could not afford the required inputs (Birner and Resnick 2010; Patel 2013).[48] The aim to secure more food for a cheap price to the increasingly urban, nonfarming, world population was nevertheless largely fulfilled. At the same time, the long-run growth of demand for foodstuffs was slowing down.[49]

Agrochemical industries, seed companies, agro-machine companies, processing industries, readymade food companies, supermarkets and retail, advanced producer service' companies (insurance, quality control, storage), traders, and a range of middlemen grew stronger out of the second food regime (Clapp 2016; Patel 2013). In addition, agribusiness firms elaborated transnational linkages between national farm sectors and global supply chains. One prominent example was the transnational animal protein complex linking soy, grain, and animal feedlots. During this regime, the national model of agricultural development of the advanced economies was universalized, at the same time as a "new international division of labor" in agriculture began to form around transnational commodity complexes (Clapp 2016; Lang and Heasman 2015; McMichael 2009, 141). The next section details how this regime was articulated in Latin America.

### 2.2.2   Latin America—The Inward-Oriented Growth Model

The export-oriented development model, centered in specialization in line with comparative advantage, was highly discredited throughout the region after the Great Depression. Real development was now believed to require a deep transformation of the productive structure, away from a reliance on primary commodities (Kay 1989; Serrano 2016). The Latin American countries expanded their production of manufactures and increased the supply of non-tradeable goods and services, using interventionist tools such as public investments, nationalizations, industrial export promotion, and ISI with preferential exchange rates for manufacturing (Bulmer-Thomas 2003, 237). In addition, many countries created central banks to compensate for the absence of foreign loans and investments.

*The United Nations Economic Commission for Latin America and the Caribbean* (CEPAL) was highly influential throughout the region in spreading research and reports that justified a new policy orientation for industrialization. The seminal economist and executive director of CEPAL, Raúl Prebish, argued that besides the long-term declining terms of trade of primary products, specialization in primary products created little backward and forward linkages, exclusively benefitting a small landholding aristocracy with little or no spill-over into the rest of the economy.[50] He also argued that capitalism worked differently in the periphery than in the core countries; thus economic development in the periphery required other tools than those used by the core countries (Prebisch 1950). The goal was to foster national capitalist growth through structural change and reduce external dependency. This, and other CEPAL studies, helped to form a school of political economy often referred to as *desarrollismo*, or Latin American structuralism. This heterodox school of thought criticized the assumptions of classical trade theory and provided important intellectual underpinnings for inward-oriented development strategies, a process of growth "from within" (Bértola et al. 2008; Ocampo 2001). The majority of the countries in the region adopted— to diverging degrees—inward-oriented models with high import tariffs on manufactures, subsidies, and favorable credit lines for industry and export taxes on many agricultural products (Flichman 1990; Grugel and Riggirozzi 2007, 88; Kay 1999; Panizza 1990).

Due to the combination of increased agrarian protectionism in the Global North and domestic bias against agriculture, through export taxes, Latin America's share of international agricultural trade declined (Kay 1989; Serrano 2016). In general, Latin America's share of world exports fell steadily, as did imports, so the region became increasingly divorced from the international trading system (Bulmer-Thomas 2003, 269–271). The industrial sector managed to increase rapidly, and its share of GDP rose throughout the period 1930–1954. Latin America reached relatively high average annual growth rates (5.5%) after World War II (Paolera et al. 2018). In addition, between 1920 and 1970, Latin America had a general drop in income inequality, even though the rates remained high in a global perspective (Álvarez and Bértola 2010, 53).

While industrialization came to the fore, Latin American governments also aimed for agricultural modernization. Public policies often supported the diffusion of green revolution technologies; both directly through for example tax deductions and credits for fertilizers and other agricultural input, and indirectly through support to research, extension, and transport. The use of fertilizers, tractors, and high-yielding varieties of mainly wheat and corn rose (Jones 1977), but the green revolution technologies also exacerbated inequality by displacing rural labor with machines and chemicals, as well as out-competing small farmers that could not afford the technology (Jones 1977; Thiesenhusen 1972, 1987).[51] Thus, agriculture absorbed fewer workers, but the urban industries were not strong enough to absorb the surplus labor and the social consequences became grave (Otero 2008, 19). Moreover, the specialization into fewer cash crops led to more monocultures, which reduced biodiversity and increased vulnerability to climate variability and pest attacks (Otero 2008). As a whole, however, the adoption of green revolution technology and subsequent yield improvement in the region was modest compared to Asia (Jones 1977). The *latifundio-minifundio* land structure curbed diffusion, as the landed elite could live well from land rents and often seemed more concerned with stability and low risk than with high productivity, whereas the poor peasants could not access the technology. In this way, the unequal distribution of land curbed new technology adoption (Jones 1977).

The question of land concentration and agrarian reform was nevertheless on the agenda all over the region.[52] Most countries in the region instituted some kind of agrarian reform, ranging from radical land redistribution based on expropriation (for example Mexico and Peru) to reforms where the state either bought land for distribution or distributed mainly public land without affecting privately owned land (for example Paraguay, Brazil, Argentina, and Uruguay). Idle private land could be expropriated if not used properly emphasizing the so-called *social function of land*, beyond the interests of the landowners (Ankersen 2006; Kay 1999). In general, however, with the exception of Cuba, no agrarian reform had lasting results on the highly unequal land structure, in spite of an increasing social critique in society against land concentration and *latifundio* (Ankersen 2006; Kay 1999). Ultimately, while the problem of underdevelopment and poverty were at the fore, the main policies aimed to increase the productive capacity and consolidate economic expansion. Natural resources were in general perceived as abundant and necessary to exploit in order to develop. Nevertheless, in these decades all countries created legal and administrative structures for natural resource management (Martinez-Alier et al. 2016, 39).

### 2.2.3    A Closer Look at Argentina, Uruguay, and Paraguay

While inward-oriented models aiming for industrialization, as *desarrollismo* from CEPAL, swept the whole region, nowhere where these policies implemented so full-heartedly as in Argentina and Uruguay. The presidency of Juan Domingo Perón (1946–1955), later referred to as *Peronismo*, was the clearest political expression of *desarrollismo* in Argentina. In Uruguay, *neobatllismo* (1947–1963) represented a fullblown *desarrollismo*. Nevertheless, both Argentina and Uruguay largely maintained industrial and welfare policies long after the heydays of

*Peronismo* and *neobatllismo*, and up until the military coups of the 1970s (Flichman 1990, 7; Ksiazenicki and Fuentes 2013). Paraguay, however, followed a slightly different path, even though it also had a period of increased state intervention in economic markets.

Argentina and Uruguay had already for decades used trade barriers to smooth fluctuations of food prices on domestic consumers through export taxes and quotas, but the postwar governments intensified interventionist and economically nationalist policies significantly. Argentine *Peronismo* combined state-led industrialization (including ISI-policies) with welfare spending, nationalism, populism, corporatism, and authoritarianism (Grugel and Riggirozzi 2007, 88). Uruguayan *neobatllismo*, represented a similar *desarrollismo*-inspired industrial impetus to that of *Peronismo*, but less authoritative and slightly less corporative.[53] Both countries strived to increase the total factor productivity through strategic state intervention, but also to redistribute the wealth and increase the levels of general welfare. For Argentina, to cast aside the traditional hostility to strong state interference in the economy and close bonds to agrarian interest groups represented a break with its political history (Flichman 1990, 6). In Uruguay, on the other hand, state-driven progressive reformism had dominated political life for the most part of the twentieth century, *Batllismo*, and was now simply taken one-step further. Self-confidence was high in Uruguay at the beginning of the period. It was the only small republic in Latin America that had established some sort of modern manufacturing, and in spite of world turmoil, it emerged into the postwar period with large currency reserves, rising living standards, and had some of the most advanced social welfare legislation in the world (Panizza 1990, 152).

One core mechanism of the new political-economic model was redistribution of income from the agrarian to the urban sectors. Through high export tariffs on primary products, agrarian rent was transferred to improve the conditions for industry and increases in wages. In addition, raised import tariffs and restriction reduced volumes of imports and shielded domestic industrial production from foreign competition in both countries. Barriers to import not only served to promote industrialization, but also to ration the use of foreign exchange (Bulmer-Thomas 2003, 264–265).[54] In Uruguay, the government even banned imports of goods competing with domestic production (Panizza 1990, 153). Another strategic instrument to promote domestic industry in both Argentina and Uruguay was the use of a multiple exchange rate policy;

state marketing boards decided on different exchange rates for each commodity, with low rates for imports considered essential and high rates for luxury consumer goods.[55] Between the mid-1940s and the mid-1950s, Uruguay's industrial production (mainly the manufacturing of consumer goods) doubled in value, and the number of industrial workers more than doubled between 1930 and 1950 (Panizza 1990, 154). Industry overtook agriculture as the main productive component of GDP in both countries (Flichman 1990; Panizza 1990).

At the same time, agricultural trade retracted significantly, diminishing supply of foreign currency. Uruguay's share of world exports declined tenfold between 1946 and 1975.[56] In Argentina, the exports of grains, particularly maize, was in the 1950s reduced into a sixth of what they were in the 1940s (Blacha and Ivickas Magallan 2013). In general, the agro-exporting activities in Argentina suffered a very long period of stagnation(1932–1970) (Flichman 1990, 7–9; Gerchunoff and Machinea 2015). The countries managed, however, to compensate their steady decline in export volumes, with favorable net barter terms of trade and some new product. In addition, in Argentina, domestic consumption became an important as growth engine. It was nevertheless hard for Uruguay to compensate for weak export performance with a rise in import-competing activities, due to its small home market (Bulmer-Thomas 2003, 213–218).[57]

Both Argentina and Uruguay largely extended state ownership and nationalizations in an explicit attempt to reduce dependence on foreign powers. For example, they nationalized the railways, which completed Britain's withdrawal from the Argentine and Uruguayan economy (Flichman 1990, 9; Jordan 2005; Panizza 1990, 153).[58] Argentina launched a program to free the country from foreign debts (converting all external debt to internal securities) and foreign ownership, by using a significant amount of its accumulated international reserves (Bulmer-Thomas 2003, 261; Flichman 1990, 9).[59] Uruguay already had a long history of construction of public enterprises and this was further completed under the postwar *neobatllismo* (Panizza 1990, 153).

The *desarrollista* development model, so fiercely adopted in Argentina and Uruguay, emphasized income distribution as a means to increase the home market for industrial goods by widening purchase capacity (Bulmer-Thomas 2003, 317). Both the Argentine and Uruguayan states took on more proactive roles in negotiating and establishing arenas for collective bargaining, with the state taking the role of arbiter between labor and capital (Flichman 1990, 7). Perón made strong bonds with the labor movement, not least through state-sponsored unions (Grugel and Riggirozzi 2007). While the Uruguayan state also made pacts with labor, the Uruguayan union movement remained autonomous, in a way that their Argentine counterparts did not (Ksiazenicki and Fuentes 2013).[60] Real wages increased, but in relation to the advanced economies in the world, both salaries and GDP/c development were weak (Álvarez and Bértola 2010, 69). Inequality, nonetheless, in terms of Gini coefficient, dropped significantly between 1930 and 1950 (Álvarez and Bértola 2010, 62). However, rural workers did not participate in the state-led wage councils.[61] Rural workers often lacked, or had weak, unions, and were tied to their *patrons* through traditional ties of loyalty, so-called *patron – client* relations. Moreover, while the welfare model adopted in Argentina and Uruguay provided in theory all citizens with a wide range of free social services (for example education and health),[62] access to these services were constrained in rural areas, and quality inferior (Ksiazenicki and Fuentes 2013). Thus, there were important exceptions to the universality of these welfare models.[63]

In Argentina, Perón nevertheless took some important steps to reduce the power of both multinational exporters and landed elites. For example, the *Juntas*—the meat and grain boards from 1933—were reformed so that the power of the state increased, while the influence of the organized landowners in SRA and private processors was reduced, to their open dissatisfaction (Basualdo 2016; Manzetti 1992, 601). The *Juntas* had been designed to protect national producers from falling international prices by setting domestic meat and grain prices above the international level, but now they instead often set prices below the international market level in order to provide finance for ISI (Blacha and Ivickas Magallan 2013, 138). Perón also established, in 1946, the *Argentine Institute for the Promotion of Trade* (IAPI)—a state agricultural purchasing and export agent, breaking the former near monopoly position of the trader *Bunge y Born* in cereal and flour exports. IAPI financed welfare and industrialization programs by using the revenues generated from buying grains and meat cheaply from rural producers while exporting them at much higher prices (Manzetti 1992, 602).

In addition, land and export taxes were increased and new legislation to improve the conditions of tenants was passed, freezing the rents of land leases and extending tenants' contracts from five to eight years (Blacha and Ivickas Magallan 2013, 141–142). This pleased the tenants and sharecroppers of FAA, but outraged SRA and the new producers' organization the *Argentine Rural Confederations* (CRA) from 1942 (Manzetti 1992, 601). Working conditions and salaries of rural wage laborers were also improved through new regulation (Blacha and Ivickas Magallan 2013, 138–139; Gomez 2008, 19). At the same time, CAN was reformed in order to support agrarian reform and on some occasions landowners were forced to sell land to the state for redistribution (Bulmer-Thomas 2003, 219–222; Flichman 1990, 5–7; Fridman 2010; Grugel and Riggirozzi 2007). However, no real land reform took place. The powerful producers' organizations SRA and CRA underlined recurrently that any land reform would be unconstitutional and economically detrimental, causing a negative impact on output (Gomez 2008, 18; Manzetti 1992, 611), and Peronism never defied private property to land. Moreover, while many of the Peronist policies aimed to support smallholders and rural workers, the industrialization of agriculture and increased mechanization under this period worked in the opposite direction (Acosta Reveles 2008). Many small farmers and seasonal workers accordingly went out of business, or lost their jobs, and moved to the cities (Grugel and Riggirozzi 2007, 88).

Uruguay remained to be a "livestock" economy. The big ranchers still concentrated the lion's share of the sparsely populated land, in spite of crop agriculture and family farming being explicit agrarian policy priorities of *neobatllismo*. High self-sufficiency targets, with investments in public agrarian infrastructure (silos, roads, and research) and prices on for example wheat and linseed well above international market price, made crop farming expand into historically unprecedented figures and peak in the mid-1950s (Panizza 1990). The amount of direct state subsidies to agriculture, in the period 1940–1952, was actually higher than subsidies to the manufacturing industry. While agricultural output increased, the yields remained relatively low and the cereal production did not manage to become competitive outside its own (protected) market.[64] Moreover, in 1948, the state created the *National Institute for Agrarian Reform* (INC) as a tool for subdivision of land, productivity raise, and promoting rural well-being.[65] INC was, however, never provided with enough resources for any substantial change in land tenure structure. By the end of the *neobatllista* period, only 154,000 ha, or about 1% of the productive land had been actually redistributed (Panizza 1990).

The land structure remained concentrated and dominated by ranching activities. While livestock production was stagnated, it remained to be

the most important export sector. The big ranchers continued to practice a livestock model mainly based on extensive natural pastures with
low amounts of improvements, such as in-sown pastures, fertilizers, or
irrigation (Garcé 2002, 66–67). In addition, increased domestic consumption diminished the surplus available for export. The consequences
of this long-term stagnation, however, were hidden by the rise in international prices of beef and wool in the postwar period, particularly in the
favorable Korean War years, 1950–1953 (Panizza 1990). The value of
Uruguay's exports doubled between 1945 and 1950. Thus, in a similar way as Argentina, the steady decline in export volumes was con-
cealed behind favorable net barter terms of trade (Bulmer-Thomas
2003, 254). However, there were some important attempts to change
this pattern. The most important emerged within the state interventionist and "developmentalist" discourse of the Agrarian Commission of
the *Technical Secretariat of the Committee on Investment and Economic
Development* (CIDE) during the 1960s (Garcé 2002).[66] This was the
most ambitious research and planning effort in Uruguayan history.
A *National Commission for the Agronomic Investigation* (CONEAT)
within CIDE, worked for ten years with capacity building of specialists,
data collection and mapping of soils of each parcel of land throughout
the territory, using both aerial photography and field studies (Former
researcher at CONEAT, Punta del Este, March 11, 2017). Uruguay
had already experienced severe problems of erosion linked to tillage
farming, but had hitherto lacked deep site-specific knowledge about its
soils (Petraglia et al. 1982). The work of CONEAT gave a new empirically based understanding of the alarming rates of erosion, threatening
the natural resource dependent economy as a whole. The commission
also created an index, CONEAT (named after the office who created
it), which effectively through a number (100 is the average productive
capacity in the country) communicated the average productivity of the
soils within one parcel of land (based on the annual potential production of beef and wool). The work also resulted in a new tax law in 1967
(Law 13.695), designed to incentivize productivity increase and punish
mere speculative use of land. Other new regulations that followed the
line of recommendations of CIDE were the 1968 new laws for fertilizers (No. 13.663), seeds (No. 13.664), and forestation (No. 13.723).
The extensive work with mapping the soils and improving the knowledge about soil quality and erosion was also reflected in the 1968 law for
conservation of Soils and Water (No. 13.667), in the institutionalization

of the CONEAT system in 1974, which is still used today (Former researcher at CONEAT, Punta del Este, March 11, 2017). Lastly, it provided the research basis for the still valid *law of water and soil conservation* from 1981 (No. 15.239), which provides the state with the right to override the interest of landowners and force them to adopt land-use and management practices that prevent soil erosion and degradation (See Chapter 5 in this book). While CIDE, and particularly the team working with soils, had an important and long-lasting impact in Uruguay, the law proposals for land reform and distribution, as well as policies to support the interests of small farmers and crop producers, that also emerged from CIDE, did not receive majority in the chambers.

As often, Paraguay—this landlocked country of great political instability, weak public institutions and of strong, often foreign, agro-exporting interest groups—followed a slightly different path. The country was ruled by a number of authoritarian military dictators, first from the Liberal Party and after 1936 from the Colorado Party, often with fascist tendencies (del Carmen Quevedo 1997, 856). However, these governments were also inspired by the *desarrollismo* from CEPAL, at least partly. While Paraguay did not adopt the inward-oriented model in any way close to its neighbors, it did adopt some of its elements; *laissez-faire* was out and national development through an interventionist state was in. Already some years before the end of the Chaco War (1935), new exchange control legislation was instituted and exporters were required to negotiate 50% of their foreign currency earnings with the Exchange Office (Abente 1989, 86). After the war, the Exchange Office was converted into a national bank with greater resources through new exchange taxes, and by 1941 the bank was given complete monopoly over foreign exchange operations (Abente 1989, 86). This made the agro-exporting interest forced to consider the state as the owner of foreign currency. The nationalization of the foreign exchange operations was soon followed by the establishment of import licenses and foreign exchange quotas. A multiple exchange rate policy was adopted to differentiate between vital and unnecessary imports, in order to protect domestic production and hinder the negative balance of trade (Nickson 2015, 384).

Paraguay also took some measures to gain more national control over the agrarian processing sectors. For example, in 1943, the state established a meat packer *Corporación Paraguaya de Carnes*, to avoid domestic shortages and keep domestic meat prices down (Nickson 2015, 384). Paraguay lacked, however, strategic promotion of the domestic manufacturing and other import-competing sectors, and did not manage to develop a strong manufacturing sector (Bulmer-Thomas 2003, 264–265). The clearest results of inward-looking policies in Paraguay were a collapse of the foreign exchange reserves, supply-side bottlenecks, and inflationary pressures (Bulmer-Thomas 2003, 277). As Alfredo Stroessner seized power in 1954, and stayed in power for 35 years, inward-orientation was abandoned and policies in favor of the export interest were back (Brezzo 2003), as will be further explored in the coming section.

### 2.2.4   The Inward-Oriented Model Erodes and Breaks—Reorientation Phase

Argentina had the most diversified economic structure and the strongest and most sophisticated industrial base in Latin America by the early 1950s. Even tiny Uruguay managed to make important progress toward industrialization, with industrial production creating economic growth in the 1940s and up until it reached its peak in 1957. However, the industrialization process was heavily constrained by its small domestic market and dependence on imported capital goods and raw materials, creating constant pressures on the balance of payments (Panizza 1990, 158). Argentina and Uruguay never managed to change their productive structures away from their original productive specialization in agriculture, and transform themselves into dynamic industrial economies (Álvarez and Bértola 2010, 70). Inward-oriented manufacturing production could not compensate for the stagnation in the livestock sector, at the same time as government expenditures were rising (Bulmer-Thomas 2003, 223–250).[67] When a steep fall in prices of beef and wool came, following the end of the Korean War (1953), the stagnated livestock sector could no longer support the rest economy.

Particularly Uruguay was extremely reliant on its beef exports. The terms of trade in external trade, and trade balance, went from positive to very negative between 1954 and 1958 (Panizza 1990, 157). At the same time, the livestock producers began to apply their economic power to press for better prices and a change in the government's economic strategy (Panizza 1990). After the collapse of beef prices in 1956, Uruguay soon entered in severe economic stagnation; only inflation and speculation soared (Jordan 2005; Panizza 1990, 149).[68] Argentina also faced rising problems of trade deficit and over-expenditure, in spite of having a more diversified and less beef-dependent economy than Uruguay.

One of the specific development constraints in the Argentine case was in addition chronic political instability. It had no less than six military interventions in less than five decades (since the coup 1930). Beginning in 1955, all coups attempted to solve the country's political crisis by proscribing or controlling *Peronism* (Fridman 2010). The producers' organizations, such as the SRA and the CRA, were active participants of the civilian coalition that supported the military coups of 1955, 1962, 1966, and 1976 (Manzetti 1992, 605). In the Uruguayan 1959 elections, the *Blancos* won on a platform favoring free trade and reforms of the agrarian sector (Panizza 1990, 159–160). ISI was gradually abandoned during the 1960s.

An increasingly number of voices argued that the inward-oriented development strategies had failed (Kay 1989; Serrano 2016). CEPAL argued that the reason for the failure was the lack of economies of scale as the manufacturing sector depended exclusively on small domestic markets. Its policy recommendations required the development of intra-Latin American trade to sustain export-levels and overcome the constraints of small markets. Several initiatives were taken to enlarge domestic markets through regional integration, but regional integration never managed to take off and significantly reduce the high tariff and nontariff barriers between the Latin American countries (Bulmer-Thomas 2003, 299–307). The failure to develop integrated regional markets was specifically problematic for Uruguay, where the domestic market was unable to sustain industrial activities on its own.

Development was also understood as constrained by the inherited unequal land structure that led to underutilization of land, migration, and poverty, as well as and powerful pressure groups representing the landed interest and concentrating wealth (Kay 1989). Attempts for land reform emerged throughout the region, but no radical change was sustained (Kay 1999; Manzetti 1992; Piñeiro et al. 1991; Riella and Mascheroni 2011). In Argentina and Uruguay, the concentrated land structure showed to be one of the most stable features over history. Some scholars stressed that the root to Latin American underdevelopment was the high levels of inequality combined with its still dependent insertion into the world system. Their solution was thus to radically break free from the world integration and radical distributive reform (Kay 1989). The increasingly dominant critique against inward-orientation, however, came from neoclassical voices. They claimed that the bias against agriculture and exports had "distorted" market signals (particularly criticizing the price controls and the deteriorated terms of trade for farmers), which had damaged exports and caused inefficiencies (Bulmer-Thomas 2003, 280–289). Moreover, the government style of *neobatllismo* and *Peronismo*—with the state acting as a central arena of negotiations, planning, and bargains for reallocation of resources between different interests and groups—was criticized to further distort incentives, boost clientelism and widespread corruption. From the 1960s, Argentina left ISI-policies behind and instead combined proactive industrial protection with export promotion of manufactures. However, the Argentine entrance in 1967 to GATT, with rules that ran counter industrial subsidies, made Argentina abandon export promotion in favor of market-oriented export substitution (Bulmer-Thomas 2003, 327–329).

The regional critique of state interventionism and ISI echoed wider shifts in the international arena. The agrofood corporations were increasingly outgrowing the setting of national regulations in which they had emerged and pushed for liberalization (Friedmann and McMichael 1989). The states were also increasingly confident in the free market as the most efficient tool for agricultural development. A strong and proactive state was no longer considered necessary for food security (McMichael 2013). With the end of the fixed exchange rates, announced by President Nixon in 1971, the Bretton Woods system and the second international food regime finally fell apart. Policies shifted rapidly toward trade liberalization, financial deregulation, and strengthened intellectual private property rights.

Both Argentina and Uruguay were throughout the 1960s in an increasingly violent cycle of deep discontent, economic crises, social protests, guerillas, increased military influence, big strikes, social unrest, guerrillas, suspended civil liberties, military interventions, human rights' violations, states of emergency, repression, and state terrorism (Panizza 1990, 163–166).[69] The military finally staged *coups d'état* and seized power completely, in Uruguay in 1973, and in Argentina in 1976.[70] Trade liberalization, fight against inflation, financial deregulation, and cuts in expenditure to combat fiscal deficit were high on the agenda. The benefits of specialization in line with comparative advantage were yet again underscored. The military regimes in Argentina and Uruguay thus shifted (back) policies in favor of ranchers' interest, but the international meat and wool prices in the early 1970s started to drop again (Panizza 1990, 166–167). The pendulum had shifted (again) toward outward-orientation. In spite of a neoliberal discourse, however, public spending and foreign debt soared during the military regimes. There were important tensions between different economic government advisors, and policies were in the end rather mixed (Bulmer-Thomas 2003, 338–340; Fridman 2010, 276).[71] The import sector remained protected by high tariffs (Fridman 2010, 288–292; Panizza 1990, 167–168).[72] The search to diversify production into nontraditional exports (shoes, rice, citrus, berries, soybeans, to mention a few), however, was starting to see some positive results for soybean cultivations in the Argentine Pampas region. Between 1969 and 1980, the soybean production in Argentina increased from 0.6 to 3.6 mmt (Martínez Alvarez 2012, 13–16; Shurtleff and Aoyagi 2009, 88, 111, 167, 173–174, 175, 242, 318, 449).[73] Argentina had launched a national program promoting research and extension services for soybeans, and yields improved significantly due to successful seed adaption (Martínez Alvarez 2012, 17–23). Uruguay also launched research and promotion programs for the development of soybean production, but did not manage to increase yields in any substantial way (Martínez Alvarez 2012, 17–23). Instead, both crop area and amount of crop producers retracted significantly during the military government, since it removed subsidies and grain self-sufficiency targets. Accordingly, most cereal production became unprofitable. In this process, almost all cultivations became concentrated to the most fertile soils along the Uruguay River, called Littoral. At the same time, crops became integrated in mixed systems with livestock, the so-called *agrícola-ganadero* model, where crop rotations serve to "boost" the pastures (Achkar et al. 2011; Díaz Rossello and Rava 2007). Uruguay also aimed for export diversification and alternative use for marginal agricultural area through promotion of the forest sector, which also resulted unsuccessful (Morales Olmos and Siry 2009, 64).

The general economic performance in both countries was deficient. Inflation was not defeated, public expenditure remained high and the balance of payments negative (Bulmer-Thomas 2003, 341). The demands for capital coincided with a huge pool of international liquidity under the control of international banks, eager to lend money and virtually free from conditions from the late 1960s onward (Felix 1990).[74] The halt to lending came with Mexican default in August 1982. The debt crisis was unleashed. The military regimes did not manage to improve the economic development in Argentina and Uruguay. On the contrary, real wages declined, unemployment rose, poverty soared (Fridman 2010,

288–292; Panizza 1990, 168–172). The turn in discourse toward neo-liberalization of the dictatorships, however, became more thoroughly implemented after the debt crisis, and even more after the re-democratization, spurred by the stabilization and adjustment programs for debt service and the so-called Washington Consensus sweeping the region (Chapter 4 in this book).

Paraguay took a different path from that of Argentina and Uruguay. After a rather brief period experimenting with inward-looking policies, followed the long and authoritarian Stroessner regime (1954–1989), when Paraguay moved back to specialization in primary products and export-orientation. There was no income tax and public spending was the smallest percentage of GDP in Latin America (del Carmen Quevedo 1997). *Stroessnerismo*, represented an eclectic blend of policy orientations including paternalism or patronage politics, with authoritarianism,[75] anti-communism, nationalism, liberal trade policies, and social reform. Free trade was combined with price interventions in agricultural markets and clientelistic support programs (Turzi 2016, 51). Stroessner's Paraguay and Nixon's US is described to have entered a cold-war marriage of convenience, where the loyalty of Paraguay to the security policies in the hemisphere of the US was paid back in financial support and military aid to enhance the army's skills in counterinsurgency warfare (Mora and Cooney 2010; Sacks 1988, 42–44).[76] Moreover, the Brazilian military regime also supported Paraguay economically (Seyler 1988, 113–116).

The return to specialization in primary products at a time of high agricultural protection in the advanced economies, and when primary product trade was growing slower than world trade, however, rendered meager results in terms of export earnings. Official trade data is, nevertheless, an unreliable guide to export performance, since contraband trade—with the close involvement of the military under Stroessner—was very widespread. Paraguay had an extreme *latifundio-minifundio* structure. According to the 1956 agrarian census, the biggest 1% of farms covered 87% of the land, while the smallest 46% of farms covered only 1% of the farmland (Seyler 1988, 113–115). The same census showed that the cattle stock was 4.5 million heads and that around 35% of the total land area was pastures for cattle raising (Nickson 2015, 384). At the same time, only 1.5% of the producers owned over two-thirds of the national herd (Nickson 2015, 384). The majority of Paraguay's rural population lived in the area around Asunción, surrounded by large *latifundio*, owned by foreign investors and members of the small political elite (Zoomers and Kleinpenning 1990).

Organized discontent with the unequal land structure, however, had been widely absent among the rural population throughout the agrarian history. This changed with the emergence in the 1960s of the *ligas agrarias cristianas* (LAC), a Christian peasant-based social movement, with support from progressive bishops and priests within the catholic church. LAC broke the silence, questioned the poor conditions of the peasantry and called for radical agrarian reform through nonviolent means (Fogel 1986, 96–97). The Stroessner administration accused the church of supporting communism. Under the 1970s government harassment against the LAC increased. LAC was violently defeated, several Jesuit bishops and priests were expelled and the government killed peasant leaders, imprisoned *campesinos* and priests, slashed down entire communities, and made raids throughout the countryside (Espínola 2008; Fogel 1986; Zoomers and Kleinpenning 1990).

At the same time, the government started to give more attention to the rural poor and the development constraints imposed by the *latifundio-minifundio* structure. Agricultural colonization became a tool to reduce political unrest (Zoomers and Kleinpenning 1990, 139). One important tool in this sense was the 1963 *Agrarian Statute*, which limited the maximum size of a single landholding to 10,000 ha in the Atlantic forest region and 20,000 ha in the Chaco region, with landholdings in excess of this size subject to taxes or possible purchase. Moreover, the state received funding from the US led *Alliance of Progress* in order to develop agriculture, modernize, and redistribute (Fogel 1986, 48–49). Small farmers in the central region were given incentives and support to establish larger, more efficient farms, particularly in the unexploited forest regions (Seyler 1988, 116–118). The Agrarian Statute also promoted settlement of farmers in the forest areas of the eastern region, and gave economic incentives to landowners who cleared their forest-land and put it under "productive use". The Rural Welfare Institute (IBR) was created to issue land titles and provide the newly settled farmers with credit, markets and technical assistance (Plant and Animal Health Specialist IICA Paraguay, Asunción, February 22, 2017).[77] The idea was to improve the situation of the poor "*minifundistas*" by offering them to leave the crowded Asunción region and start a new life as legal owners of larger holdings in the Atlantic forest region (Zoomers and Kleinpenning 1990). From 1963 to the late 1980s, the IBR titled millions of hectares of land and created hundreds of colonies (settlements), directly affecting the circumstances of roughly one-quarter of the population (PNUD 2014, 164–165). Moreover, the government issued 15–20 ha land grants to military personnel upon completion of their service, if the land would be used for farming purposes (del Carmen Quevedo 1997). In this way, several thousands of colonists were resettled in the unexploited eastern region. Through this state-led horizontal expansion of the land frontier, incorporating "new" land, the government managed to reduce social tensions, without threatening the interests of the landlords. Unsurprisingly, the producers' organizations—representing the landed elite, ARP, and SNA—strongly favored these policies (Zoomers and Kleinpenning 1990). While the land titling programs of IBR explicitly aimed to redistribute land to the landless and small producers, most "new" land that was incorporated under agriculture ended up in the hands of medium and large producers (PNUD 2014, 165). IBR was thoroughly corrupt and is described to have turned into the primary office for developing rural patronage networks for the Colorado Party (Hetherington 2009).

Forests were considered unproductive—an untapped resource that should be transformed into uses that are more "productive"—and deforestation was seen as a sign of progress. Under the doctrine of the *social function of land*, it was established in the Agrarian Statute that unproductive *latifundio* land, such as forests, could be expropriated and distributed among small farmers, which further increased the incentives of forest landowners to clear the land. As late as in 2001, the definitions of "unproductive" land changed in the Agrarian Statute, so that forest conservation no longer was seen as "unproductive" (Piera Valdés 2016, 30). Many small farmers from Paraguay's central zone also moved spontaneously to the eastern border region, without any support of IBR, and simply grabbed some of the abundant, fertile, but forested land of the eastern departments. The *Forest Law* (422/73) from 1973 put some protection on the forest by stating that 25% of all land should remain under forest cover and by establishing fiscal incentives for reforestation. However, loopholes in the law still made it possible for landowners to be able to clear 75% of the remaining 25%, by transferring it to other owners (Chapters 4 and 5, in this book for further discussion about deforestation and forestry regulation in Paraguay).

While a large number of farmers settled in the areas of colonization, increasing the agricultural production, the majority of the *campesinos* did not receive definitive property titles and did not succeed in establishing modern commercial farms and raising their standard of living substantially. Instead, many farmers became a reserve of cheap labor to be used for the expanding large-scale farming, led by Brazilian farmers, and forest exploitation (Zoomers and Kleinpenning 1990, 142–143). While the government promoted settlement in the eastern regions and made some investments in highways and electricity, there was an almost complete lack of public services—schools, roads, health care—in the majority of the newly settled areas (Director of DCEA-MAG, Asunción, February 21, 2017).

The most significant land-use change in Paraguay was nevertheless not driven by peasant farmers, but by the soybean expansion and more capitalized farmers. During the 1960s, Stroessner launched a national plan for self-sufficiency in wheat. Within the realm of this plan, the soybean was suggested as an ideal supplementary summer crop (wheat being the winter one), and was thus promoted with financial incentives, technical support, and fiscal stimuli (OEA 2009). After soybean prices increased, however, much of the land slated for wheat was sown with soybeans instead (Fogel 1986, 84). The soybean expanded rapidly in the Atlantic forest region—eastern Paraguay, were Brazilian farmers rapidly became to dominate most of the production. Brazilian farmers arrived in Paraguay attracted by its lower land prices (Hopewell 2014).[78] In 1970, Paraguay produced 75,000 t of soybeans, in 1978 it produced 375,000 t in 1980 it produced 0.54 mmt and in 1985 it already produced 1.2 mmt (Shurtleff and Aoyagi 2009, 374–375, 449). As the lucrative nature of soybean cultivation and processing became apparent, several large agribusiness firms arrived, mainly from Brazil, but also from the US and other countries, engaging in large-scale, commercial production of soybeans and soybean meal/oil. Foreign agribusiness firms also dominated the economic activities derived from agriculture; industry, commerce, and services. Stroessner also supported the development of the cotton sector, which was Paraguay's number one export crop in the 1970s and early 1980s, and the sector in which most small farmers were engaged. The rising world prices on both cotton and soybeans from the early 1970s to the early 1980s, in combination with Paraguay's free-market economy, triggered rapid expansion of both these cash crops (Seyler 1988, 113–116).

The penetration of the dense Atlantic forest was not only driven by expansion of cash crops, but made possible through large investments in infrastructure, not least roads. Moreover, Brazil financed and constructed the largest hydroelectric dam in the world, the *Itaipú Dam*, between 1973 and 1982. Argentina built another huge hydroelectric dam (*Yacreta*) in the same region. All these activities brought even more capital, infrastructure and colonists to the area (Seyler 1988, 115–116). This created a lot of economic dynamism. During most of the 1970s, Paraguay's economy grew at the fastest pace of all the Latin American countries. Paraguay also benefitted from not having received many international loans and was thus less affected when the debt crisis came. Low export tariffs and low government spending was the main tune played by the Paraguayan authorities. As both public and private investments in infrastructure reduced transport costs, the economic rationale for further agrarian frontier expansion into the forest region increased.[79] Rapid deforestation of the Atlantic rain forest led to a dramatic loss of biodiversity and wildlife habitat, at the same time as many indigenous groups living in the forest were displaced (Hetherington 2009).

While crops were expanding in the eastern regions, the government also had livestock fund *fondo ganadero*, providing soft loans for productivity raising measures, such as artificial pastures and improved breeding (Nickson 2015). Meat products represented a share of 35% of total exports by 1972 (Nickson 2015, 384–385).[80] The government also provided fixed prices on meat to protect farmers from market fluctuations (Plant and Animal Health Specialist IICA Paraguay, Asunción, February 22, 2017). The state, with financial support from the World Bank and the Inter-American Development Bank, also launched a rather successful program with technical assistance to farmers to raise productivity and end the foot-and-mouth disease (Nickson 2015, 384–386). While these reforms raised productivity, it mainly benefitted the big ranchers and the handful of foreign meat packers. At the same time, protectionist agricultural policies in the advanced economies yielded rather low prices on the meat exports. The cattle herd had risen to 8 million heads by 1989, and pastureland area was 20 Mha, almost half the productive land (Nickson 2015, 384–385). The government aimed for cattle production to expand into the unexploited Chaco forest region, but did not succeed to get any major influx of people into the remote Chaco region, almost completely devoid of public services (health, schools, electricity, roads). The region remained instead for a long time (until the recent "boom") to be covered by immense intact forests. While sparsely populated, it was still the home of 19 different native groups, as well as scattered Mennonite colonies engaged in beef and dairy production.[81]

The great majority of peasant farmers, however, continued to use traditional farming methods and live on small parcels of land at a subsistence level. Despite the abundance of land, the distribution of the country's farmlands remained highly skewed, favoring large farms (Seyler 1988, 113–115; Zoomers and Kleinpenning 1990). According to the 1981 agricultural census, 1% of the nation's more than 273,000 farms covered 79% of the nation's farmland in use. By contrast, the smallest farms, which made up 35% of all farms, covered only 1% of the land, making the average size of a *minifundio* 1.7 ha, or less than was necessary for one family's subsistence (Seyler 1988, 113–115). In addition, more than half of the small farmers did not have formal land titles to the plots they held (Zoomers and Kleinpenning 1990, 144). Stroessner managed, however, to form a rather strong bloc with the peasant organizations (after his dismantlement of the previously mentioned LAC). Probably some of the leaders of the organizations truly believed that eventually the reforms of IBR would bring about a real land reform and bring prosperity to the poor rural population, while others remained loyal because of personal benefits and corruption (Hetherington 2009).

## 2.3    Conclusion

This chapter has presented a short historical review over how the post-independent Latin American states in shifting ways, handled the multiple and evolving opportunities, constraints, and internal tensions that emerged during the period 1870 to the 1970s in relation to agrarian and national development. The focus has been on the shifting positions and roles taken by Argentina, Paraguay, and Uruguay, but these have also been situated in the broader regional context, as well as in the wider world capitalist agrofood relations, as conceptualized by the Food Regimes approach.

The chapter has shown both long lines of continuity and abrupt game-changing shifts. One of the central continuities is the tension between export-oriented interests and other interests, as well as the tension between large landholders and peasant farmers, established already under colonial rule. While the exact relations of force fluctuated, it is clear that Latin America early became, and throughout the period remained, the most unequal region in the world (Bértola et al. 2008). Distribution of land was even more unequal than the distribution of income (Bulmer-Thomas 2003, 313). The early agrarian export bonanza reinforced inequality as it increased the wealth of landowners.

In all three countries, powerful landed livestock producers' organizations emerged in the late nineteenth century, which still today constitute important voices for the ranchers' interests. The crop farming model developed in the Southern Cone was not in any way similar to the patterns of settlement developed in the North American Corn Belt. The *farmer's way* in the US was a model of medium-size property holders, based on mainly family labor (Balsa 2004). They invested in the land, and local communities with churches, schools, commerce, and light manufacture emerged and flourished around them. This did not happen in the underdeveloped countryside of the *Río de la Plata*. Domestic landowning ranchers controlled the actual production stage in Argentina and Uruguay. The concentrated land pattern resulted to be extremely resilient, as Argentina and Uruguay did never have any land reform. Small Uruguay, where the land frontier was exhausted already 1830, makes an extreme case of continuity. The land structure observed in the agrarian census of 1908, in which around 65% of all productive land was controlled by 10% of the farming units, showed to remain more or less the same throughout the periods here considered until the 1980s (Fernández 2007; Pérez Arrarte 1984a, 81–83; 1984b). Paraguay, however, experienced a path of more "dependent development", characterized by a high degree of enclave commodity exports completely in the hands of foreign groups and interests. The war-ridden Paraguayan state was badly equipped to build up an independent state as it started out the period by selling off huge tracts of land to both big foreign and national property owners in order to repay war debts after the war of the triple alliance. By contrast, Argentina and Uruguay managed the consolidation of a relatively strong and autonomous state apparatus. Nonetheless, the power of the states was always, in some way or another, conditioned by a combination of landed and foreign interest. While land was mainly concentrated in national hands allowing for capital accumulation, foreign capital groups typically controlled export and processing stages of the livestock chain also in Argentina and Uruguay. The role of "London capital" was early problematized in the *Río de la Plata*, but the ruling politicians in Argentina were in general loyal with the British interests, whereas the Uruguayan rulers lacked the resources to fulfill their plans for enhanced national independency.

One of the most vibrant discussions throughout the region and the period, and still going on today, was about whether high reliance on natural resources is a viable and desirable strategy for long-term economic development. In most Latin American countries, the early trade integration based on commodity specialization created short cycles of booms and busts, but poor long-term growth performance. Argentina and Uruguay were among the selected few countries in the region where agrarian-based exports under the first food regime actually managed sustained high growth rates and general rise in well-being. However, the vulnerability inherent to the insertion into the world market as a highly specialized commodity provider became evident, as the international meat and grain markets started falling. Instead, the idea that real development required industrialization and inward-oriented growth started to spread, and became dominant throughout the region after the Great Depression. Argentina and Uruguay in particular, full-heartedly adopted this new development orientation of *desarrollismo*. Under the inward-looking model, the states aimed to reduce both agrarian and foreign dependence through national industrialization. While the manufacture sector grew and inequality dropped slightly, they did not manage to change their productive structures away from their original productive specialization in agriculture and transform themselves into dynamic industrial economies. Capital constraints (economic and human capital) and powerful domestic groups in alliance with foreign interests, however, made the region remain highly dependent on agrarian commodities, foreign markets, finance, traders and technology. The concentrated land structure also remained largely intact. When international beef prices fell after 1954, both Argentina and Uruguay entered in recession. The negative balance of payments, inflation and high public expenditures resulted in a deep economic crisis.

Paraguay did not follow the path of Argentina and Uruguay during the second food regime. It went from another devastating war (Chaco), over to a short period of inward-oriented interventionism, and into a fifty year long authoritarian, but mainly "modernization" and free trade oriented rule. The extreme levels of foreign control that Paraguay had suffered under the first period were nevertheless partially reduced under the strong and authoritative hand of Stroessner, although his hand would have been less strong if it was not from the support from the US and later from the military regime in Brazil. In spite of a dramatic expansion of the land frontier into the Atlantic forest, and in spite of many decades of "agrarian reform", Paraguay remained as a country of extreme inequality and a large amount of poor rural peasants.

At the international level, the state-centered second food regime finally fell apart together with the Bretton Woods system in 1973, and pressures toward agrofood liberalization and deregulation became increasingly influential. Argentina and Uruguay fell under authoritarian right-wing military rule in the 1970s, which expressed a renewed faith in export-oriented growth in line with "comparative advantage". While the economic discourse of the military dictatorships in both Argentina and Uruguay were celebrating macroeconomic stability, privatization, and liberalization, it was not until after re-democratization, and particularly under the 1990s, that the *Washington Consensus* recipes became more firmly adopted (Grugel and Riggirozzi 2007, 89). The general push in the region toward neoliberalism was also somewhat exempted under the long authoritarian rule of Stroessner in Paraguay, which lasted until 1989. The rules and power relations of this recent history, which Philip McMichael refers to as the third (corporate) food regime, will be thoroughly presented in the remaining chapters of the book. The historical backdrop presented in this chapter will be vital for the thorough understanding and analysis of today's agrarian change in Argentina, Paraguay, and Uruguay.

## NOTES

1. Friedmann, acknowledges the increased power of corporations, but she argues that this period involves too many challengers, resulting in too much contestation and contradictory trends, to qualify as a proper food regime (Friedmann 2016). According to Friedmann, the contemporary moment of agrofood history is still a period of reorganization, and the eventual emergence of a third regime is yet to be seen. She thus agrees with McMichael in the analysis of increasingly powerful corporations relative nation-states since the 1980s, but still finds that the "regime" require more stability, legitimacy, and acceptance than what is currently to be observed in the food system (Friedmann 2016).
2. Commodity exchanges, such as the Chicago Board of Trade (CBoT) and London Corn Trade Association (LCTA) began in the late nineteenth century to develop tools to measure and quality grade agro-commodities, which reduced uncertainties and spurred international trade.
3. Food regime research has notably focused on how wheat emerged as the first international price setting market for a basic food item.

4. The focus here is on Latin America, but clearly agricultural exports from the region depressed European agricultural prices, which, at least in Britain, stimulated the process of capitalist accumulation and urbanization to the detriment of the old landholding interests.

5. Merchant capitalism refers to the earliest phase in the development of capitalism as an economic and social system, which Latin America was inserted to, and formed a constituting part of, since colonization (post-1492).

6. As a whole, Latin America provided the world by the turn of the century, with coffee, cacao, bananas, sugar, rubber, copper, tin, nitrate, gold, and silver, while the temperate regions exported mainly cattle and its by-products, as well as wool (Bulmer-Thomas 2003, 59–66).

7. There are several examples from Latin American history of natural resource export booms that lacked social-ecological sustainability. For example, the guano boom in Peru led to its depletion and the expansion of grazing and agricultural crops as well as monocultures of trees have fragmented and destroyed many forests throughout the region (Martinez-Alier et al. 2016).

8. In 1910, GDP/c in Argentina was USD 3822, and in Uruguay USD 3136. This can be compared to the GDP/c in some European countries in 1910: USD 1895 in Spain, USD 3348 in Germany, USD 4611 in the UK and USD 2965 in France (Bolt et al. 2014)

9. Argentina also protected processed products through a differentiated tariff system and manufactures were spurred by the big domestic market (Bulmer-Thomas 2003, 141–142). In addition, many of the Argentine landed ranches diversified their activities by investing in agro-industrial activities (Manzetti 1992, 591).

10. One of the largest industrial complexes in the region, with more than 5000 employees, was the meat packer *Liebig's Extract of Meat Company*. Established in the port city *Fray Bentos* in Uruguay in 1873 it produced mainly corned beef for the English working class.

11. Argentina alone accounted for almost one half of British Foreign Direct Investment for the whole of Latin America in 1913 (Jones 2000, 62). From the turn of the century, France, Germany, and the USA also became increasingly involved in business and trading activities, including the establishment of meat packaging industries, but Britain remained the most important foreign direct investor, trading partner, and capital provider throughout the period (Jones 2000, 257).

12. Between 1870 and 1913, the total factor productivity of the Uruguayan livestock sector grew annually by 2% (Moraes 2008, 85)

13. The SRA was dominated by the ranchers considering themselves the true national elite because of the pivotal role that beef and grain exports

played in the economic development of the country (Manzetti 1992, 590). In practice, the members of SRA constituted the backbone of the National Autonomist Party (PAN), which dominated Argentine political life until 1912 (Manzetti 1992, 592).

14. Proposals to tax income and land were discussed many times, but in the end almost all revenue came from trade taxes (Bulmer-Thomas 2003, 181). The standard 20% tariff on imports was the mere result of the difficulties to collect public revenue from anything else, but it still had some protective effects on domestic production (Scobie 1964, 131).

15. Federal, provincial, and municipal governors nevertheless almost completely neglected the road building to the delight of the British railway operators (Scobie 1964, 90).

16. While Argentine public policies for the most part did not proactively support the development of the grain sector, it nevertheless adopted in 1877 a protectionist tariff on wheat and flour to preempt external competition, turning Argentina into a major net exporter in a few years (Bulmer-Thomas 2003, 141–142). Argentine exports represented nearly 30% of Latin America's total export earnings in 1913 (while only 9.5% of the population). The main products were wheat, corn, chilled and frozen beef, lamb, wool, linseed, rye, barley, and hides (Bulmer-Thomas 2003, 170).

17. In fact, the percentage of the harvest that sharecroppers turned over to the landowners rose from an average of 10–12% in the 1890s to 20–30% in 1910 (Scobie 1964, 50–52).

18. The commercial houses, the local millers, merchants, and storekeepers bought the wheat from the farmers and resold it to the highly concentrated multinational wheat-exporting firms.

19. The lion's share of the arriving European immigrants never became farmers in the new land, but settled in the city. Some immigrants who came to work the land became seasonal harvest workers. Harvest labor was either traveling back and forth between Southern Europe and Argentina (*golondrinas*), or between cities and countryside, to work the harvest seasons (Scobie 1964, 80).

20. The ranchers were understood as backward and depopulating the countryside, while the British investors were understood to stand in the way for independent national development, taking advantage of their almost monopoly position and sending back all profits to London, and in the end threatening sovereignty (Barrán and Nahum 1981).

21. Some if the established public corporations were the Bank of the Republic (BROU), the National Mortgage Bank, the State Insurance Bank and the Port of Montevideo (ANP). These were later followed by public corporations for electricity (UTE), telecommunications (ANTEL) and railways

(AFE) (Panizza 1990, 153). The agrarian Census of 1908, and the creation of the faculties of agronomy (FAGRO) and of veterinary (FVET) in 1907 and 1905, had the explicit aim to use the science and knowledge for the transformation of the agrarian system into higher productivity, more crop production, and the subdivision of land (Baraibar 2014, 135).

22. Moreover, *Batllismo* was only one branch of the Colorado party, and it was in minority in the legislative chamber (Maubrigades 2009).

23. The rural sector was largely exempted from the social advances that benefited the urban laborers, for example the introduction of 8-hour workday and the introduction of wage councils.

24. Nationalist economic policies, redistribution through taxes, relative high salaries, and social protection provided some conditions for the development of a manufacturing sector for domestic consumption (Bulmer-Thomas 2003, 181–185).

25. The following presidents had ruled: José Gaspar Rodriguez de Francia (1814–1840), Carlos Antonio López (1841–1862), and Francisco Solano López (1862–1870).

26. All figures are broad-brush estimations. Unfortunately, there are no reliable national accounting data before World War II in Paraguay. This includes for example statistical information on trade, population size, land tenancy, and GDP per capita (Bulmer-Thomas 2003, 308).

27. The Liberal and Colorado parties emerged in the late nineteenth century. For long periods, the Colorado Party was the only political party allowed. The power within the Colorado Party was nevertheless disputed, and under several decades around the turn of the century, there was a series of coups and presidents that were rapidly overthrown and replaced with other Colorado leaders (Sacks 1988, 31–32).

28. It was first up until 1938 called the Ranchers' Society of Paraguay. While ARP had the most influential voice and represented the ranchers had the greatest land holdings, the *Sociedad Nacional de Agricultura* (SNA) from 1902, mainly representing commercial crop farmers, have also been an important national producers' organization and still is.

29. An illustrative example was when the government attempted to raise the export tax by about 20% in 1917. The exporting companies threatened to retaliate by closing down their factories and moving them to Argentina, which made the government back down.

30. During the period 1870–1916 twenty-one governments passed. During the period 1904–1922, Paraguay had fifteen presidents.

31. The decision of UK and US to abandon the gold standard made Argentina and Uruguay adopt a dual exchange rate system (Bulmer-Thomas 2003, 203–204).

32. Besides strategic tariffs, the US supported its agricultural sector through huge amounts of public funds in agricultural development (agricultural research labs, experiment stations and land grant colleges), through extension services, to farmers through investments in infrastructure (not least public irrigation projects), and through subsidized credit provision to farmers (Chang 2009, 490–495).

33. Argentina, Brazil, Paraguay, and Uruguay, in contrast to the rest of the region, continuously depended more on Europe as market and foreign investor, than on the US (Gerchunoff and Machinea 2015, 108).

34. For example, lower exchange rate for British imports, guaranteed purchases of British coal, lower export tariffs on all exports to Britain, repayment of current and future debts to Britain at a privileged exchange rate, high fees to the British railroads and guaranteed monopoly positions for some British owned sectors (Gerchunoff and Machinea 2015, 115–117).

35. These tripartite boards, with representatives from the government, the producers' organization, and the processors; meat packers and milling.

36. Moreover, in 1934, the Argentine Meat Producers Corporation (CAP) was created to end the oligopolistic situation of the meat packers and to enhance the relative power of the producers by establishing proper meat-packaging plants (Basualdo 2016; León and Rossi 2003).

37. The first president of this "conservative reaction" was Gabriel Terra (1931–1938) and his successor was Alfredo Baldomir (1938–1942).

38. For example, in 1931 the state-owned industrial corporation for oil refining and distribution and alcohol and cement production (ANCAP), was established and given monopoly on the sales of fuels and alcohol.

39. Since their formation as independent republics in the nineteenth century, neither Bolivia nor Paraguay had been able to agree on a common border in the Chaco region. Although Paraguay had held the Chaco for as long as anyone could remember, the country did little to develop the area. Aside from scattered Mennonite colonies and nomadic Indian tribes, few people lived there. Bolivia's claim to the Chaco became more urgent after it lost its sea coast (the Atacama region) to Chile during the 1879–1984 War of the Pacific.

40. Paraguay lost approximately 3.5% (36,000 dead) of its population.

41. In 1944, the Food and Agriculture Organization (FAO) was established and "food security"—understood as enough calories for all citizens—became a fundamental part of the development agenda all over the world (McKeon 2015, 13–15).

42. Because agriculture was excluded from GATT, trade in primary products grew less rapidly than the rest of world trade (Bulmer-Thomas 2003, 269–271).

43. This was the rationale behind the US' farm bills (first appearing under Roosevelt's New Deal).
44. These aims were still in mind when the European Economic Community (later the European Union) designed its Common Agricultural Policy.
45. Many farmers in the receiving countries had no chance to compete with the subsidized grains (Friedmann 1982).
46. The green revolutions were spurred by the Rockefeller and Ford Foundations, the International Wheat and Maize Improvement Center, the International Center for Tropical Agriculture, agronomic universities, and states.
47. The new "high-input-high output" productive paradigm reduced self-sufficiency of farmers, and increased instead the use of synthetic fertilizers and pesticides. In 1961, the International Union for the Protection of New Varieties of Plants (UPOV) was created, marking an important step toward strengthened property rights over plant varieties (McKeon 2015, 124–125).
48. The specialization brought by the green revolution separated animal/livestock farming from grain, and specialized fields of maize and soybeans became devoted specifically to feeding animals (Friedmann 1982).
49. Due to falling birthrates and Engel's law; the percentage of income allocated for food purchases decreases as income rises.
50. Both Raúl Prebisch and Hans Singer presented studies showing that the terms of trade of the primary products exported by the Latin American countries had in the long-run been unfavorable vis-à-vis the manufactured products imported. According to the economic historian Victor Bulmer-Thomas, however, any suggestion of a secular decline in net barter terms of trade for Latin America before 1913, needs to be regarded with extreme caution because of unreliable data and extremely fluctuating data (Bulmer-Thomas 2003, 78–80).
51. The diffusion of green revolution technologies was highest in Mexico and Colombia, followed by Ecuador, Peru, Venezuela, and Chile.
52. The US initiative "The Alliance for Progress" for capital transfer to Latin America, not least to curb an expansion of the Cuban revolution, also proclaimed land reform and subdivision of large estates.
53. During Neobatllismo 1947–1963 the Colorado Party (led by Luis Batlle Berres) was mostly in government, but the last term was led by the Blanco Party (1959–1963).
54. Credit policies were mainly decided by the state and not left to the market. Interest rates were controlled by the state and maintained below inflation rates. Access to foreign currency was limited to foreign export and tourism (Grugel and Riggirozzi 2007, 89).
55. This allowed for the provision of cheap foreign currency for raw materials, capital goods for industry and selected consumer goods, as well as to collect fiscal revenue, and to redistribute income (Bulmer-Thomas 2003, 265; Panizza 1990, 153).

56. Uruguay's share of world exports fell from 0.45% in 1946 to 0.08% in 1970 and 0.047% in 1975. In absolute volumes exports fell only in the 1950s (Bulmer-Thomas 2003, 284), but since trade in the rest of the world increased rapidly in this period, the relative decline was far greater, and the gap with the advanced economies widened.

57. The depression had effected domestic demand less than the export sector, since the rates of return of capital had fallen more sharply than real wages (Bulmer-Thomas 2003, 216).

58. Uruguay also nationalized the Montevideo water supply from the British. It should be mentioned however, that the British government was in favor of the nationalizations as a way of paying their wartime debt and the price is now considered to have been excessive (Bulmer-Thomas 2003, 261; Panizza 1990, 153).

59. For example, the telephone and gas services were nationalized. The state created a water and electric energy enterprise to operate in the hinterland (remaining in private hands in Buenos Aires) and a national airline *Aerolineas Argentinas.*

60. The main labor negotiations in Uruguay took place in tripartite commissions (state-employers-employees) called wage councils, created in 1943. The Wage Councils decided wage claims by branches of industry. The vote of the state representative was decisive in the outcome of the claim.

61. Agriculture was at the same time absorbing a small and rapidly declining share of labor force. Labor in agriculture even fell in absolute terms after 1960 in both Argentina and Uruguay (Bulmer-Thomas 2003, 316), while the number of public employees increased substantially during the period.

62. Public policies in Argentina and Uruguay, between 1930 and 1960, emphasized public education levels, diminishing the gap with the leading economies in the world at the end of the period (Álvarez and Bértola 2010, 69–70). In 1913, the level of education of the population in relation to the leading countries in the world, reached 52% in Argentina and 42% in Uruguay. By 1960, the level of education in these countries had catch up significantly and reached 80% of the level of the most advanced countries (Álvarez and Bértola 2010, 69–70).

63. This welfare model is described as "universalist, but stratified" (Ksiazenicki and Fuentes 2013).

64. Continuous tillage in crop farming systems in the vicinity of Montevideo created severe problems of soil degradation and erosion. Consequently, crop producers gradually moved to the fertile soils along the littoral (Bertino and Bucheli 2000, 8).

65. See INC's official web page: https://www.colonizacion.com.uy/antecedentes. Accessed 23 July 2018.

66. One of the most influential and long-lasting efforts came from the *National Commission for the Agronomic Investigation* (CONEAT) of

the Land. This was launched to map and measure the quality of the soils of each parcel of land and to develop criteria for the definition of their productivity. The commission worked several years with capacity building and recollection of data from producers, technicians, aerial photography and also from field studies, laboratories, and institutions (Former researcher at CONEAT, Punta del Este, March 11, 2017). The main objective of the program was to design a land tax system that considered the productivity potential of the soils, in order to create economic incentives for productivity increase and hamper land speculation, as well as for land distribution. This never happened, and instead Uruguay entered a period of military regime and faith in liberalism, but the CONEAT system became vital for many purposes beyond taxation.

67. Many consumer goods that had received support for decades never managed to raise productivity as expected or become competitive outside their own protected domestic markets (Bulmer-Thomas 2003, 276–299). The economies remained largely dependent on the imports of capital and intermediate goods, which created balance-of-payments problems. In addition, the high nominal tariff rates created a gap between world and domestic prices and imposed a heavy burden on consumers (Bulmer-Thomas 2003, 279).

68. Uruguay was one of the few countries in the world where the growth of GDP was below the rate of growth of population (Panizza 1990, 149).

69. In the 1971 elections, the left party-coalition *Frente Amplio* received almost 20% of the votes, challenging the long political history of bipartisan oligopoly of *Blancos* and *Colorados* (Panizza 1990, 156).

70. While the core of the military rule was authoritarian politics against "the threat" of leftist movements and parties, it also relied on professional neoclassical trained economists more than any government had before.

71. There were tensions between traditional liberals, monetarists and neoconservatives, between advocates of gradualism and shock therapy, and between statist and nationalist traditions and neoliberal tenets.

72. The military regimes took a series of measures to control the inflationary episodes, but the remedies seemed to make the situation worse (Fridman 2010, 281–292).

73. While soybeans expanded, the success of this strategy was still rather limited due to the reintroduction of export taxes on agricultural commodities in 1969 for fiscal revenue (Director of Agricultural Affairs CARI, Buenos Aires, March 1, 2017).

74. Between 1975 and 1982, Latin American external debt quadrupled, from USD 75 billion (B) to 315B, which represented 50% of the region's GDP. Surprisingly, the banks and other creditors continued to supply

Latin American countries with new loans even after the second oil crises, in the wake of the Iranian revolution in 1979 (Bulmer-Thomas 2003, 364). As global interest rates surged, the debt service grew even faster than debt, from USD 12B in 1975 to 66B in 1982 (Felix 1990).

75. Stroessner declared a state of siege which allowed him to suspend civil liberties, and arrest and detain anyone indefinitely without trial (del Carmen Quevedo 1997). Stroessner dedicated large proportions of the Paraguayan national budget to the military and police apparatus, both fundamental to the maintenance of the regime (Sacks 1988, 42–44).

76. From 1947 until 1977, the US supplied about USD 750,000 worth of military hardware each year and trained more than 2000 Paraguayan military officers in counterintelligence and counterinsurgency (Sacks 1988, 44–46). Between 1954 and 1961, Paraguay received USD 53.2 million in aid and loans from the US, representing 2.7% of national GDP (Mora and Cooney 2010).

77. The respondent has more than 30 years' experience in the area of strategic agrarian planning. During the 1980s, he was the director of the planning office for control of agricultural, livestock, and forestry products at MAG. He has been international consultant at various international organizations and authored various publications on production, animal health, and management of rural development projects.

78. Brazil had since the 1960s been in a process of dramatic agricultural expansion, mechanization, and intensification. Brazil had adopted a system of differential export taxes on soybeans, with high export tariffs on whole beans and low on processed agricultural products, to foster agro-industry and developing processing stages and intensive meat farming systems (Shurtleff and Aoyagi 2009, 318). Brazil had also increased significantly its stock of grazing cattle, and developed a competitive and cost-effective broiler and pork industry, in which crushed soybean meal played a pivotal role (Lapitz et al. 2004, 15).

79. The soybean thus expanded rapidly in eastern departments of Itapúa, Alto Paraná, Canendiyú, and Amambay.

80. By the mid-1970s, the livestock sector represented 12% of national GDP, and employed 5.5% of the labor force.

81. The Paraguayan congress had passed a law in 1921 allowing the Mennonites in Paraguay to create a state within the department of Boqueron in the Paraguayan Chaco. Through trial and error, they developed and adapted techniques of water storage and managed to develop a highly productive farming system in the Chaco, and accordingly turned the region into the main producer of dairy products (Cartes et al. 2015; Mereles and Rodas 2014)

REFERENCES

Abente, Diego. 1989. "Foreign Capital, Economic Elites and the State in Paraguay During the Liberal Republic (1870–1936)." *Journal of Latin American Studies* 21 (1–2): 61–88.

Achkar, M., A. Dominguez, I. Díaz, and F. Pesce. 2011. "La intensificación del uso agrícola del suelo en el litoral oeste del Uruguay en la última década." *PAMPA, Revista Interuniversitaria de Estudios Territoriales* 7 (Suplemento especial temático: Impactos territoriales asociados a la reconfiguración del sistema productivo primario): 143–157.

Acosta Reveles, Irma L. 2008. "Capitalismo agrario y sojización en la pampa Argentina. Las razones del desalojo laboral." *Lavboratorio, Cambio Estructural y Desigualdad Social (CEyDS)/Facultad de Ciencias Soicales, UBA* 10 (22): 8–12.

Alonso, José M 1981. *El Proceso Histórico de la agricultura uruguaya.* Vol. 3. Temas Nacionales. Montevideo: Fundación de Cultura Universitaria & Centro Interdisciplinario de Estudios sobre el Desarrollo Uruguay (CIEDUR).

Álvarez, Jorge, and Luis Bértola. 2010. "Desarrollo y desigualdad: miradas desde la historia económica." In *Pobreza y (des) igualdad en Uruguay: una relación en debate,* 55–57.

Álvarez, Jorge, Luis Bértola, and G. Porcile. 2007. *Primos ricos y empobrecidos: crecimiento, distribución del ingreso e instituciones en Australia-Nueva Zealand vs Argentina-Uruguay.* Montevideo: Editorial Fin de Siglo.

Ankersen, Tom, and Ruppert, Thomas. 2006. "Tierra y Libertad: The Social Function Doctrine and Land Reform in Latin America." *Tulane Environmental Law Journal*, University of Florida Levin College of Law Research Paper 19 (69): 52.

Astori, D. 1984. "Principales Interpretaciones sobre la Problemática Agraria Uruguaya." In *La cuestión agraria en el Uruguay,* edited by M. Buxedas, R. Jakob, D. Astori, C. Perez Arrarte, L. Sierra, R. Irigoyen, C. Paolino, and J. M. Alonso. Montevideo: Ciedur.

Balsa, Juan Javier. 2004. "Consolidación y desvanecimiento del mundo chacarero: Transformaciones de la estructura agraria, las formas sociales de producción y los modos de vida en la agricultura bonaerense, 1937–1988." Ph.D. in History, Universidad Nacional de la Plata, Facultad de Humanidades y Ciencias de la Educación (153).

Baraibar, Matilda. 2014. "Green Deserts or New Opportunities? Competing and Complementary Views on the Soybean Expansion in Uruguay, 2002–2013." Stockholm Studies in Economic History, 64. Doctoral thesis Monograph, Economic History, Stockholm University.

Barrán, José Pedro, and Benjamín Nahum. 1977. *La civilización ganadera bajo Batlle: 1905–1914.* Edited by J. P. Barrán and B. Nahum. Vol. 6. Historia rural del Uruguay moderno. Montevideo: Banda Oriental.

Barrán, José Pedro, and Benjamín Nahum. 1981. *Un dialogo difícil 1903–1910*. Vol. II. Batlle, los estancieros y el imperio Britanico. Montevideo: Ediciones de la Banda Oriental.

Barrán, José Pedro, and Benjamín Nahum. 1984. "Uruguayan Rural History." *The Hispanic American Historical Review* 64 (4): 655–673.

Basualdo, Marcelo Ernesto. 2016. "La historia de la CAP, el frigorífico exportador de los ganaderos argentinos, en resumen." XXV JORNADAS DE HISTORIA ECONOMICA, Salta, 21–23 September.

Benvenuto, Luis C. 1969. *La quiebra del modelo*. Vol. 48. Enciclopedia Uruguaya. Montevideo: Arca.

Bertino, M., and H. Tajam. 1999. *El PBI de Uruguay 1900–1955*. Montevideo, Uruguay: Instituto de Economía, Facultad de Ciencias Económicas y de Administración, Universidad de la República.

Bertino, M., and Gabriel Bucheli. 2000. "La agricultura en el Uruguay, 1911–1930." Serie Documentos de Trabajo/FCEA-IE; DT08/00.

Bertino, M., and H. Tajam. 2000. La ganadería en el Uruguay 1911–1943. In *Serie documentos de trabajo*. Instituto de Economía, Facultad de ciencias económicas, Udelar.

Bértola, Luis. 2000. *Ensayos de Historia Económica: Uruguay y la región en la economía mundial, 1870–1990*. Montevideo: Ediciones Trilce.

Bértola, Luis, and J. G. Williamson. 2003. "Globalization in Latin America Before 1940." NBER Working Paper Series.

Bértola, Luis, Cecilia Castelnovo, Javier Rodriguez, and H. Willebald. 2008. Income Distribution in the Latin American Southern Cone During the First Globalization Boom, ca: 1870–1920. In *Working Papers in Economic History*, edited by Instituto Figuerola de historia económica, Universidad Carlos III de Madrid.

Birner, Regina, and Danielle Resnick. 2010. "The Political Economy of Policies for Smallholder Agriculture." *World Development* 38 (10): 1442–1452.

Blacha, Luis Ernesto, and Maximiliano Ivickas Magallan. 2013. "El Consejo Agrario Nacional y la acción colectiva: Reflexiones socio históricas sobre las estrategias estatales peronistas (1946–1949)." *Estudios Interdisciplinarios de América Latina y el Caribe* 24 (2): 131–155.

Bolt, Jutta, Marcel Timmer, and Jan Luiten van Zanden. 2014. "GDP Per Capita Since 1820." In *How Was Life?*, edited by Jan Luiten van Zanden, Joerg Baten, Marco Mira d'Ercole, Auke Rijpma, Conal Smith, and Marcel Timmer, 57–72. Uutrecht: OECD Publishing.

Bonino-Gayoso, Nicolás, Antonio Tena-Junguito, and Henry Willebald. 2015. "Uruguay and the First Globalization: On the Accuracy of Export Performance, 1870–1913." *Revista de Historia Economica-Journal of Iberian and Latin American Economic History* 33 (2): 287–320.

Brezzo, Liliana M. 2003. "La historiografía paraguaya: del aislamiento a la superación de la mediterraneidad." *Diálogos* 7 (1): 157–175.

Bulmer-Thomas, Victor. 2003. *The Economic History of Latin America Since Independence*. Cambridge: Cambridge University Press.

Burch, David, and Geoffrey Lawrence. 2009. "Towards a Third Food Regime: Behind the Transformation." *Agriculture and Human Values* 26 (4): 267.

Cartes, José L., J. J. Thompson, and A. Yanosky. 2015. "El Chaco paraguayo como uno de los últimos refugios para los mamíferos amenazados del Cono Sur." *Paraquaria Natural* 3 (2): 37–47.

Chang, Ha-Joon 2006. "Policy Space in Historical Perspective with Special Reference to Trade and Industrial Policies." *Economic and Political Weekly* 41 (7): 627–633.

Chang, Ha-Joon. 2009. "Rethinking Public Policy in Agriculture: Lessons from History, Distant and Recent." *The Journal of Peasant Studies* 36 (3): 477–515. https://doi.org/10.1080/03066150903142741.

Clapp, Jennifer. 2016. *Food*. 2nd ed. Cambridge: Polity Press.

Coordinadora de Derechos Humanos del Paraguay. 2007. *Informe Chokokue: Informe al Relator Especialsobre las ejecuciones extrajudiciales, sumarias o arbitrarias del Consejo de Derechos Humanos de Naciones Unidas sobre las violaciones al derecho a la vida en contra de miembors y dirigentes de las organizaciones campesinas en el contexto de reforma agaria en Paraguay (1989–2005)*. Asunción: Codehupy.

Cote, Stephen. 2013. "A War for Oil in the Chaco, 1932–1935." *Environmental History* 18 (4): 738–758. https://doi.org/10.1093/envhis/emt066.

de la Torre, Nelson, Julio C. Rodriguez, and Lucia Sala de Touron. 1971. *Artigas: Tierra y revolución*. Montevideo: Ediciones por Uruguay (EPU).

del Carmen Quevedo, Oscar. 1997. *Crónica histórica ilustrada del Paraguay: Paraguay siglo XX*. Vol. 3. Buenos Aires: Distribuidora Quevedo de Ediciones.

Díaz Rossello, R., and C. Rava. 2007. "Aportes de la ciencia y la tecnología al manejo productivo y sustentable de los suelos del Cono Sur." In *Programa Cooperativo para el Desarrollo Tecnológico Agroalimentario y Agroindustrial del Cono Sur*, edited by IICA INIA La Estanzuela and PROCISUR. Montevideo: Instituto Interamericano de Cooperación para la Agricultura.

Ernst, O., and G. Siri-Prieto. 2011. "La Agricultura en Uruguay: su trayectoria y consecuencias." II Simposio Nacional de Agricultura, Montevideo, Paysandú, Uruguay.

Espínola, Julio. 2008. "Ligas Agrarias Cristianas, un movimiento contrahegemónico en Paraguay." *Revista de la Facultad* 14: 121–145.

Felix, David. 1990. "Latin America's Debt Crisis." *World Policy Journal* 7 (4): 733–771.

Fernández, Emilio 2007. "La evolución económica y social del sector agropecuario." In *El Sector agropecuario en el Uruguay: Una mirada desde la sociología rural*, edited by Marta Chiappe, Matías Carámbula, and Emilio

Fernández. Montevideo: Departamento de ciencias sociales, Facultad de la Agronomía, Universidad de la República.

Finch, Martin Henry John. 1981. *A Political Economy of Uruguay Since 1870*. New York: St. Martin's Press.

Flichman, Guillermo. 1990. "The State and Capital Accumulation in Argentina." In *The State and Capital Accumulation in Latin America*, edited by Christian Anglade and Carlos Fortin, 1–32. London: Macmillan.

Fogel, Ramón. 1986. *Movimientos campesinos en el Paraguay, Serie Estudios Agrarios*. Asunción: Centro Paraguayo de Estudios Sociológico.

Fridman, Daniel. 2010. "A New Mentality for a New Economy: Performing the Homo Economicus in Argentina (1976–83)." *Economy and Society* 39 (2): 271–302.

Friedmann, Harriet. 1982. "The Political Economy of Food: The Rise and Fall of the Postwar International Food Order." *American Journal of Sociology* 88: S248–S286.

Friedmann, Harriet. 2000. "What on Earth Is the Modern World-System? Foodgetting and Territory in the Modern Era and Beyond." *Journal of World-System Research* XI (2): 480–515.

Friedmann, Harriet. 2005. "From Colonialism to Green Capitalism: Social Movements and Emergence of Food Regimes." In *New Directions in the Sociology of Global Development*, edited by Frederick H. Buttel and Philip McMichael, 227–264. Bradford: Emerald Group.

Friedmann, Harriet. 2009. "Discussion: Moving Food Regimes Forward: Reflections on Symposium Essays." *Agriculture and Human Values* 26 (4): 335.

Friedmann, Harriet. 2016. "Commentary: Food Regime Analysis and Agrarian Questions: Widening the Conversation." *The Journal of Peasant Studies* 43 (3): 671–692.

Friedmann, Harriet. 2017. "Towards a Natural History of Foodgetting." *Sociologia Ruralis* 57 (2): 245–264. https://doi.org/10.1111/soru.12144.

Friedmann, Harriet, and Philip. McMichael. 1989. "Agriculture and the State System: The Rise and Fall of National Agricultures, 1870 to the Present." *Sociologia Ruralis* 29 (2): 93–117.

Garcé, Adolfo. 2002. *Ideas y competencia política en Uruguay (1960–1973): Revisando el "fracaso" de la CIDE, Ciencia Política*. Montevideo: Ediciones Trilce.

Gerchunoff, Pablo, and José Luis Machinea. 2015. "Going Through the Labyrinth: The Political Economy of Argentina's Abandonment of the Gold Standard (1929–1933)." *CEPAL Review* LC/G.2652-P (117): 118.

Gomez, Marcelo. 2008. "La soja de la discordia. Los sentidos y estrategias en la movilización de la pequeña buruesía." *Lavboratorio, Cambio Estructural y Desigualdad Social (CEyDS)/Facultad de Ciencias Soicales, UBA*. 10 (22): 22–35.

Gordon, Peter, and Juan José Morales. 2017. *The Silver Way: China, Spanish America and the Birth of Globalisation, 1565–1815.* London: Penguin Books.

Grugel, Jean, and Maria Pia Riggirozzi. 2007. "The Return of the State in Argentina." *International Affairs* 83 (1): 87–107.

Hetherington, Kregg. 2009. "Privatizing the Private in Rural Paraguay: Precarious Lots and the Materiality of Rights." *American Ethnologist* 36 (2): 224–241.

Holt Giménez, Eric, and Annie Shattuck. 2011. "Food Crises, Food Regimes and Food Movements: Rumblings of Reform or Tides of Transformation?" *The Journal of Peasant Studies* 38 (1): 109–144.

Hopewell, Kristen. 2014. "The Transformation of State-Business Relations in an Emerging Economy: The Case of Brazilian Agribusiness." *Critical Perspectives on International Business* 10 (4): 291–309. https://doi.org/10.1108/cpoib-03-2014-0019.

Howard, Philip. 2016. *Concentration and Power in the Food System: Who Controls What We Eat?* Edited by D. Goodman and M. K. Goodman. Vol. 3. Contemporary Food Studies: Economy. Culture and Politics. London: Bloomsbury.

Hughes, Matthew. 2005. "Logistics and the Chaco War: Bolivia Versus Paraguay, 1932–1935." *The Journal of Military History* 69 (2): 411–437.

Jacob, Raúl, Martín Buxedas, Danilo Astori, Carlos Pérez Arrarte, Lilian Sierra, Rodolfo Irigoyen, and José M. Alonso. 1984. "Los principales modelos históricos." In *La cuestion agraria en el Uruguay*, edited by M. Buxedas, R. Jakob, D. Astori, C. Perez Arrarte, L. Sierra, R. Irigoyen, C. Paolino, and J. M. Alonso. Vol. 13. Temas Nacionales. Montevideo: Fundación cultura Universitaria, CIEDUR.

Jones, David M. 1977. "The Green Revolution in Latin America: Success or Failure?" Publication Series (Conference of Latin Americanist Geographers).

Jones, Geoffrey. 2000. *Merchants to Multinationals: British Trading Companies in the Nineteenth and Twentieth Centuries.* Oxford: Oxford University Press on Demand.

Jordan, Paul R. 2005. "From Bureaucratic Alienation to Political Exile: Evolving Views of Uruguayan Identity in the Work of Mario Benedetti." *The Modern Language Review* 100 (2): 383–395.

Kay, Cristóbal. 1999. "Agrarian Reforms: The Latin American Experience." *Revista Envío* (216).

Kay, Cristóbal. 1989. *Latin American Theories of Development and Underdevelopment.* Routledge Library Editions, Development. London: Routledge.

Ksiazenicki, Inés, and Guillermo Fuentes. 2013. "Permanencia y rupturas. Una mirada en torno a las matrices de bienestar social en Argentina y Uruguay a comienzos del siglo XXI." *Papel Político* 18 (1): 83–114.

Lang, Tim, and Michael Heasman. 2015. *Food Wars: The Global Battle for Mouths, Minds and Markets.* 2nd ed. London: Routledge.

Lapitz, Rocío, Gerardo Evia, and Eduardo Gudynas. 2004. *Soja y carne en el Mercosur: comercio, ambiente y desarrollo agropecuario*. Montevideo: Coscoroba Ediciones.

León, Carlos Alberto, and Carlos Alberto Rossi. 2003. "Aportes para la historia de las instituciones agrarias de la Argentina (I): la Junta Nacional de Granos." *Realidad económica* 196: 84–101.

Manzetti, Luigi. 1992. "The Evolution of Agricultural Interest Groups in Argentina." *Journal of Latin American Studies* 24 (3): 585–616.

Margulis, Matias E. 2014. "Trading Out of the Global Food Crisis? The World Trade Organization and the Geopolitics of Food Security." *Geopolitics* 19 (2):322–350.

Martinez-Alier, Joan, Michiel Baud, and Héctor Sejenovich. 2016. "Origins and Perspectives of Latin American Environmentalism." In *Environmental Governance in Latin America*, 29–57. Dordrecht: Springer.

Martínez Alvarez, Diego Leonardo. 2012. Historia de la soja en la Argentina: Introducción y adopción del cultivo. In *El cultivo de soja en Argentina*, edited by Baigorri and Salado Navarro, 1. Buenos Aires: Agroeditorial.

McKeon, Nora. 2015. *Food Security Governance: Empowering Communities, Regulating Corporations*. Oxon: Routledge.

McMichael, Philip. 2009. "A Food Regime Genealogy." *The Journal of Peasant Studies* 36 (1): 139–169. https://doi.org/10.1080/03066150902820354.

McMichael, Philip. 2013. *Food Regimes and Agrarian Questions*. Edited by Kate Kennedy. Agrarian Change & Peasant Studies. Winnipeg: Fernwood Publishing.

Mederos Porto, Leticia. 2014. "La fiebre aftosa como problema para la producción ganadera en Uruguay y la demanda de ciencia, tecnología e innovación endógenas 1870–2001." Master of Science in Economic History, Facultad de Ciencias Sociales, Unidad Multidisciplinaria, Montevideo: Udelar.

Mereles, María Fátima, and Oscar Rodas. 2014. "Assessment of Rates of Deforestation Classes in the Paraguayan Chaco (Great South American Chaco) with Comments on the Vulnerability of Forests Fragments to Climate Change." *Climatic Change* 127 (1): 55–71. https://doi.org/10.1007/s10584-014-1256-3.

Mora, Frank O., and Jerry Wilson Cooney. 2010. *Paraguay and the United States: Distant Allies*. London: University of Georgia Press.

Moraes, María Inés. 2008. *La pradera perdida: historia y economía del agro uruguayo: una visión de largo plazo 1760–1970*. Montevideo: Linardi y Risso.

Morales Olmos, Virginia, and Jacek P Siry. 2009. "Economic Impact Evaluation of Uruguay Forest Sector Development Policy." *Journal of Forestry* 107 (2): 63–68.

Morgan, Kevin, Terry Marsden, and Jonathan Murdoch. 2006. *Worlds of Food— Place, Power and Provenance in the Food Chain*. Oxford: Oxford University Press.

Maubrigades, Silvana. 2009. "Intentos de Reforma Agraria, el latifundio como héroe o martir. Uruguay entre 1870 y 1915." Fundação de Economía e Estatistica, Porto Alegre, Brazil.

Nickson, A. 2015. *Historical Dictionary of Paraguay.* Edited by J. Woronoff. 3rd ed. Historical Dictionary of the Americas. Lanham, MD: Rowman & Littlefield.

Ocampo, José Antonio. 2001. "Raúl Prebish and the Development Agenda at the Dawn of the Twenty-First Century." *CEPAL Review* 75: 23–37.

OEA. 2009. Evaluación regional del impacto de la sostentabilidad de la cadena productiva de la soja: Argentina - Paraguay - Uruguay. In *OAS Official Records,* edited by Secretaría General de la Organización de los Estados Americanos Departamento de Desarrollo Sostenible. Asunción. Paraguay: Organización de los Estados Americanos (OEA - OAS).

Otero, Gerardo. 2008. "Neoliberal Globalism and the Biotechnology Revolution: Economic and Historical Context." In *Food for the Few: Neoliberal Globalism and Biotechnology in Latin America,* edited by G. Otero, 1–29. Austin, TX: University of Texas Press.

Otero, Gerardo. 2012. "The Neoliberal Food Regime in Latin America: State, Agribusiness Transnational Corporations and Biotechnology." *Canadian Journal of Development Studies* 33 (3): 282–294.

Panizza, Francisco. 1990. "Accumulation and Consensus in Post-war Uruguay." In *The State and Capital Accumulation in Latin America,* edited by Christian Anglade and Carlos Fortin, 149–181. London: Macmillan.

Paolera, Gerardo della, Xavier H. Duran Amorocho, and Aldo Musacchio. 2018. "The Industrialization of South America Revisited: Evidence from Argentina, Brazil, Chile and Colombia, 1890–2010." NBER Working Papers 24345, National Bureau of Economic Research Inc.

Patel, Raj. 2013. "The Long Green Revolution." *The Journal of Peasant Studies* 40 (1): 1–63.

Pérez Arrarte, Carlos. 1982. *El agro Uruguayo y el mercado internacional.* Vol. 4. Temas Nacionales. Montevideo: Fundación de Cultura Universitaria; Centro Interdisciplinario de Estudios sobre el Desarrollo Uruguay, CIEDUR.

Pérez Arrarte, Carlos 1984a. "La estructura agraria en Uruguay." In *La cuestión agraria en el Uruguay,* edited by Ciedur. Montevideo: Fundación de cultura universitaria, CIEDUR.

Pérez Arrarte, Carlos. 1984b. "El Complejo Productor de Carne Vacuna en Uruguay." In *La cuestión agraria en el Uruguay,* 83–106. Montevideo: Fundación de Cultura Universitaria, CIEDUR.

Petraglia, Cecilia, Ruben Puentes, Ricardo Cayssials, José Barrios, and José P. Lucas. 1982. *Avances en conservación de suelos en el Uruguay.* Montevideo: Departamento de Uso, Manejo y Conservación de Suelos, Ministerio de Agricultura y Pesca, Instituto Interamericano de Cooperación para la Agricultura.

Piera Valdés, A. 2016. Consultoría Nacional: Análisis del Marco Legal e Institucional Vigente para la Implementación de REDD+ en Paraguay. In *Program ONU-REDD+ Paraguay*. Asunción: FAO, PNUD, PUMA, INFONA, SEAM, FAPI.

Piñeiro, Diego, Carmen Améndola, Rodolfo Irigoyen, Gonzalo Kmaid Ricetto, Raúl Latorre, and Alberto Riella. 1991. *Nuevos y no tanto: Los actores sociales para la modernización del agro uruguayo*. Edited by CIESU. Montevideo: Centro de Información y Estudios del Uruguay.

Pinilla, Vicente, and Agustina Rayes. 2017. "Why did Argentina Become a Super-Exporter of Agricultural and Food Products During the Belle Époque (1880–1929)?" Working Papers 0107. European Historical Economics Society (EHES).

PNUD and OIT. 2014. Informe Nacional sobre Desarrollo Humano Paraguay 2013 Trabajo Decente y Desarrollo Humano. In *Informe Nacional sobre Desarrollo Humano*. Asunción: UNDP/ILO.

Prebisch, Raúl. 1950. "The Economic Development of Latin America and Its Principal Problems." *American Economic Review* 40 (6): 473–485.

Riella, Alberto, and Paola Mascheroni. 2011. "Desigualdades sociales y territorios en Uruguay." *PAMPA, Revista Interuniversitaria de Estudios Territoriales* 7 (Suplemento especial temático: Impactos territoriales asociados a la reconfiguración del sistema productivo primario): 39–63.

Sacks, Richard. 1988. "Historical Setting." In *Area Handbook Series: Paraguay: A Country Study*, edited by Dennis M. Hanratty and Sandra W. Meditz, 1–50. Washington, DC: Library of Congress; Washington, DC: Federal Research Division.

Scobie, J.R. 1964. *Revolution on the Pampas: A Social History of Argentine Wheat, 1860–1910, Latin American Monographs*. Austin, TX: University of Texas Press.

Serrano, R., and V. Pinilla. 2016. "The Declining Role of Latin America in the Global Agricultural Trade, 1963–2000." *Journal of Latin American Studies* 48 (1): 115–146.

Seyler, Daniel. 1988. "The Economy." In *Area Handbook Series: Paraguay: A Country Study*, edited by Dennis M. Hanratty and Sandra W. Meditz, 97–156. Washington, DC: Library of Congress; Washington, DC: Federal Research Division.

Shurtleff, William, and Akiko Aoyagi. 2009. *History of Soybeans and Soyfoods in South America (1882–2009): Extensively Annotated Bibliography and Sourcebook*. Lafayette: Soyinfo Center.

Teubal, Miguel. 2008. "Genetically Modified Soybeams and the Crisis of Argentina's Agriculture Model." In *Food for the Few—Neoliberal Flobalism and Biotechnology in Latin America*, edited by G. Otero, 189–216. Austin: University of Texas Press.

Thiesenhusen, William C. 1972. "Green Revolution in Latin America: Income Effects, Policy Decisions." *Monthly Labor Review* 95: 20.

Thiesenhusen, William C. 1987. "Review: Rural Development Questions in Latin America." *Latin American Research Review* 22 (1): 171–203.

Turzi, Mariano. 2016. *The Political Economy of Agricultural Booms: Managing Soybean Production in Argentina, Brazil, and Paraguay*. Cham, Switzerland: Springer.

Wallerstein, Immanuel. 1974. "The Rise and Future Demise of the World Capitalist System: Concepts for Comparative Analysis." *Comparative Studies in Society and History* 16 (4): 387–415.

Velkar, Aashish. 2010. "'Deep' Integration of 19th Century Grain Markets: Coordination and Standardisation in a Global Value Chain." Economic History Working Papers, London School of Economics 145 (10): 67.

Williamson, Jeffrey. 2002. "Land, Labour and Globalization in the Third World, 1870–1940." *Journal of Economic History* 62: 55–85.

Zoomers, Elisabeth B., and Johan M. G. Kleinpenning. 1990. "Elites, the Rural Masses and Land in Paraguay: A Case Study of the Subordination of the Rural Masses to the Interests and Power of the Ruling Class." *Revista Geográfica* (111): 129–148.

Zum-Felde, Alberto. 1987. *Proceso histórico del Uruguay*. 10th ed. Montevideo: Arca.

# CHAPTER 3

# Agrofood Globalization: The Global Soybean and Beef Commodity Chains

Since the late 1970s, inter- and transnational trade of food and agricultural commodities has increased in an ever accelerating pace. Population growth, urbanization, dietary changes, economic growth, and financialization of land and food markets, have created an unprecedented market demand for food, water, fiber, and energy (Godfray et al. 2010).[1] Food has increasingly become transported over greater and greater distances in order to ensure year-round availability, overcome natural restrictions, and consolidate global linkages (Morgan et al. 2006, 10). The main economic agents of the agrofood system are big transnational corporations (TNCs), increasingly vertically and horizontally integrated. While the expansion strategies of these corporations are important drivers behind contemporary agrarian globalization and frontier expansion, it is also supported by global shifts in political structures and geopolitical arrangements. In broad terms, the global political-economic governance regarding food production and security has shifted from a "national developmentalist" model—adopted to varying degrees in most part of the world after WWII—to an increasingly institutionalized form of globally integrated market, marked by liberalization, de-regularization, and financialization of food and land markets (Stiglitz 2008).

The restructuring process of the global food system can be seen to represent a new phase of agrarian capitalist expansion, which also involves spatial transformations. Asia, Russia, and the Middle East emerge for example as increasingly reliant on foreign land and water

© The Author(s) 2020
M. Baraibar Norberg, *The Political Economy of Agrarian Change in Latin America*, Governance, Development, and Social Inclusion in Latin America, https://doi.org/10.1007/978-3-030-24586-3_3

for food security as a result of changing dietary and demographic patterns, loss of domestic productive land due to environmental degradation and urbanization, as well as economic growth and policy change (OECD/FAO 2016a). In this process of spatial reconfigurations, Latin America has become the largest net exporter of food in the world, surpassing North America (Egas and De Salvo 2018).[2]

Agropastoral area in Latin America has enlarged faster than in any other place in the world during the twenty-first century. Between 2001 and 2013, pastureland expanded with 96.9 Million hectares (Mha), and cropland with 44.27 Mha (Graesser et al. 2015). Argentina, Paraguay, and Uruguay were at the center of this expansion. Both these types of LULCC are driven by snowballing global appetite for animal products, which in turn has increased global demand for land for animal source foods; pastureland for livestock grazing and cropland for the production of high-protein vegetable feed grains. The main protagonists of current agrarian change are soybeans expanding into former grazing lands or forests, and cattle expanding into savannah and forestland. The rising competition for land further pushes for land-use intensification. Since the lion's share of recent agrarian change in Latin America is then to be explained by growing international demand of soybeans and beef, it becomes relevant to ask how these products are embedded in wider chains, and how the structure of these chains looks like.

This chapter presents in detail the evolution and articulation of these intertwined commodity complexes, as well as the wider context of contemporary agrofood globalization. Inspired in a historically informed Global Commodity Chain (GCC) approach, shifts in the main places, actors, assets, and activities of the global soybean and beef commodity chains are addressed (see the Introduction chapter for a more detailed presentation of the GCC approach). Besides mapping out the overtime shifting main geographies, drivers, producers, sellers, and buyers involved in soybeans and beef flows, the chapter provides particular emphasis on those firms that are identified as chain drivers, playing a lead role in constructing and managing the configuration of the chain (Bair 2009, 7–9, 11). In this way, this chapter highlights dynamics and relations at the international level, which co-shapes recent agrarian change in Latin America.

The chapter is organized the following way: Section 3.1 presents the evolution of the soybean and beef production and trade. This includes a short world historical backdrop (Sect. 3.1.1), an exploration into how the soybean was converted into a main ingredient in the meat production systems (Sect. 3.1.2), and a presentation of the shifting geographies of trade the past fifty years (Sect. 3.1.3). This is followed by a section on the concentration and vertical integration of the driving transnational business actors involved in these chains, and the role of new technologies in these processes (Sect. 3.2). This section is divided into one subsection about the upstream stages of seeds, biotechnology, and agrochemicals (Sect. 3.2.1) and one subsection about the downstream stages of processing, storage, and trading (Sect. 3.2.2). The chapter ends with concluding remarks and a discussion about the main tenants of the current agrofood system, but also about some of its contradictory trends and elements.

## 3.1 THE EVOLUTION OF SOYBEAN AND MEAT PRODUCTION AND TRADE

Planning, organization, reproduction, and distribution of crop and livestock production have been at the heart of human society, since the Neolithic revolution, some 11,000 years ago. Both soybeans and livestock have been important sources to nutritious food for thousands of years. The soybean has historically been very important in an almost completely vegetarian diet in Asia, while meat has been highly appreciated (while often scarce) in the cuisines of Europe. However, as the soybean was discovered as a cheap source of feed, the relation between meat and soybeans dramatically shifted. This section presents the evolution of soybean and meat production and trade, including a world-historical backdrop, a short exploration into how soybeans became integrated with the meat industries, and finally a presentation of the shifting geographies of production and consumption the past decades, with emphasis on the growing appetite for soybeans and beef in China.

### 3.1.1 A Short World Historical Backdrop of Soybeans and Meat

The soybean was domesticated in Asia, in the eastern part of today's northern China, around eleventh century B.C. (Mintz et al. 2008). It developed gradually into the cornerstone of East Asian nutrition, taking a variety of forms—cooked, fermented, milled beans—and foods—miso soup, tofu, tempeh, and soy sauce (Hymowitz 1990). While the Asian cuisine and diet remained mainly vegetarian over millennia (with an exception for the Mongols), the European cuisine, particularly in some areas, developed around meat and dairy as high status, desirable food. In preindustrial European agricultural systems, however, farm animals were mostly fed either with crop residues (hay) and by-products from food processing, or by grazing land not suitable for cropping, and were accordingly not a major competitor for land with crops (Krausmann 2004).

Europe "discovered" the soybean in the seventeenth century and trade from east to west with soya sauce became common. After international bulk trade took off in the late nineteenth century, whole soybeans begun to be traded to Europe. During the first three decades of the twentieth century, Manchuria was the heart of soybean trade, providing Europe and Japan with an increasing amount of soybeans (whole beans, as well as crushed to meal and oil) to the burgeoning dairy and margarine industries. Soybeans became an increasingly important vegetable protein meat extender, particularly for the dairy sector both in Europe and the US. Most animal fodder, however, was still produced on-farm. While Manchuria was the most important soybean exporter, China remained the leading world soybean producer until 1950, when it was surpassed by the US (High Quest Partners and Soyatech 2008, 5).[3]

Cattle were brought to the Western hemisphere with colonization, and they reproduced rapidly in vast areas of the Southern Cone and North America. Newly industrialized Europe, particularly Britain, was experiencing a chronic shortage of meat. The British started accordingly to invest heavily in creating an efficient meat exporting industry in the New World. Extensive cattle grazing, often on vast *latifundio* land, became the dominant agrarian activity in the region, and remained so for centuries (Chapter 2 in this book). Foreign (mainly English) investments and inputs in breeding, railways, meatpacking factories, shipping, and trading opened the way for massive beef trade. Thanks to large amounts of imports of corned, and later frozen and chilled beef, the British rapidly increased its per capita consumption of meat at the beginning of the twentieth century.

### 3.1.2   Soybeans Become Part of Meat Production

Soybeans had been known in the US since the late eighteenth century, but had not moved outside the experimental research fields to any significant degree. In the 1920s and 1930s, however, the *United States Department of Agriculture* (USDA) put a significant amount of effort and money on investigation on in the versatile and highly nutritional soybean for enhancing effectivity of animal farming; bringing the animals to slaughter weight the cheapest and fastest way possible. The soybean contains 40% protein and 20% oil, along with 8 essential amino acids. By crushing whole soybeans into soybean meal and oil, the meal can be used as a protein-rich ingredient in feed and the oil can be used in a wide range of commodity chains, from industrial to edible purposes. In 1917, two US researchers demonstrated that properly heated soybean meal is superior in nutritional quality to unheated soybean meal (Hymowitz 1990). This increased the potential high value of processed soybeans a cheap and nutritious input for the expanding dairy and meat industries. The USDA also developed new high yielding soybean cultivars based on new Chinese germplasm, and started to massively promote and subsidize soybean production among farmers. In this way, the soybean finally become popular among commercial producers (Hymowitz 1990).[4] The soybean cultivation in the US expanded from an area of 20,235 ha in 1907, to 2.226 Mha in 1935 (Shurtleff and Aoyagi 2009, 47). Between 1941 and 1942, soybean production doubled. From 1930 to 1942, the US' share of the world production increased from 3 to 46.5%.[5]

Within the efforts to develop technologies to raise production per animal at a facile price for consumers, the soybean meal became increasingly important as a low-cost, high protein feed ingredient. By feeding animals with feed concentrates, rather than allowing them to graze freely, productivity increased substantially. Soybean meal became the primary protein feed source for livestock kept in confined quarters. The US has led the world in large-scale factory-type animal farming, pioneering the use of intensive livestock rearing of pigs, cattle, sheep, and poultry. Notably, a wide array of meat produced in the US became to an increasing, albeit varying, degree a soybean-based product.[6] Soybean oil was highly valued and demanded for industrial purposes, and after World War II, it also became increasingly used for edible purposes, for example as the main ingredient in margarine and as a key ingredient in ready-made frozen foods.

*Soybeans and Meat—From Engagement to Marriage*
The US integrated intensive soybean-meat model started to spread the world. The most intensive animal farming is typically poultry and pork, where animals are raised indoors under factory-like conditions and given protein-rich cakes to fatten quickly (Lapitz et al. 2004, 59). According

to the *Food and Agriculture Organization of the United Nations* (FAO), intensive systems account for about 60% of global pork and chicken meat (FAO 2018).[7] Moreover, artificial insemination and hybridization in the poultry and swine sectors have raised feed conversion and growth rates. The use of soybeans has also increased dramatically in dairy production. Protein-rich feed together with other new intensifying technologies has led the average milk cow to produce six to seven times as much milk today as a century ago (Howard 2016, 95).

Notwithstanding the diversity of bovine productive systems, in most cases the calves are typically born in the spring, weaned by the fall and sold to producers specialized in fattening the cattle to slaughter weight.[8] Different operators then follow different development paths. Some have the ruminants kept in concentrated confinement quarters (feedlots), while others allow the cattle to graze, in permanent or in-sown pastures. Grazing can be supplemented with dry forage, silage, legumes, and grain. Before being sent to the slaughter, almost all cattle (least in the US and Europe) spend some time in the feedlot—eating high-energy feed rations, including soybean meal (USDA 2016).[9] Beef is nevertheless still a very heterogeneous productive activity, ranging from cattle raised in very extensive grazing systems to feedlots (including the most intensive and large-scale factories—*Concentrated Animal Feeding Operations*—CAFOs).[10] The overall trend is nevertheless that an increasing number of livestock that were once raised on pastures are now increasingly raised on feed formulated on the basis of vegetable protein meals to speed up their growth while minimizing costs.

Processed soybeans (soybean meal) are now the largest source of livestock protein feed in the world (Davis and D'Odorico 2015; de Waroux et al. 2017; MacDonald et al. 2015).[11] Soybean meal has also become the most widely used plant protein source in intensive fish farming, increasingly replacing fishmeal (Craig et al. 2017). While soy-based foods, such as tofu, soymilk, and soy sauce have gained popularity outside Asia in recent years, only a small portion of the rapidly increasing soybean harvest is used for such foods, the lion's share of the soybean harvest ends up as meat. Consumers are, nevertheless, not aware that meat, margarine, dairy, and cultivated fish are largely soybean-products, since they are not marketed as such. The soybean's oil (a "byproduct" from the crushing of whole beans to meal) is also valued for its multiple uses, ranging from cooking oil, to margarine and biodiesel (Davis and D'Odorico 2015; de Waroux et al. 2017; MacDonald et al. 2015; OECD/FAO 2016b).

*Skyrocketing Demand*

In response to the soaring soybean meal demand, the soybean world production increased a remarkably 22-fold between 1950 and 2017, from 16 million metric tons (mmt), to 350.8 mmt in 2017/2018 (Shurtleff and Aoyagi 2009, 233; USDA 2018c). More than 80% of world production is processed into soybean meal and oil (USDA 2018c). This is not only driven by the higher share of soybeans in animal feed (intensified animal production), but also by the dramatic absolute increase in meat and dairy production (OECD/FAO 2016b). The average yearly meat consumption per capita has risen from 23 kg in 1961 to 43 kg in 2013.[12] During the same period, the world population has almost doubled, exacerbating demand further (Roser and Ortiz-Ospina 2018). Only in a decade, between 2005 and 2015, the international meat production increased almost 20% (OECD/FAO 2016b, 107–108).

The rapid rise in meat consumption thus resulted in higher demand for feed crops. Livestock consumes around 6B tons of feed (dry matter) annually—representing approximately one-third of global cereal production (about 40% of global arable land). Soybean cakes represent around 4% of the global livestock feed intake (Mottet et al. 2017).[13] The growth of meat production the past decades has mainly been met by large-scale, specialized feedlot farms, and this trend can be expected to continue (Mottet et al. 2017). However, demand has also risen for grazing resources for ruminant animals, such as permanent pastures and meadows (Mottet et al. 2017). Livestock uses 2 Bha of grasslands worldwide today, of which about 700 Mha could be used as cropland, while the rest cannot be cropped because of slope, elevation, type of soil, or climate (Mottet et al. 2017).

Grass-fed meat, previously the norm, has now become a niche market; consumers can pay a little more to eat meat from cows that have been exclusively grazing. In order to receive this price premium, however, sophisticated and cost enhancing traceability systems need to be adopted so that consumers know that they get what they pay for. The meat and dairy production complex is the world's largest user of land resources, with grazing land and cropland dedicated to the production of feed representing almost 80% of all agricultural land.

### 3.1.3    The Shifting Geographies of Production and Trade

Soybean cultivation skyrocketed in the US after 1940, while the Asian soybean production remained at more or less the same level between 1922 and 1979 (Shurtleff and Aoyagi 2009, 112–113). In 1950, the US passed China as the leading soybean producer in the world. In 1960, the US produced 60%, and China 33% of the 26.08 mmt of the world's total soybean production (Shurtleff and Aoyagi 2009, 112–113). Almost 90% of total soybean export the same year came from the US, since China consumed all of its production domestically (Shurtleff and Aoyagi 2009, 81). Since then, the global production of soybeans is more than 13 times bigger and an increasing share is traded. The US soybean production and export have continued to rise in absolute terms, but the relative share has been falling, whereas the share of Latin America has been rising (Howard 2016, 92–93; Interagency Agricultural Projections Committee 2011, 34; USDA 2011). While there has been a general rise in the world production and trade of the several agricultural products since the 1960s, e.g., maize, wheat, rice, and cotton, none has grown so steep and fast as the soybean and derives.

*Latin America as an Increasingly Important Soy Producer*
Latin America was an important world agrofood provider already by the nineteenth century, as presented in detail in Chapter 2. It is not surprising that in particular Brazil, Argentina, Paraguay, and Uruguay have a long history of high economic reliance on agricultural products, since around 90% of the land area in this region is apt for agrarian production (Lapitz et al. 2004, 10–11). Latin America's participation in the international agrofood market, however, declined 1930s–1970s, partially explained by the inward-looking development strategies adopted in the region (Chapter 2 in this book). Since then, the region has become an increasingly specialized agrofood provider again, and its share of total agricultural trade has risen (Clapp 2016, 10).

Soybean cultivations started to expand in Brazil in the 1970s, resulting from public promotion of agricultural expansion and "modernization" through subsidized credits, bank debt write-offs, exports subsidies, investments in infrastructure and processing (Lapitz et al. 2004, 11; Shurtleff and Aoyagi 2009, 174, 183, 184, 242).[14] This production "spilled-over" rapidly to Paraguay, as Brazilians expanded into the Eastern parts of the country, but also due to agricultural programs of the Stroessner regime.[15] Argentina also promoted the soybean through research and extension and had rapid increase of production.[16] The pace and scope of the soybean cultivation expansion in Latin America accelerated significantly after the turn of the century. The soybean area in Brazil, Argentina, Paraguay, Bolivia, and Uruguay soybeans expanded from 30 Mha in 2000, to 60 Mha in 2017 (USDA 2018c).[17]

The soybean became the main driver or land-use change, which in turn was driven by meat demand (Mottet et al. 2017). The increased demand was in part translated into the soybean prices. Global soybean prices (based on Chicago Soybean Futures Contracts traded in CBoT) in March 2001 were USD 164/t, in March 2007 prices were USD 330/t. Prices went down after the financial crisis in 2008, but recovered rapidly. In March 2014, the soybean price stood at USD 522/t (OECD/FAO 2016a, b; Sandoval 2016).[18] The economic margins of soybean production could be even higher than suggested by the rising prices, since yields per ha in Latin America were rising and since costs on labor and fuel decreased by the adoption of the genetically modified soybeans (*Roundup Ready*—RR). Advances in container trading further increased margins since transport costs were reduced. Moreover, the economic margins of soybean business were also boosted, at least for the transnational agribusiness firms, by regulative shifts in the international arena during the 1990s in favor of liberalization. Worth to mention here are the *Agreement of Agriculture* (AoA) and the agreement of *Trade-Related Aspects of Intellectual Property Rights* (TRIPS), both of the World Trade Organization (WTO). The deregulations of financial activities on the commodity markets and the new trade agreements with dispute settlement mechanisms protecting private firms vis-à-vis states in third countries (Chang 2009, 480; Clapp 2016, 64–67; Wolford et al. 2013, 2), further strengthened the flows of international capital and firms moving into the Latin American soybean business.

As explored in detail in the next chapter (Chapter 4), however, the expansion of soybeans in Latin America cannot exclusively be explained by exogenous factors, but was also made possible by previous public funded research and extension during the 1960s and 1970s, and perhaps more importantly by the neoliberal turn in the region in the 1980s and 1990s. Export-led growth in line with "comparative advantage" became the new development receipt for resource-rich developing countries (Clapp 2016, 64–67; Wolford et al. 2013, 2). The low land prices in Latin America was its main competitive advantage for soybean production (Lapitz et al. 2004, 162). The economic margins of soybean production in Latin America was higher than in the US, despite the significant amount of farm subsidies still provided to the soybean sector in the US (Lapitz et al. 2004, 59).[19]

Brazil alone is since 2017 on par with the US in soybean production (representing around 33% each of total world production), and estimated to pass it for the next harvest (USDA 2018c).[20] The US's relative participation in the soybean trade is smaller than its production, but it is nevertheless still the second largest exporter in the world of whole soybeans (after Brazil), mainly to China (USDA 2018c).[21] The third biggest soybean producer in the world is Argentina (and largest exporter of soybean meal and oil). These three giant producers grow 86% of today's global soybean production—cultivated over more than 100 Mha of land (Service/USDA 2018; USDA 2018c).[22] While soybeans are very important in Paraguay and Uruguay, they are not big world exporters compared to their giant neighbors. Here it is important to bear in mind that these are small countries. Argentina is for example more than 6 times bigger than that of Paraguay and more than 15 times bigger than that of Uruguay. In spite of their small size, they figure among the world's top ten exporters.

Chart 3.1 show the share of countries that exported whole soybeans, soybean meal and soybean oil in 2017. Total value of whole soybeans was USD 58.1B, value of soybean meal exports were USD 23B, and value of soybean oil USD 9.3B. Areas are proportional to the export value in dollars per country. Source: Observatory of Economic Complexity, data from Comtrade FAO statistics. Interactive versions are available (open access) at atlas.media.mit.edu.

*Beef Trade*

As outlined in Chapter 2, beef has been important in the Southern Cone for the past hundred and fifty years, but the sector was largely stagnated for the most part of the twentieth century. Brazil and Paraguay started, however, a process of both rise in area and productivity in the 1960s.[23]

Compared to the soybean, the livestock sector appears still as an overall national sector. In 2017, less than 15% of global meat, milk, and egg production were traded internationally (FAO 2018). International trade on meat is more restricted, taxed, and regulated in protectionist ways, than the "free" soybean of no quota and low tariffs. Thus, while the soybean is heavily traded to become cheap animal feed the meat

**Whole soybean Exports per country, 2017:**

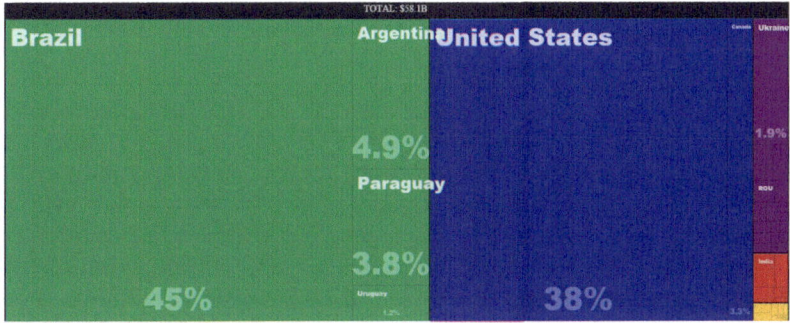

**Soybean Meal Exports, per country, 2017:**

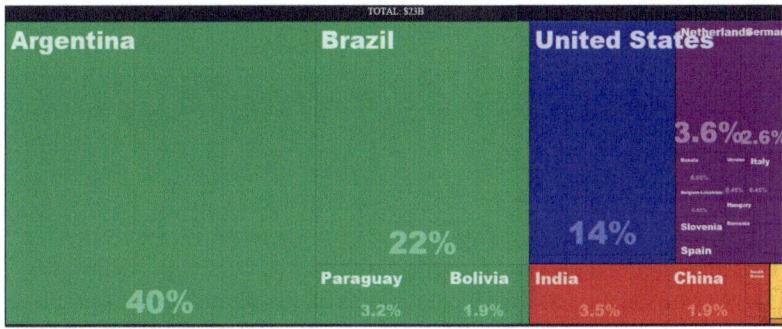

**Soybean Oil Exports, per country, 2017**

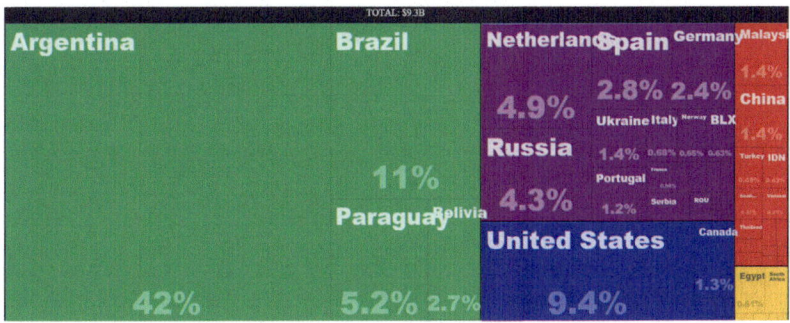

**Chart 3.1** Soybean exports per country, 2017

products are still mostly consumed nationally. However, behind the label of "national meat" there is an increasing amount of imported soybeans. On the other hand, meat has a far higher value than soybeans. Accordingly, of the total value of international agrofood trade between 2000 and 2009, almost half was attributable to meat and animal products, while wheat, soybean, and maize only represented 21% of the value (MacDonald et al. 2015). In the past decades, however, meat trade has increased, particularly beef trade (USDA 2018b). Latin America accounts for more than 25% of world bovine production, and more than 20% of global poultry production (Cepal, FAO, and IICA 2015, 14). Between 2005 and 2015, exports of bovine meat from the region doubled (Cepal, FAO, and IICA 2015, 14).

In 2018, global beef production was 63.0 mmt. The same year, 10.5 mmt of beef was internationally traded (USDA 2018b). The top-four beef exporting countries; Brazil, India, Australia, and the US concentrate together the lion's share of total beef exports (USDA 2018b).[24] Some of the constraints imposed on beef imports will be presented in the next section with focus on the importing side, but it is important to bear in mind that also beef producing countries can put constraints to trade. For example, Argentina is the third top beef producer in the world, but it is only the tenth top exporter with 0.35 mmt of beef exported in 2018 (USDA 2018b). This is mostly the result of national regulations in favor of domestic beef consumption at low prices (Chapter 4 in this book). Both Paraguay (0.4 mmt) and Uruguay (0.42 mmt) actually exported

**Chart 3.2** Frozen bovine meat exports (including beef, steak, red meat) per country, 2017

more beef than Argentina, as the eighth and seventh biggest beef exporters respectively in 2018 (USDA 2018b).[25] If considering export value
per country, instead of volume, the export share of each country and
region in 2017, can be visualized as follows:

Chart 3.2 shows the share of countries that exported Frozen Bovine
Meat in 2017. Total value of soybean meal exports were USD 22.4B.
Areas are proportional to the export value in dollars per country. Source:
Observatory of Economic Complexity, data from Comtrade FAO statis-
tics. Interactive versions are available (open access) at atlas.media.mit.edu.

*China's Growing Appetite for Soy and Meat*
If the most important shift in the soybean and meat producing side is
the relative rise of Latin America as the most important producer, and
even more important exporter, the biggest shift on the consumption
side is the exploding rise of Asia, particularly China (ECLAC 2017;
Hairong et al. 2016; OECD/FAO 2016a, 107; Zhang and Liu 2014).
This is due to the snowballing global meat and dairy consumption per
capita in China, and many other parts of Asia. Soybeans have been a fundamental ingredient in the mainly vegetarian Chinese diet for millennia, but the unprecedented increased appetite (and purchase capacity)
for meat, dairy, and egg, have created an exponential growth in soybean
demand from its burgeoning animal farming industry. As recently as
1990, the US soybean consumption of soybean meal was quadruple that
of China. Today China consumes the double of that of the US.

While China has become the biggest soybean meal consumer in
the world, the EU is actually the biggest buyer. The EU is the world's
third-biggest soybean meal consumer (after China and the US), but
more than half of the soybean meal it consumes is imported (USDA
2018c).[26] By contrast, China imports whole soybeans and makes the
crushing into soybean meal and oil domestically. China, as many other
countries, use a differentiated tariff-system with higher import taxes
on processed agricultural products, such as soybean meal and oil, than
unprocessed primary products, such as whole beans (Lapitz et al. 2004,
33; USDA 2018c).[27] The Value-added Tax (VAT) rate applied to
imports of processed agricultural products in China has recently been
lowered to 16% in 2018. It is, nevertheless, still higher than the 10%
VAT rate on unprocessed agricultural products (Ward 2018).[28] These
policies are designed to protect the activities of more value-added, while
at the same time access inputs as cheap as possible. China, thus, supplies most of its poultry and pork demand domestically, albeit based on

imported whole soybeans. The 12.9 mmt of domestically grown soybeans in China (GM-free) are instead mainly used for direct human consumption, such as miso soup and other vegetarian dishes, soya sauce, and kitchen oil (USDA 2018c).

The cradle of the soybean—and still the world's fourth-largest soybean producer—has in this way become the world's most important buyer of whole soybeans (He et al. 2016). China accounted for almost two-thirds of global whole soybean imports; 93.495 mmt out of 152.9 mmt in 2016/2017 (USDA 2018c).[29] This development is in stark contrast to high self-sufficiency targets that marked Chinese agricultural and food security policies in the past. China represented for a long time an important exception to the general rule toward agro-food trade liberalization,

**Whole Soybean Imports per country, 2017:**

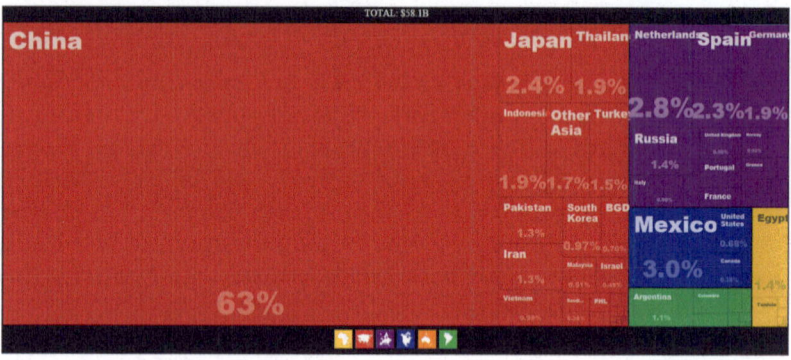

**Soybean Meal Imports per country, 2017:**

**Chart 3.3**    Soybean imports per country

through its national food security plans based on self-sufficiency. After joining the WTO in the 1990s, China nevertheless relaxed import restrictions on grains slightly (Clever and Xinping 2016; Hairong et al. 2016; MacDonald et al. 2015, 275). Since then, the imports of whole soybeans have been further liberalized in several steps (Zhang and Liu 2014).

It is hardly impossible to exaggerate the role of China in the international soybean market. China's appetite for whole soybeans is the main motor by the soybean expansion in Latin America the past decades (CAS 2018). More than 70% of soybeans exported from Latin America, in 2016, ended up in China (ECLAC 2017; Hairong et al. 2016; OECD/FAO 2016a, 107; Sandoval 2016). China has also been an important market for the US (USDA 2018c). The future of US soybean exports to China is nevertheless highly uncertain. In response to the Trump administration's duties on a wide range of Chinese goods (US 301 Investigation), China imposed in July 2018, an additional 25% tariff on Imported US Agricultural Products, including soybeans and pork (Ward 2018). China argued to be forced to take these "necessary countermeasures" in response to the US launching "the largest trade war in economic history".[30] While there is a high degree of uncertainty about the future, China is in the short-term expected to largely replace the US imports with more imports from South America. However, the South American supplies fall to seasonal lows in the first and fourth quarters, so Chinese companies are expected to still import small amounts from the US when South American supplies run down.[31]

Chart 3.3 show the share of countries that imported whole soybeans and soybean meal respectively in 2017. Total value of whole soybeans were USD 58.1B and of soybean meal imports were USD 23B. Areas are proportional to the import value in dollars per country. As seen, China is extremely important as buyer of whole soybeans, but not of soybean meal. Europe is still important in this market. Source: Observatory of Economic Complexity, data from Comtrade FAO statistics. Interactive versions are available (open access) at atlas.media.mit.edu.

While the global soybean trade now is facing more uncertainty and tariffs, it is still much more open and "free" than the meat trade. Almost all countries in the world use protectionist measures for the domestic meat sector, limiting market access on imported meat products through high import tariffs and restricted quotas (Lapitz et al. 2004, 69–70). For example, soybeans face almost no restrictions from the European market, but the meat products are subjected to extremely high tariffs.[32] However, the EU also has tariff rate quotas with duty preference inside the quota. One important tariff quota for the Mercosur countries is the so-called "Hilton quota" for high-quality fresh, chilled, and frozen beef (No. 593/2013). In 2018, the quota was 66,826 MT, divided among the suppliers Argentina, Brazil, Uruguay, Paraguay, US, Canada, Australia, and New Zealand.[33] Inside the Hilton quota, beef enjoys a duty preference charged 20% (USDA 2018a).[34] Another EU quota is the autonomous beef quota covering grain fed hormone-free beef, sometimes known as the "EU 481 grain fed quota", which consists of 48,200 MT/yr, with zero quota.[35] The suppliers are Australia, New Zealand, Uruguay, Argentina, Canada, and the US, following the allocation principle is "first come, first served".

Total external imports to the EU of fresh/frozen beef stood at 196,000 MT in 2017 and 70% came from the Mercosur countries (142,500 MT).[36] The high beef tariffs outside the quotas reduce economic margins for exporters. The EU and the four founding members of Mercosur (Argentina, Brazil, Paraguay, and Uruguay) are since many years negotiating a trade agreement as part of a bi-regional *Association Agreement*.[37] This involves talks about an increase of reduced-tariff quota on beef and other agricultural products, which will give the Mercosur countries unprecedented access to Europe's agricultural markets in exchange for buying industrial products from the EU. The negotiations have encountered several setbacks and one of the major obstacles is the beef trade.[38] This issue is especially sensitive in EU member states with strong domestic farmers' organizations opposing foreign competition on meat, like France, Poland, and Ireland. However, on June 29, 2019, after two decades of negotiations, it was finally announced that the parties had reached an agreement. The Mercosur countries will win a new 99,000 MT quota of beef at a 7.5% tariff, phased in over five years, along with tariff-free 180,000-tonne quotas each for sugar and of poultry, as well as decreased tariffs on rice, meat, honey, hides, and many other agrarian based products. Meanwhile, the EU will get sharp tariff reductions on industrial products. The companies from the EU will also have access to bid on public procurements in the Mercosur countries. It is too early to say in detail what the effects of this historical trade deal will be, but it is clear that it will benefit the agribusiness sector in Mercosur and the industrial sector in Europe, while harming the industrial sector in Mercosur and European farmers. However, the agreement still needs to be ratified by the national parliaments of all member countries of both blocs, as well as by the European Parliament and EU Council. No timeline has been given (as of July 2019).[39]

The rapid rise in beef demand in Asia has in this way been important for the beef exporting countries in Latin America. China's rise in beef imports alone has by far been able to replace the highly protected European meat market. China has become the second biggest beef consumer in the world, after the US, but before Brazil and the EU (USDA 2018b). China's beef consumption has had an annual growth rate of 7.8% in the past decade and became the top net buyer of beef (USDA 2018b). While China still has high self-sufficiency targets, it faces severe constraints to supply the soaring meat demand, not least because of loss of productive land due to degradation, erosion, and urbanization. Thus, Chinese beef production is estimated to 7110 mmt of beef, while consumption is estimated to 8140 mmt in 2018 (USDA 2018b). In this way, Chinese beef imports are around five times bigger than those of the EU.

In addition, in the wake of environmental and sanitary scandals, a new purchase-strong sector of urban citizens prefers to buy imported food, particularly luxury food such as beef. Robust demand in China and Hong Kong is expected to continue growing, as stagnant domestic production is unable to fulfill rising consumption. Besides China, Japan, South Korea, Egypt, and Malaysia have had a rapid increase in beef imports the past ten years (USDA 2018b) (Chart 3.4).

**Frozen Bovine Meat Imports (including beef, steak, red meat) per country, 2017:**

**Chart 3.4**  World bovine meat imports, 2017

Areas are proportional to the export value in dollars per country. Besides the clear dominant of China, it is clear that Asia, as a region, has become extremely important international beef buyer, and represents around two-thirds of the total imported amount of USD 22.4B of frozen bovine meat. Source: Observatory of Economic Complexity, data from Comtrade FAO statistics. Interactive versions are available (open access) at atlas.media.mit.edu.

## 3.2   Concentration and Vertical Integration in the Soybean and Livestock Chains

The past sections presented how the soybean and meat production and trade have developed over the years. This narrative is based on official data from sources such as USDA and FAO, and their state-centric accounting systems. However, the main agents involved in the actual trading and processing are not states but giant agribusiness firms. International trade is increasingly intra-firm trade, and the dominating commodity trading firms are enormous in size, although their precise scale is hard to determine, because of lack of official statistical procedures and reliable data. When investigating into power relations and market shares of the agribusiness firms, there is no digital databases to dig into from FAO, OECD, USDA or the World Bank, but mostly business magazine and newspapers, writing about spectacular mergers and acquisitions, all the company's annual reports and websites. This section is an attempt to present the soybean and meat chains beyond the state-centric narrative, and concerning the upstream (input) and downstream (trading/processing) stages.

### 3.2.1   Upstream: GM, Seed, and Agrochemical Stages

Since the development of industrial agriculture, and the worldwide diffusion of the same through the green revolution, food production is increasingly intensive in "external" input. Thus, in the same way as meat production increasingly relies on feed supplements (soybeans and grains) coming from far away from the animal farm (and often supplemented with antibiotics, breeded sperms and hormones to maximize their weight gain), the actual crop cultivation relies more and more on irrigation, agrochemicals, and seeds from outside the farm (in other words, farmers save less seeds, and the monocultarization of agriculture require more pest and weed control, since the pest and weed community get specialized). The latest step along this path is bioengineering, which has emerged under an increasingly strong international intellectual property rights regime (IPR) with patent rights on seed genome and other immaterial innovations, strengthened by the adoption of the TRIPS-agreement. As intellectual property rights have been strengthened, the concentration rates of these markets have become rather extreme. These regulative shifts created huge economic incentives for private investments in Research and Development for biotechnology (Clapp 2016, 78–79). At the same time, biotechnology has increasingly linked previously separated input markets—biotech, seeds, and pesticides—under the umbrella of the same technological package, which has spurred vertical integration.

*The Gene Revolution*

Seeds and animal breeds have been selected in ways that develop specific characteristics (traits) over millennia. Deliberate, intentional, scientific selective breeding to produce desired results have occurred in agriculture since the eighteenth century. Until the late twentieth century, however, farmers and firms could only breed new traits of a plant or organism from closely related species, In the wake of technological innovations and enforced intellectual property protection emerged the so-called "gene revolution". According to the *International Service Acquisition* (ISAAA), biotech crops are the fastest adopted crop technology in the history of modern kind. From 1996 to 2016, the world biotech crop area increased from 1.7 Mha to 185.1 Mha.[40] In contrast to the diffusion of the green revolution technologies, the gene revolution has not been diffused by public organizations with social intentions, but by an ever more concentrated group of giant private firms.

The first biotech, or genetically modified, crop was the Monsanto soybean (GTS 40-3-2) commercialized in 1996. The history behind this "success" is both illustrative and widely known; the big agro-chemical firm Monsanto from Iowa, US (owner of the glyphosate-based weed-killer brand Roundup), patented, in 1996, an herbicide-tolerant (HT) trait, with the brand name RR.[41] When this genetically modified trait is inserted into a soybean seed, this becomes tolerant to glyphosate, which allows for effective and cheap weed control through glyphosate spraying throughout the growing season, without harming the soybean plant. In this way, the agrochemical and seed stages started to integrate intensively. The capacity of the soybean to adapt to different climate and soil conditions was already developed under the system of conventional soybean farming, but the diffusion of the new technological package—centered in herbicide tolerant soybeans and no-tillage farming—further enhanced the capacity of good yields in less perfect soils and standardized the production forms. Moreover, the combination of no tillage with glyphosate allows to spare both labor and gasoil (Barbazán et al. 2011; Souto and Ferenczi 2010; Souto 2013).

This technology was rapidly diffused through Monsanto's strategy of strategic mergers and acquisitions of seed companies and widespread licensing, which rendered the company an almost monopoly position in the soybean market (Moss 2009). Farmers had no longer right to save their own seeds and Monsanto aggressively and successfully filed lawsuits against farmers that used the technology in saved seeds, without paying patent. The practice of saving commodity seeds accordingly dropped—in the US, the amount of saved seeds went from 63% in the mid-1990s to 10% in 2001 (Howard 2016, 108–110).

Besides the actual biotech events, the seed companies became extremely concentrated in the wake of the gene revolution. The concentration rate of the four top seed companies (CR4) in the soybean seed market is estimated to 70% (Howard 2016, 107).[42] In addition, by calling the new soybean trait for RR, Monsanto managed to maintain an 80% market share of the glyphosate market, in spite that several years had passed since Monsanto's patent on glyphosate had expired (Howard 2016, 109).

The patent of soybeans RR (GTS-40-3-2) expired in 2016, but Monsanto successfully made the producers adopt new varieties. In Latin America, for example, it launched soybean RR2Y (MON89788) in 2013, which is marketed to particularly fit Latin American soils and climate conditions.[43] One reason to the fast adoption of new varieties with new biotech traits, is that the—now generic—soybean RR has become less effective. This is due to rising levels of resistance to glyphosate within the weed community, decreasing its efficiency as total herbicide. Besides causing weed resistance, the effects of the massive use of glyphosate in agricultural communities (both directly and indirectly through glyphosate residues in groundwater and drinking water) are highly disputed. Accordingly, new genetically modified events, to be combined with other herbicides, have emerged, for example with glufosinate 2,4-Dichlorophenoxyacetic acid (usually called 2,4-D), and dicamba. The herbicide glufosinate is about to be phased out of EU fields.[44] However, EU has at the same time allowed the imports of soybeans designed to be combined with these herbicides.

The majority of the 35 biotech soybeans events (approved for cultivation at least somewhere in the world) is still HT—designed to be combined with specific herbicides. There are, nevertheless, also insect resistant (IR) traits, "quality" enhancing traits (such as a soybean with higher levels of oleic acid) and stacked traits (combining HT with IR) in the biotech soybean portfolio. Drought resistant traits are in the pipeline to be commercialized in the coming years. The lion's share of new traits still comes from either the US or Europe (ISAAA 2017; ISGA 2017; Waltz 2015).[45]

While there nowadays exist many crops with commercialized biotechnological traits (maize, cotton, canola, rice, alfalfa, potato, sugar beet, squash, papaya, eggplant), the soybean, with 91.4 Mha of biotech area in 2016, is still accounting for around half of total biotech area. It is also the crop with the highest share of biotechnology use (the 91.4 Mha of biotech soybean represents 78% of the total soybean area of 117 Mha, in 2016). In Argentina and Uruguay, all soybean is biotech, while in Paraguay 95% of soybeans are estimated to be biotech (ISAAA 2017; ISGA 2017).

While the biotech soybean area is huge, it is actually concentrated to only eleven countries of which ten are in the Western hemisphere (ISAAA 2017).[46] China's domestic production of soybeans, which is mainly used for direct human consumption, is GM-free. The EU does not allow genetically modified soybean varieties for planting within the EU, but it does allow many GM varieties for imports for animal feed and for food. Soybean meal is the primary source of protein for the EU animal feed market, representing more than 60% of vegetable protein (European Parliament 2015). The EU is actually the world's biggest soybean meal importer (USDA 2018c). Around 90% of the imports originate from countries in which almost all cultivated soybeans are genetically modified, mainly Latin America (European Parliament 2015, USDA 2018c).[47] There are, on the other hand, hardly any imports of GM soybeans for food on the EU market. This is probably due to the compulsory labeling of food containing GM organisms, and the expectation that consumers prefer to buy food that is not declared GM. The consumers' aversion toward GM-food does not, however, translate into the soybean meal market, since animals can be fed with GM soybeans without meat, eggs, and milk being labeled as GM-derived products.

Before a GM event is approved for commercialization, it has to pass a relative severe, long, and complex procedure of risk assessment. One side effect has been that today only a small range of powerful private corporations can afford those costly and long processes. The biotech approval procedures of the EU are even stricter and slower than elsewhere. This is heavily criticized by the biotech industry (European Parliament 2015; ISGA 2017). For example, the *International Soybean Growers Alliance* (ISGA), representing the soy industry in Argentina, Brazil, Canada, Paraguay, Uruguay, and US, and thus more than 90% of global soy-exports, criticizes the EU policies in regards to GM. They claim that the EU-decisions rest on politics and emotions rather than science, leading to unsound non-tariff barriers to trade (ISGA 2017).[48] Europe's livestock and feed manufacturing industries have also argued in favor of EU authorizing more soybean imports since they depend heavily on shipments of soy products (beans for crushing in Europe and already crushed soybean meal) as a source of protein-rich and high-quality feed. At the same time, NGOs, the media and the *European Parliament* have expressed criticism about EU being too soft in regards to the biotech companies and that the policies are too GM-friendly.

Animal genetics differ slightly from the world of crops. There are about 40 domesticated species and 9000 breeds of livestock (the products of both natural selection and human-controlled selection, as well as crossbreeding). Locally adapted livestock breeds have characteristics that help them cope with local ecosystems, including grazing patterns, climates, parasites, and diseases. Business-led selection and crossbreeding have focused on a limited number of breeds and concentrated on production traits leading to standardization. These are often internationally traded from developed to developing countries and tend to replace or be crossbred with local animals.

The control of the genetic breeds of livestock in general is extremely concentrated. For example, in the poultry breeding market, three firms control 94% of sales, and in dairy cattle breeding one firm makes up for more than 85% of the milking cows in the US (Howard 2016, 117–119). Animal genetics is nevertheless a much smaller and less important economic sector than plant breeding. The commercial breeds lead to standardization, but the intention is to improve feed conversion rates of livestock to boost growth rates, yields, and reproductive efficiency. There are also important controversies involved in this technology. Together with a handful of other big pharmaceutical and chemical companies, Monsanto developed and hold patents on the controversial *recombinant bovine somatotropin* (rBST)—a synthetic hormone for dairy cows that enhance milk productivity (Otero et al. 2008, 159–165). One of the controversies of rBST is the problems of surplus production of milk, chronical in the dairy business since the US begun with price supports and government purchases in the 1930s (Howard 2016, 93–94).

*The Political Economy of GM*

The "gene-revolution" made the soybean business even more profitable for seed companies, agrochemical firms, and traders alike, while it also increased the incentives for further integration with the meat chains. The alma mater of biotech soybeans, Monsanto (who first developed and patented the soybeans RR, GTS 40-3-2), has largely managed to remain its market leading positions and was until 2018 the world's biggest supplier of genetically modified seeds. Monsanto is the company with the most approved soybean events, followed by Bayer, with the market brand *LibertyLink*. The net sales of Monsanto's soybean seeds and traits totaled almost USD 2.7B in fiscal 2017. Gross profits from soybean products climbed 35% over 2016, according to the *Securities and Exchange Commission*, of the US government.[49]

Monsanto can be described as a powerful chain driver, playing one of the lead roles in constructing and managing an increasingly powerful segment of the agrofood system. However, in June 2018, the company name Monsanto, after 117 years, ceased to exist. The German agrochemical and biotech company Bayer bought Monsanto for USD 62.5B—the largest cash-deal on record.[50] Soybeans, corn, cotton, and wheat were a major emphasis of both Bayer and Monsanto. To satisfy the requirements for approval of the Bayer-Monsanto deal by the US *Justice Department* (DOJ), Bayer sold off some of its soybean businesses (for example Bayer's *LibertyLink* soybean to be combined with glufosinate, was sold to German BASF in an USD 8B deal) as well as its weed killer glufosinate. Bayer is now the largest seed and Agrochemical Company in the world, with more than a quarter of the world's markets on seeds and pesticides.[51] An executive briefing on trade from the *United States International Trade Commission* expressed in April 2018 that the Monsanto-Bayer merger can increase seed prices on soybean (and maize and cotton) for farmers (DeCarlo 2018). The decision to drop the brand name, Monsanto, is probably due to the company being one of the most hated companies in the world.[52]

The Bayer-Monsanto deal has received a lot of attention, but should be seen in the light of a wider movement of *mergers and acquisitions* (M&A) in the global agrochemical and seeds industry. Some fifty years ago, there were approximately seventy agrochemical companies only in the US. By the turn of the century, roughly eight major multinational firms controlled the majority of the domestic market and large part of the international market. Until 2015, the global agrochemicals and seeds market was commonly referred to as led by the "Big 6" (Bayer,

BASF, Dow, DuPont, Monsanto, and Syngenta). Together they controlled around 75% of the agrochemical sector and 61% of the seeds sectors (DeCarlo 2018). In 2017, the Chinese state-owned company ChemChina's acquired the seed and agrochemical giant Syngenta for USD 43B.[53] ChemChina had already in 2011 acquired the Israeli generic agricultural chemical company, *Makhteshim Agan Group* (now ADAMA Agricultural Solutions), for around USD 2.4B. This marks a new era with China becoming a global giant in the field of agrochemicals and seeds. In 2017, there was also a merger between, the US companies Dow and DuPont (the 4th and 5th largest biotechnology and seed companies in the world respective), now DowDuPont. Thus, from 2015 the seven companies that dominated the global markets of pesticides and seeds have been reduced to four (Bayer-Monsanto; DowDupont, ChemChina-Syngenta, BASF).[54] The three newly formed conglomerates alone are estimated to dominate more than 60% of the market for commercial seed and agricultural chemicals.[55]

These giant mergers and acquisitions can be seen as the logical result of an increasingly strong IPR, combined with high costs and long time-frames involved in taking new agrarian products to market.[56] The private firms have also benefitted from the relative withdrawal of the state from agrarian research, development, and innovation. In addition, the technology of genetic modification allows for altered seeds to be combined with specific herbicides, insecticides, and pesticides, which further paved the way for extreme market concentration and vertical integration (DeCarlo 2018). In this way, it is clear how public regulation has acted in the benefit of the big agribusiness firms. At the same time, public regulation can sometimes act in counter-veiling ways to corporate power. One prominent example is the controversy around the weed-killer glyphosate (widely known under the brand name Roundup of Monsanto, now Bayer, but also sold by other agrochemical companies under other brand names). France decided to ban glyphosate domestically, and is currently within the EU pushing for making the whole Union to follow. The *World Health Organization* (WHO) declared in 2015 that glyphosate was "probably carcinogenic". However, both the US *Environmental Protection Agency* (EPA) and the EU have still considered glyphosate safe to use.[57] A surprising turning point in this debate emerged on August 10, 2018, as a San Francisco jury in a superior court chocked the world and ruled that Monsanto has to pay USD 289M in redemption to a cancer sick man, Dewayne Johnson. The court asserted that the company was responsible for "negligent failure", since it knew or should have known that Roundup (glyphosate) was "dangerous", but still failed to warn him of the health hazards from exposure. The court decision could result in Bayer ending up paying billions to persons that have cancer, as several thousands of similar cases from across the US are currently going through the national legal system.[58] According to Bayer, who saw shares abruptly fall by 11% in response to the court ruling, the ruling will probably be overturned by an appeal court, as it was "at odds with the weight of scientific evidence, decades of real-world experience and the conclusions of regulators around the world".[59] The future will show, but a regulatory crackdown on glyphosate would not only hit sales of pesticides and weed-killers, but also the seeds that firms have developed to be used with them. That would be a major game-changer in the soybean business, but it is beyond the scope of this book to speculate in.

### 3.2.2 Downstream: Trading and Processes Stages

While the lead firms of the input stages, as seen in the past section, clearly hold a lot of power through their technologies (with high degree of asset specificity through the patents), partly determining downstream activities and processes, the giant transnational traders also govern the chain by linking together land, input, infrastructure, market information, access to finance, transport, and consumption in complex webs.

*The ABCD-Companies and the Challenge from China*
The private grain trading multinational firms *Archer Daniels Midland* (ADM), *Bunge, Cargill,* and *Louis Dreyfus Commodities* (LDC), dominate global grain trade flows, including soybeans.[60] These four (often referred to the ABCD-companies) emerged under the first globalization wave, or the *first food-regime,* in the late 1800s or early 1900s. Their power in the agrofood system has increased exponentially along with the massive increase in trade. Today, the four ABCD mega-firms control around 75% of soy exports (de LT Oliveira and Hecht 2017), and around 70% of the world trade of corn, wheat, and soy taken together (Clapp 2016, 105). The process of market concentration, driven by mergers and acquisitions, will probably continue and further enhance market power of the selected few at the top. ADM has for example recently announced interest in buying Bunge.[61] There are, however, new-coming challengers to the ABCD-dominance. *Glencore,* a global

commodity firm that historically focused on nonagricultural commodities has now increasingly moved into agriculture, capitalizing on its knowledge of a range of markets of relevance for agriculture. Glencore has also announced interest in buying Bunge.[62]

The most spectacular rapid advancement in the international grain markets comes, nevertheless, not from yet another private transnational firm, but from the Chinese state-owned trading company *China National Cereals, Oils and Foodstuffs Corporation* (COFCO). COFCO is rapidly increasing its market share. The company presents itself in the following way, on its international website:

> We are industry leaders in rice, wheat, corn, oil and oilseeds, sugar, and cotton in over 140 countries and regions of the world. We have a global coverage of grain and oil producing areas and own a sophisticated global production & procurement platform and trade network. The company already earns more than 50% of its operating income from overseas business. With its access and strong planning, COFCO can ensure a stable supply for two markets, Domestic and International, and be the foundation for food security.[63]

COFCO restructured in the 1990s into an internationally operating entity, but it is only in recent years that it started to invest heavily abroad. In 2018, COFCO passed Cargill in Paraguay as the most important soybean trader, exporting 8.27 mmt of soybeans, which represents 15% of total soybeans exported.[64] It has moved significantly into infrastructure, such as storage, processing, transportation, and port facilities, creating a whole value chain that stretches around the world (Gaudreau 2015). The most symbolically value-laden overseas investment of COFCO was the purchase of the two "classical" agribusiness firms Nidera and Noble Agri, in 2014. This acquisition meant that COFCO considerably strengthened its position in processing, storage, and shipping and infrastructure covering corn, soybean, rice, and wheat among other commodities and fertilizer inputs. In 2018, COFCO sold the seed business part of Nidera (Nidera Seeds) to Syngenta. Until 2017, Syngenta was a Swiss seed and agrochemical company, but it is now owned by another Chinese state-owned enterprise, ChemChina—the previously mentioned upcoming giant in the international agrochemical and seed business.[65]

*The Business Model*
"You can arbitrage the soybean sourcing from any origin," said Soren Schroder, chief executive of the global grain merchant Bunge, to Reuters

May 16, 2018.[66] This quote provides important insights into the business model of soybean trade. The outstanding flexibility and geographical adaptability of the soybean—enhanced by the biotechnology—together with the decreasing costs and increased capacities for storage and container shipping, have allowed for the soybean to become the ultimate commodity; a standardized product that can be sourced from anywhere, or nowhere, for the handful of transnational grain traders to make profits on arbitrage. Besides buying and selling grain, the traders also undertake a range of activities from finance to production, to processing and distribution. In the words of the family owned food-giant Cargill:

> Our team of 155,000 professionals in 70 countries draws together the worlds of food, agriculture, nutrition and risk management. For more than 150 years, we have helped farmers grow more, connecting them to broader markets. We are continuously developing products that give consumers just what they're seeking, advancing nutrition, food safety and sustainability. And we help all of our partners innovate and manage risk, so they can nourish the world again tomorrow. (Cargill 2018)[67]

Cargill—the world's largest processor and trader—engages thus not only in the world's grain trade (where it controls 45% of the market), but also in seeds and fertilizers, contracts with grain farmers, livestock rearing contracts, animal feed production, beef processing, meat packaging, financial advice, financial speculation, elevator/storaging, and shipping. Cargill is not alone; all big traders are but also increasingly vertically integrated with various stages in a variety of commodity chains, such as with agrochemical, seed, crushing, fodder production, livestock rearing, and meat-packaging markets.

The power of the big traders has thus also increased because of strategies of vertical integration adopted by the proper firms themselves (Clapp 2015; Lapitz et al. 2004, 35). The traders have bought export infrastructure (silos, elevators, port terminals, transport fleets, and so forth) as well as big processing plants. The big traders control most of the soybean crushing and processing in the world. In the US, Bunge ADM and Cargill alone control almost 70% of the US soybean crushing (Howard 2016, 74). Accordingly, they have been able to take benefit from the falling prices and often overproduction brought by the development of the model of industrial agriculture (co-created and subsidized by many governments), since it allows them to get input—soybeans—at a low price (Howard 2016, 94–95).[68]

Moreover, the traders take positions at the future market at for example the, *Chicago Board of Trade* (CBoT) and offer farmers to commit their future harvest at a fixed price (for both spot and future markets) to minimize risk and to be able to cover production costs before harvest. Usually, they discount cost for logistic services from the final price when accounts are settled. Thus, the traders make money by buying for example soybeans at one discount rate and sell them at another (Baraibar 2014). Along with the deregulation and liberalization impetus of the 1980s and 1990s, the rules surrounding position limits and financial intermediaries have been relaxed, and the transnational traders have engaged more heavily in financial transactions linked to soybean futures market and other financial investment products linked to agriculture (Clapp 2014, 797–799). The traders use their information advantage to take positions in the commodity market, in particular in commodity index funds such as trade with futures contracts, based on speculation of price movements, and can thus profit from both predicted price rises or falls (Clapp 2014, 797–799; 2016, 18, 133–135; Gonzaga Belluzo 2015). At the same time, the traders increase their influence over prices. For many traders today, the financial instruments offer higher profits than the physical activities (Howard 2016, 74). Financial capital has in this way expanded in terms of both scope and depth. At the same time, there has been a substantial increase in price volatility as well as co-integrated price moves between previously independent markets (Clapp 2014, 799–806; McMichael 2009, 160).

Through their size and their involvement in various stages of the commodity chain—from financial markets and trade to infrastructure and food processing—the trading companies are uniquely positioned in the commodity chains. Their coordinating role, yielding important information and market advantages, make the traders increasingly powerful chain coordinators.[69] As noted by the GCC scholar, Gary Gerefi, information and the mechanisms for delivering it are the unifying force that holds together the structure of business in increasingly distant and complex commodity chains.[70]

*Processing Stages of Meat and Dairy*
The meat sector is increasingly integrated with the soybean chain through the feed industry. In this way, while most meat is consumed in the same country as where the animals grew up and became slaughtered, it is increasingly fed on inputs cultivated elsewhere. It is also

increasingly concentrated and vertically integrated in trading and processing stage. In addition and as already mentioned, traders such as Cargill have moved into livestock rearing contracts, animal feed production, beef processing, and meat packaging. Mega food companies such as Nestlé and Dean dominate the dairy business. The dairy processors have benefitted from the chronic oversupply of milk in the world, created by governmental subsidies (direct and indirect), particularly in the US (Howard 2016, 93–94). The low market price on milk hit hard, however, on the milk producers.

While there is an important move toward concentration and internationalization among the top international meat processing firms, the majority are still predominately regional and national in scope (Clapp 2016, 107). One exception is the largest meat processing company in the world JBS S.A., originally from Brazil.[71] Through an aggressive strategy of acquisitions, JBS became the world's largest company in the beef sector and a leading top company in chicken and pork production and processing.[72] It currently operates in five continents with slaughter, cold storage and meat processing facilities and produces beef, pork, and chicken, including fat, feed, meat by-products, and derivate products. It is also involved in tanning of leather, management of industrial waste, production of aluminum cans, soap, glycerin, and biodiesel, among others.[73] Its products are distributed under various brand names, such as Friboi, Swift, and Bertin.

However, in 2017 JBS was involved in a large corruption scandal. The Batista family, who control the JBS, admitted to paying bribes to Brazilian politicians in exchange for favors in a scandal (which threatened to topple former President Michel Temer, 2016–2018), which rendered the company a record fine of USD 316B. The fine forced JBS to sell many of its subsidiaries in neighboring Argentina, Uruguay, and Paraguay to its competitor Minerva for a total of USD 300M.[74] See Chapter 4 in this book for further details of the articulation of the meat complex in Argentina, Paraguay, and Uruguay.

## 3.3 Concluding Remarks

The historical evolution of the GCC of soybeans and beef illustrates wider shifts in the international agrofood system. For example, technological and organizational shifts in the US around World War II allowed for the transformation of soybeans into cheap feed for the animal industry. This model formed part of the industrialization of agriculture, resting on increasing food supply through the reliance on sophisticated (and well-protected) technology and inputs from outside the farm. This brought profound changes into the agrofood system, as the model and collateral dietary patterns spread the world. The so-called "gene revolution" has reinforced the model. New technologies, and the property-rights regime in which they emerged, involve a significant amount of lock-ins, and even path dependency. For example, as GM soybeans have become the norm and interchangeability is central to keep costs down in the soybean supply chain, the costs to produce for a non-GM market rise considerably since it has to bear the costs of establishing a completely parallel storage and logistics system. Another example of technological lock-in is the rising resistance to glyphosate in the weed-community, in the wake of the massive use of biotechnology designed to be combined with it. The problem tends to require "solutions" that ends up strengthening the same paradigm; new biotech traits (with new patents), in a new type of technological treadmill.

The historical evolution of soy/meat GCC also illustrates how the overall organization of the production and distribution systems of food has become increasingly complex with diverse constellations of economic actors linked together in long-distant supply chains, ranging from inputs to food processing. The increasingly powerful corporate actors are engaged in strategies of both vertical integration and widespread outsourcing of productive functions (Buttel 1987, 173; Clapp 2016, 9; Lang and Heasman 2015; McKay and Colque 2016; McMichael 2013). The advancement of transnational agribusiness firms was also spurred by a technological package (soybeans RR) allowing for good yields in less perfect soils. As business actors have advanced their positions, they have taken over large tracts of space previously occupied by nonmarket arrangements, including the state (Clapp 2016, 102–103; FAO 2012, 22). States have nevertheless played a vital role in allowing the business actors to advance through shifts in regulations (liberalization of agricultural trade, deregulation of financial markets, and stronger intellectual property rights), not least through the WTO. Decreasing transport costs have also fueled more trade. Particularly the soybean chain seems to encapsulate the tenants and trends of what has been described as the corporate or the *third food regime* by many food regime scholars; extreme market concentration, transnationalization, vertical integration, strengthened intellectual property rights, and financialization (Burch and

Lawrence 2009; Mcmichael 2009, 2013). The meat chain, in the wake of snowballing meat consumption, has also become increasingly transnational and integrated with the soybean business, but it still involves many elements of *the second food regime* (see Chapter 2); highly regulated and protectionist markets. While the international beef market is regulated by the advanced economies in order to protect domestic farmers, soybeans are among the most "freely" traded agricultural commodities.

The meat chain constitutes and important reminder of that (rich) states still have significant amount of power in the agrofood system. Besides agricultural subsidies and trade restrictions, state intervention in the agrofood system takes the form of investments in ground research, education, infrastructure, strong settlement mechanisms, and so on. Regulations thus still matters. Therefore, agribusiness firms spend annually billions of dollars on lobbying activities. There is also a well-known rotation practice—the so-called revolving door—where the same persons go from high positions within the agribusiness firms to USDA, to the US congress, to negotiate trade agreements on behalf of states, and back to business again (Clapp 2009, 2016, 126–128; Howard 2016, 76–78). While strong states often have acted in the interest of agribusiness, the current court-trials in the US against glyphosate also shows the potential counter-veiling power of the states. The current "trade-war" between China and the US with huge implications on the soybean trade, also shows that states do not only act in the interest of transnational agribusiness firms in favor of more trade. The EU's demand for GM-labeling on food is also one example of public regulations that can go against agribusiness interest, but the importance of EU in the global agrofood system has nevertheless declined significantly.

While the US has had an almost hegemonic position in articulating the GCC of soybeans and beef since the mid-twentieth century, these chains have recently come to reflect an emerging new economic geography of food, in which China is taking an increasingly dominant position (OECD/FAO 2016a; Ray and Gallagher 2016). China is not only advancing rapidly as food importer, but also as transnational trader, as owner of seed, and agrochemical companies, and lately even as patent-holder of new genetically modified traits. In Latin America, China has become the biggest agrofood market and it is financing projects in infrastructure and setting new quality norms for agrofood products. The rise of China as a mega food power also puts a big question mark to the notion of ever increasing corporate power and control. Everything indicates that the increasingly transnational Eastern giant will be even more powerful within the international food system, and capable of reshaping global rules and institutions from "Washington Consensus" to its unique brand of state capitalism. So far, however, the state-owned Chinese enterprises are acting in similar ways as the transnational private agribusiness firms; increasingly concentrated, vertically integrated, and with longer geographical distance, as well as in accordance with the logics of industrial agriculture (high input—high output, monoculture).

Indeed, there are uncertainties as well as inconsistent trends that can be identified through the analysis of the soybean and meat chains, albeit the main path undoubtedly has been moving in the direction of increased concentration, distance, and geographical productive specialization. As mentioned in Chapter 2, the food regime scholar Harriet Friedmann refutes the idea of a *third food regime*. She acknowledges the increased power of corporations, but argues that the current period involves too much contestation and contradictory trends, to qualify as a proper food regime (Friedmann 2016). The rising demand for organic products, farmers' markets (empowering farmers and cutting intermediaries), and "fair trade" can for example be conceptualized as challengers to "the corporate model" and agriculture form "nowhere". There has also emerged a rather radical critique against corporate agriculture, both within academia and among social movements. Within this vein, the global soybean complex stands as the ultimate symbol for a food system dominated by corporate interest, while destroying livelihoods, peasant communities, food security, and health. Under the banners of agro-ecology, slow food, Via *Campesina*, *food sovereignty*, and *Sumak Kawsay*, several radically different models for food production and distribution are proposed (Altieri and Toledo 2011; Patel 2009; Shiva 2009; Stahler-Sholk and Vanden 2011). The impacts on the wider system of these movements are, nevertheless, as of yet negligible.

In the midst of dramatic geopolitical shifts, the Mercosur bloc (Argentina, Brazil, Uruguay, and Paraguay) has become the largest soybean and beef exporter in the world (Bureau et al. 2005; Lapitz et al. 2004, 162; USDA 2018b). An important part of agrarian change in the Southern Cone countries can ultimately be seen as a mere response to increased global meat consumption, which requires ever more land for both soybeans and pastures (USDA 2018b). In other words, Latin America had the geographic conditions—availability of relatively cheap land, ports, and already installed export-infrastructure—to become an increasingly important global agrofood provider (Clapp 2016, 61; ECLAC 2017; Morgan et al. 2006, 10; OECD/FAO 2016a; Oosterveer and Sonnenfeld 2012, 61, 73–77). While the region is important to the world, agrofood exports are even more important for the economies of the region. In Argentina, agrarian-based exports represent 60% of total exports, in Paraguay, agrarian-based exports represent 63%, and in Uruguay they represented 73.4% of total exports (CAS 2018). In turn, soybeans and beef weight heavily in the basket of agrarian-based exports in these countries. Soybeans and beef alone represent 74% of the Paraguayan agrarian-based exports, 52% of Argentine agrarian-based exports, and 48% of Uruguayan agrarian-based exports (CAS 2018).[75] However, as explored at length in the coming chapters of this book, pressures from on-going agrofood globalization cannot exclusively explain agrarian change in Latin America, but state regulations and historical legacies have also been important co-shapers of recent agrarian change.

In short, this chapter has shown how the beef and soybean export-booms in South America should be understood in the light of a general process of liberalization of international agricultural trade. While these features of the GCC shape the ways the soybean and beef complexes are articulated in the region, they are not the only relevant factors. Historically formed institutions and local political-economical arrangements also matter. The region has been an important global meat provider since the first international food regime (as outlined in Chapter 2) with strong historical legacies still relevant for today's articulation. The soybean became an important export-crop as late as in the 1970s, and it was not until after the "gene-revolution" in the 1990s that production formerly exploded in the region. However, national agricultural policies promoting soybean cultivation also played an important role, particularly in the beginning of the soybean expansion. The next chapter provides a presentation of the articulation of the soybean and beef complexes in Latin America during the late twentieth century, with specific emphasis on the national regulations in Argentina, Paraguay, and Uruguay that paved the way for "boom" of the twenty-first century.

# Notes

1. The total value of the international food and agriculture market has grown from USD 15B in 1990 to 15 trillion in 2014 (Clapp 2016, 61).
2. The last decade (2006–2016), total agricultural exports from Latin America increased by almost 90%, from USD 107B to 195B, with the largest increases in vegetable products (49B), food products (25B), and animal products (13B) (Egas and De Salvo 2018).
3. Between 1909 and 1913, the estimated share of China of the world's soybean production was 71.5%—followed by Manchuria (16.5%), Japan (5.9%), Korea (5.5%), and Indonesia (Dutch East Indies, less than 1%)—but China was never an important exporter of soybeans due to its high domestic consumption.
4. The soybean had been cultivated in field trials in both the U.S. and Europe since the early eighteenth century.
5. The advent of World War II in 1940 disrupted soybean trade between Manchuria and Europe, and it never resumed after the war (Shurtleff and Aoyagi 2009, 112–113).
6. The U.S. government had since the 1930s been involved in developing all types of technologies to enhance agricultural productivity in livestock systems. Besides the increased use of animal feed, other technologies were developed, such as selective breeding, artificial insemination, antibiotics, and growth hormones (Chang 2009).
7. Considering all types of livestock, crop residues and agro-industrial by-products represent close to 30% of the livestock feed intake (FAO 2018).
8. The calves born in the spring are weaned by the fall, when their digestive system allows them to process whole feed.
9. Poor livestock producers in the South are less specialized and engage less in intensive production, but are having an increasingly hard time to compete with the more intense and capital-intensive systems.
10. Around 13% of global beef output is currently estimated to be completely raised in feedlots (Mottet et al. 2017).
11. The cost of feed typically accounts for 50–80% of total operating costs in intensive animal farming systems.
12. Source: FAOSTAT Database, "Food Supply—Livestock and Fish Primary Equivalent". Accessed 11 April 2018.
13. Producing 1 kg of boneless meat requires an average of 2.8 kg human-edible feed in ruminant systems and 3.2 kg in monogastric systems (Mottet et al. 2017).
14. Giant Brazil tripled its production between 1971 and 1974; from 2.1 to 7 mmt, and then doubled it again between 1974 and 1980; from 7 to 15.2 mmt. Brazil adopted differentiated export taxes on soybeans in the 1970s

(higher on whole beans than on meal and oil) in order to support the development of a domestic processing industry, and ultimately the development of domestic chicken and pork export industries. This strategy proved to fulfill the government's intention to add more value to its soybean production, by selling broilers and pork instead of whole soybeans (Shurtleff and Aoyagi 2009, 318).

15. Paraguay went from producing 0.075 mmt of soybeans in 1970, to 1.2 mmt in 1985 (Shurtleff and Aoyagi 2009, 374–375, 449).

16. The Argentine production increased six-fold between 1969 and 1980, from 0.6 mmt to 3.6 mmt (Martínez Alvarez 2012, 13–16; Shurtleff and Aoyagi 2009, 88, 111, 167, 173–174, 175, 242, 318, 449).

17. The soybean-area in Argentina jumped from 6.7 Mha in 1997 to 18.3 Mha in 2017. In Brazil, the area increased from 18.4 Mha in 2003 to 33.9 Mha in 2017. In Paraguay, from 1.2 Mha in 2000 to 3.4 Mha in 2017. In Bolivia, from 0.1 Mha in 1990 to 1.1 Mha in 2017. Uruguay had hardly participated at all in the soybean complex during the 1970s, 1980s and 1990s, but now the soybean area skyrocketed from 0.03 Mha in 2001, to 1.1 Mha in 2017.

18. In March 2016, prices were down at USD 375, but then started to recover, albeit with important fluctuations. In March 2018, the price stood at USD 443. See Trading Economics "Soybeans: Historical Chart", https://tradingeconomics.com/commodity/soybeans. Accessed 5 September 2018.

19. From 1995 to 2016, the U.S. soybean subsidies totaled USD 35.6B, and the biggest soybean farmers received the lion's share of it. See The Environmental Working Group, EWG's farm subsidy database, https://farm.ewg.org/progdetail.php?fips=00000&progcode=soybean. Accessed 17 July 2018.

20. Brazil is the world's biggest exporter of whole soybeans. In 2016/2017, it exported 74.65 mmt, mainly to China (USDA 2018c). Brazil crushed, in 2016/2017, 43 mmt of soybeans into 31.66 mmt soybean meal, of which 13.76 mmt soybean meal was exported (USDA 2018c). Brazil exports an important share of its production, but over time an increasingly amount of soybean meal is instead directly used in its expanding and intensifying animal farming industry. The crushing also produced 8.25 mmt of soybean oil, of which 1.5 mmt was exported (USDA 2018c).

21. In 2016/2017, 54.84 mmt soybeans in the US were crushed into 43.24 mmt soybean meal and 10.04 mmt soybean oil. The same year, the US exported 59.157 mmt of whole soybeans, 10.52 mmt of soybean meal and 1.16 mmt of soybean oil (USDA 2018c).

22. Out of the globally produced 350.8 mmt of soybeans, 116.9 mmt were produced in the US, 114.6 mmt from Brazil and 57.8 mmt from Argentina (USDA 2018c). After these soybean-giants, comes the following soybean producing countries in descending order: China, India, Paraguay, Canada, Ukraine, Uruguay, Russia, the EU, Bolivia, South Africa and Nigeria. Together, they represent more than 99% of world soybean production (USDA 2018c).

23. Brazil had 78.5M heads of cattle in 1960 and 169.9M heads of cattle 2000.

24. Beef exports in 2018: Brazil (2.03 mmt), India (1.85 mmt), Australia (1.61 mmt), and the U.S. (1.37 mmt).

25. World exporters after the top-four in order of size: New Zealand (0.56 mmt), Canada (0.48 mmt), Uruguay (0.42 mmt), Paraguay (0.4 mmt), EU (0.37 mmt), Argentina (0.35 mmt), Mexico (0.31 mmt) (USDA 2018b).

26. The participation of soybean meal in animal feed in EU has also risen significantly since the World War II, and is expected to continue growing (DeCarlo 2018). This trend was spurred by the outbreak of *bovine spongiform encephalopathy* (BSE) or "mad cow disease", associated with the practice to feed cattle with bone meal in the early 2000s. EU consumed in 2017, 30.34 mmt of soybean meal, of which 18.92 mmt was imported, and 11.38 mmt domestically produced from the crushing of 14.9 mmt whole soybeans within EU (USDA 2018c). The EU is the most important market for Argentine soybean meal and soybean oil (Service/USDA 2018). Soybean oil is also important in both the European margarine industry and for industrial uses. Since EU target blends of so-called biofuels, as part of the Kyoto agreement, the soybean oil is additionally appreciated as a source of biodiesel (Davis and D'Odorico 2015).

27. China thus "only" imported 0.57 mmt of soybean meal and 1617 mmt of soybean oil in 2016/2017, while it the same year bought 93.495 mmt of whole soybeans.

28. In 2017, the VAT rate for unprocessed agricultural products was lowered from 13 to 11%, and in May 2018, the VAT was further lowered from 11 to 10%.

29. After China comes the following countries as buyers of whole soybeans: EU (14.1 mmt), Mexico (4.4 mmt), Japan (3.25 mmt), Thailand (3.15 mmt), Egypt (2.8 mmt), Indonesia (2.7 mmt), Taiwan (2.55 mmt), Turkey (2.3 mmt), and Russia (2.3 mmt). The rest of the importing countries together import 18.36 mmt (USDA 2018c). There is a much bigger spread among soybean importers than among exporters.

30. The Chinese Ministry of Commerce Spokesperson said in an official discourse 7 July 2017, translated by USDA: "The United States violated the World Trade Organization's rules and launched the largest trade war in economic history to date. This kind of taxation is typical trade bullying, which is: seriously jeopardizing the security of the global industrial supply chain and the value-chain; hindering the pace of global economic recovery; triggering global market turmoil; and affecting more innocent multinational corporations, general enterprises, and ordinary consumers. This action will not help, but harm the interests of American businesses and people. The Chinese side promised not to fire the first shot. But to defend the core interests of China and the interests of its people, we were forced to take necessary countermeasures [...] We will promptly inform the WTO about the situation and work with countries around the world to jointly safeguard free trade and the multilateral system. At the same time, China reiterates that we will unswervingly deepen reform, expand opening up, protect entrepreneurship, strengthen property rights protection, and create a good business environment for companies from around the world in China. We will continue to assess the impacts on affected companies and work to take effective measures to help their business" (Ward 2018).

31. See *South China Morning Post*, "Companies set to cancel outstanding US soybean order when tariff bites", published 3 July 2017. https://www.scmp.com/news/china/diplomacy-defence/article/2153557/china-expected-cancel-11-million-tonnes-soybeans-us-new. Accessed 19 July 2018.

32. The Mercosur countries, except Paraguay, no longer benefit from the EU *Generalised Scheme of Preferences* due to their classification as high middle-income countries. While Paraguay still forms part of the preferential scheme, beef is excluded from the list of the products covered by it. See European Commission, "Mercosur", 2018, http://ec.europa.eu/trade/policy/countries-and-regions/regions/mercosur/. Accessed 19 July 2018.

33. See EUR Lex, Tariff quotas for high-quality fresh, chilled, and frozen beef and for frozen buffalo meat: https://eur-lex.europa.eu/legal-content/EN/TXT/?qid=1527588274070&uri=CELEX:02013R0593-20170921. Accessed 26 September 2018.

34. Since 21 September 2017 when the CETA-agreement went into effect, Canada can export beef with zero percent duty.

35. See EUR Lex, Tariff quota for high-quality beef: https://eur-lex.europa.eu/legal-content/EN/TXT/?qid=1527601011534&uri=CELEX:02012R0481-20170921. Accessed 26 September 2018.

36. See Oborne, Rebecka, "EU Beef Quota Back in the Limelight", published Thursday, 22 March 2018, http://beefandlamb.ahdb.org.uk/market-intelligence-news/eu-beef-quotas-back-limelight/. Accessed 19 July 2018.

37. Venezuela has been a member of Mercosur since 2012 and is an observer in the trade negotiations.
38. See information from the European Commission: http://ec.europa.eu/trade/policy/countries-and-regions/regions/mercosur/. Accessed 25 June 2019, as well as from the Farming Independent, "Trade deal with Mercosur is No. 1 priority: EU trade chief" https://www.independent.ie/business/farming/beef/trade-deal-with-mercosur-is-no-1-priority-eu-trade-chief-38216379.html. Accessed 25 June 2019.
39. Mercosur's exports to the EU were €40.6B in 2016, mostly agricultural products such as foodstuffs, beverages, and tobacco, followed by vegetable products including soya and coffee, and meats and other animal products. The EU's exports to the four Mercosur countries totaled €41.5B in 2016, mostly manufactures such as machinery, transport equipment, followed by chemicals and pharmaceutical products. See Euronews, "What's in the historic EU-Mercosur free-trade deal?" https://www.euronews.com/2019/06/28/eu-and-mercosur-strike-historic-free-trade-deal-after-20-years-of-talks; Financial Times, "EU-Mercosur trade deal: what it all means"; European Commission, http://trade.ec.europa.eu/doclib/docs/2017/december/tradoc_156465.pdf. Accessed 1 July 2019.
40. See the official website of ISAAA: http://www.isaaa.org/resources/publications/pocketk/16/. Accessed 12 May 2018.
41. The GM event contains the gene coding the enzyme 5-enolpyruvylshikimate-3-phosphate synthase (EPSPS) that gives the plant resistance to glyphosate, the active ingredient in the herbicide Roundup.
42. The CR4 of the agrochemical firms in the pesticide market is 62%. Both the seed and pesticide markets had Monsanto, DuPont Pioneer and Syngenta among the top four, in 2011 (Howard 2016, 107).
43. See ISAAA, "GM approval database", http://www.isaaa.org/gmapprovaldatabase/advsearch/default.asp?CropID=19&TraitTypeID=Any&DeveloperID=Any&CountryID=Any&ApprovalTypeID=3. Accessed 14 June 2018.
44. See the European Food Safety Authority (EFSA) Scientific Report (2005), 27, 1–81, "Conclusion on the peer review of Glufosinate", https://www.efsa.europa.eu/en/efsajournal/pub/2609. Accessed 18 July 2018.
45. One exception is a drought tolerant soybean trait developed by a joint venture (called Verdeca) of a Rosario, Argentina–based firm (Bioceres Davis) and a California–based firm (Arcadia Biosciences), so far only approved for cultivation in Argentina.
46. Biotech soybean was only planted in 11 countries in 2016. Countries in the world with biotech soybean area: Brazil (32.7 Mha), USA (31.8 Mha), Argentina (18.7 Mha), Paraguay (3.2 Mha), Canada (2.1 Mha), Uruguay (1.2 Mha), Bolivia (1.2 Mha), and smaller areas in South Africa, Mexico, Chile, and Costa Rica (ISAAA 2017).

47. EU is currently the third biggest soybean meal consumer in the world (after China and the U.S.).
48. See the ISGA's website: https://globalsoygrowers.org/. Accessed 1 July 2018.
49. See the United States Securities and Exchange Commission, FORM 10-K Monsanto Company Cf, https://www.sec.gov/Archives/edgar/data/1110783/000111078317000l8-7/mon-20170831x10k.htm#s-24080DADAFAC5B10B7F0BD740F7A630B. Accessed 22 August 2018.
50. See *Reuters*, "Battle of the beans", www.reuters.com/article/us-usa-pesticides-soybeans-insight/battle-of-the-beans-monsanto-faces-a-fight-for-soy-market-idUSKBN1FD0G2. American Antitrust Institute (AAI), "White Paper Monsanto—Bayer", www.antitrustinstitute.org/sites/default/files/White%20Paper_Monsanto%20Bayer_7.26.17_0.pdf. Accessed 20 April 2018.
51. Ibid.—Reuters and AAI, as well as the European Commission, "Press release Monsanto", http://europa.eu/rapid/press-release_IP-18-2282_en.htm; Market Watch, "Monsanto reports improved earnings driven by Bayer merger", www.marketwatch.com/story/monsanto-reports-improved-q1-earnings-driven-by-bayer-merger-soybean-sales; Handelsblatt, "BASF is the winner of Bayer-Monsanto merger", https://global.handelsblatt.com/companies/basf-winner-bayer-monsanto-merger-901801. All websites accessed 10 July 2018.
52. See *The Washington Post*, "Why Monsanto is no more?", https://www.washingtonpost.com/news/wonk/wp/2018/06/04/why-monsanto-is-no-more/?noredirect=on&utm_term=.9698ad801516; *The Guardian*, "Monsanto to ditch its infamous name after sale to Bayer", https://www.theguardian.com/business/2018/jun/04/monsanto-to-ditch-its-infamous-name-after-sale-to-bayer. Accessed 10 July 2018.
53. From the companies' websites: http://www.xinhuanet.com/english/2018-04/08/c_137094136.htm and https://www.syngenta.com/site-services/chemchina-transaction. Accessed 10 July 2018.
54. See *Handelsblatt Global*, "BASF: The unexpected winner in the Bayer-Monsanto merger", https://global.handelsblatt.com/companies/basf-winner-bayer-monsanto-merger-901801. Accessed 10 July 2018.
55. See *Atlasmanufaktur/Heinrich Böll Foundation*, "Monsanto and co: From seven to four, growing by shrinking", https://www.boell.de/en/2017/10/31/monsanto-and-co-from-seven-to-four-growing-by-shrinking. Accessed 16 July 2018.
56. To develop a new genetically modified (GM) seed trait takes an average of 13 years and costs about USD 136M (DeCarlo 2018).
57. See *The Economist*, "A shock court verdict against Monsanto's Roundup", 18 August 2018.

58. See *The Guardian*, "Monsanto ordered to pay $289m as jury rules weed killer caused man's cancer", 11 August 2018, https://www.theguardian.com/business/2018/aug/10/monsanto-trial-cancer-dewayne-johnson-ruling. Accessed 12 August 2018.

59. See *Reuters*, "Bayer says can begin integration with Monsanto", 16 August 2018, https://www.reuters.com/article/us-monsanto-m-a-bayer-integration/bayer-says-can-begin-integration-of-monsanto-idUSKB-N1L11G8. Accessed 17 August 2018.

60. See *Reuters*, "New Titans on the block", https://www.reuters.com/article/us-brazil-grains/new-titans-on-the-block-abcds-lose-top-brazil-grains-spot-to-asian-rivals-idUSKCN0WP19V. Accessed 24 May 2018.

61. See *Financial Times*, https://www.ft.com/content/250af818-a1c1-11e2-8971-00144feabdc0. Accessed 24 May 2018.

62. See *Bloomberg*, "Bunge takes soybeans loss after wrong-way bet on trade war", published 1 August 2018, https://www.bloomberg.com/news/articles/2018-08-01/bunge-profit-trails-estimates-on-loss-from-soybean-contracts. Accessed 22 August 2018. See also Glencore website: http://www.glencore.com/what-we-do/agriculture. Both accessed 22 August 2018.

63. See the official website of COFCO: http://www.cofco.com/en/AboutCOFCO/. Accessed 18 May 2018.

64. See *ADN*, "Empresa china lidera ranking de las exportadoras de soja", published 26 September 2018, http://www.adndigital.com.py/empresa-china-lidera-ranking-las-exportadoras-soja/. Accessed 29 October 2018.

65. In a press release at Syngenta's homepage, it was stated that Nidera was acquired because of its importance in Brazil, Argentina, Uruguay, and Paraguay, and because it owns a pool of proprietary germplasm, which would enhance Syngenta's ability to be competitive in seeds: https://www.syngenta.com/media/media-releases/yr-2018/07-02-2018. Accessed 16 July 2018. Before Cofco sold Nidera Seeds to Syngenta, in response to the commitments demanded by the European Commission in order to accept ChemChina's acquisition of Syngenta, Syngenta sold off agrochemicals to other companies. See for example *Reuters*, "China's Cofco International sells Nidera seeds business to Syngenta", published 6 November 2017, http://www.reuters.com/article/china-cofco-syngenta/chinas-cofco-international-sells-nidera-seeds-business-to-syngenta-idUSL3N1NC2U2. *Syngenta website*, "Syngenta finalizes sale of remedy assets in the EEA in relation to the ChemChina acquisition", published 16 March 2018, https://www.syngenta.com/media/media-releases/yr-2018/16-03-2018. Both accessed 16 July 2018.

66. See *Reuters*, "ADM, Bunge say they can navigate U.S.—China trade tensions", published 16 May 2018, www.reuters.com/article/us-bunge-ceo/adm-bunge-say-they-can-navigate-us-china-trade-tensions-idUSKCN1IH1SW. Accessed 10 July 2018.

67. From Cargill's annual report 2017: "We reached $3.04 billion in adjusted operating earnings in fiscal 2017, an 85 percent increase year-on-year with gains across all four business segments. Net earnings on a U.S. GAAP basis rose 19 percent to $2.84 billion. Revenues grew 2 percent to $109.7 billion on higher sales of grain, oilseeds and metals. Cash flow from operations climbed 38 percent to $4.69 billion" (Cargill 2018).

68. Farmers, on the other hand, are harmed by the low prices, particularly in the countries with less governmental protection for agriculture.

69. The giant agribusiness company, expressed this in the following way, in its annual report from 2007: "ADM is uniquely positioned at the intersection of three global trends: growing demand for food to feed the growing and more prosperous global population; increasing demand for energy agriculture and, transportation fuels from renewable resources; and growing desire for environmental improvement" (Shurtleff and Aoyagi 2009, 545).

70. A traditional distinction in GCC analysis has been between producer-driven and buyer-driven commodity chains (Bair 2009, 10; Gereffi 2001, 10). This denotes whether power in the chain is mainly captured by leading firms (or, driving firms) within the production end of the commodity chain who own capital-intensive industries (high degree of asset specificity and high barriers to entry), often involving several stages of vertically organized suppliers—producer-driven—or if the chain is governed by leading firms coordinating far-reaching networks through sub-contractation; buyer-driven (Bair 2009, 19–21).

71. The Brazilian rancher Jose Batista Sobrinho founded JBS (after the initials of the owner) as a slaughtering business, in 1953. Sobrinho expanded rapidly throughout the country in the 1960s. By the 1980s, the company began purchasing other meat processing companies. In 2007, JBS became a publicly held company on Brazil's Sao Paulo Stock Exchange. From this year, the company—now JBS S.A.—started to receive a major investment from the Brazilian Development Bank (BNDES). Between 2007 and 2010, JBS received around USD 2.5B in investments from BNDES. The strategy of the company became more international and started to have units on the five continents.

72. In 2005, JBS bought the Argentine meat processor Swift Armour for USD 200M and changed name to JBS Swift. In 2007, JBS bought the third largest US meat processor, Swift and Company. See *Reuters*, "Brazil's JBS-Friboi to buy Swift for $225 million", published 29 May 2007, https://www.reuters.com/article/us-swift-friboi-idUSN2930167420070529. Accessed 30 June 2018.

73. See *Reuters*, "Profile JBS S.A.": https://www.reuters.com/finance/stocks/companyProfile/JBSS3.SA. Accessed 29 June 2018.
74. See *Financial Times*, "JBS sells Argentina, Paraguay, Uruguay businesses for $300m", published 6 June 2017, https://www.ft.com/content/dd330d94-0ed0-3aaa-8b84-bbcf439c98b3. Accessed 30 June 2018; *Reuters*, "Minerva expects Brazil to resume beef exports to U.S. in first quarter", published 28 November 2017, https://www.reuters.com/article/us-minerva-outlook/minerva-expects-brazil-to-resume-beef-exports-to-u-s-in-first-quarter-idUSKBN1DS1N0. Accessed 30 June 2018.
75. In Uruguay, Soybeans, beef and pulp represent 75% of total Uruguayan agrarian exports, in turn representing 73% of total exports (CAS 2018, 48–51).

## REFERENCES

Altieri, Miguel A., and Victor Manuel Toledo. 2011. "The Agroecological Revolution in Latin America: Rescuing Nature, Ensuring Food Sovereignty and Empowering Peasants." *Journal of Peasant Studies* 38 (3): 587–612.

Bair, Jennifer. 2009. "Global Commodity Chains: Genealogy and Review." In *Frontiers of Commodity Chain Research*, edited by Jennifer Bair, 1–281. Stanford: Stanford University Press.

Baraibar, Matilda. 2014. "Green Deserts or New Opportunities?: Competing and Complementary Views on the Soybean Expansion in Uruguay, 2002–2013." Doctoral thesis Monograph, Economic History, Stockholm University (Stockholm Studies in Economic History 64).

Barbazán, Mónica, Carlos Bautes, Licy Beux, Martín Bordoli, Juan Diego Cano, Oswaldo Ernst, ... & Andrés Quincke. 2011. "Fertilización potásica en cultivos de secano sin laboreo en Uruguay: rendimiento según análisis de suelos." *Agrociencia Uruguay* 15 (2), 93–99.

Burch, David, and Geoffrey Lawrence. 2009. "Towards a Third Food Regime: Behind the Transformation." *Agriculture and Human Values* 26 (4): 267.

Bureau, Jean-christophe, Maria Priscila Ramos, and Luca Salvatici. 2005. "Tariffs, TRQs and Import Composition: The Case of Beef Trade Between the EU and Mercosur." European Trade Study Group, Seventh Annual Conference.

Buttel, Frederick H. 1987. "New Directions in Environmental Sociology." *Annual Review of Sociology* 13 (1): 465–488.

Cargill. 2018. "Momentum—2017 Annual Report". In *Annual Report*. Cargill, Incorporated.

CAS. 2018. *Anuario de Comercio Exterior de base agraria de los países CAS 2013–2017*. Edited by M. Ackermann and L. Gorga. Montevideo: Consejo Agropecuario del Sur.

Cepal, FAO, and IICA. 2015. "Perspectivas de la agricultura y del desarrollo rural en las Américas: Una mirada hacia América Latina y el Caribe 2015–2016". In *Perspectivas de la agricultura y del desarrollo rural en las Américas*. San José: Comisión Económica para América Latina y el Caribe (CEPAL), Organización de las Naciones Unidas para la Agricultura y la Alimentación (FAO), Instituto Interamericano de Cooperación para la Agricultura (IICA).

Chang, Ha-Joon. 2009. "Rethinking Public Policy in Agriculture: Lessons from History, Distant and Recent." *The Journal of Peasant Studies* 36 (3): 477–515. https://doi.org/10.1080/03066150903142741.

Clapp, Jennifer. 2009. "Corporate Interests in US Food Aid Policy: Global Implications of Resistance to Reform." In *Corporate Power in Global Agrifood Governance*, edited by Jennifer Clapp and Doris Fuchs, 125–152. Cambridge, MA: The MIT Press.

Clapp, Jennifer. 2014. "Financialization, Distance and Global Food Politics." *The Journal of Peasant Studies* 41 (5): 797–814. https://doi.org/10.1080/03066150.2013.875536.

Clapp, Jennifer. 2015. "ABCD and Beyond: From Grain Merchants to Agricultural Value Chain Managers." *Canadian Food Studies/La Revue canadienne des études sur l'alimentation* 2 (2): 126–135.

Clapp, Jennifer. 2016. *Food*, 2nd ed. Cambridge: Polity Press.

Clever, Jennifer, and WU Xinping. 2016. "China: Oilseeds and Products Update." In *GAIN Report*, edited by USDA Foreign Agricultural Service. USDA.

Craig, Steven, Louis Anthony Helfrich, David Kuhn, and Michael H. Schwarz. 2017. "Understanding Fish Nutrition, Feeds, and Feeding." *Virginia Cooperative Extension, 420–256*. Virinia Tech and Virginia State University.

Davis, Kyle F., and Paolo D'Odorico. 2015. "Livestock Intensification and the Influence of Dietary Change: A Calorie-Based Assessment of Competition for Crop Production." *Science of the Total Environment* 538: 817–823. https://doi.org/10.1016/j.scitotenv.2015.08.126.

DeCarlo, S. 2018. Executive Briefings on Trade April 2018 United States International Trade Commission. USITC Office of Industries.

de LT Oliveira, Gustavo, and Susanna B. Hecht. 2017. *Soy, Globalization, and Environmental Politics in South America*. New York: Routledge.

de Waroux, Yann le Polain, Rachael D. Garrett, Jordan Graesser, Christoph Nolte, Christopher White, and Eric F. Lambin. 2017. "The Restructuring of South American Soy and Beef Production and Trade Under Changing Environmental Regulations." *World Development* 121: 188–202. https://doi.org/10.1016/j.worlddev.2017.05.034.

ECLAC, FAO, and IICA. 2017. *The Outlook for Agriculture and Rural Development in the Americas: A Perspective on Latin America and the Caribbean 2017–2018*. San José, Costa Rica: Economic Commission for Latin America and the Caribbean (ECLAC), Food and Agriculture Organization of the United Nations (FAO), Inter-American Institute for Cooperation on Agriculture (IICA).

Egas, Juan José, and Carmine Paolo De Salvo. 2018. "Agricultural Support Policies in Latin America and the Caribbean: 2018 Review." *Agricultural Policy Reports.*

European Parliament. 2015. "Imports of GM Food and Feed Right of Member States to Opt Out." In *Briefing EU Legislation in Progress.* European Union.

FAO. 2012. "The State of Food and Agriculture 2012: Investing in Agriculture for a Better Future." In *The State of Food and Agriculture.* Rome: Food and Agriculture Organization of the United Nations.

FAO. 2018. *Livestock and Agroecology—Shaping the Future of Livestock.* Rome: Food and Agriculture Organization of the United Nations.

Friedmann, Harriet. 2016. "Commentary: Food Regime Analysis and Agrarian Questions: Widening the Conversation." *The Journal of Peasant Studies* 43 (3): 671–692.

Gaudreau, Matthew. 2015. "Seeds, Grain Trade, and Power Off-land: Chinese Agribusiness in Global Agrarian Change." Conference Paper.

Godfray, H. Charles J., Ian R. Crute, Lawrence Haddad, David Lawrence, James F. Muir, Nicholas Nisbett, Jules Pretty, Sherman Robinson, Camilla Toulmin, and Rosalind Whiteley. 2010. "The Future of the Global Food System." *Philosophical Transactions of the Royal Society B: Biological Sciences* 365 (1554): 2769–2777. https://doi.org/10.1098/rstb.2010.0180.

Gonzaga Belluzo, Luis. 2015. "La reciente internacionalización del régimen del capital." In *Neoestructuralismo y corrientes heterodoxas en América Latina y el Caribe a inicios del siglo XXI*, edited by Alicia Bárcena and Antonio Prado, 112–125. Santiago de Chile: CEPAL.

Graesser, Jordan, T. Mitchell Aide, H. Ricardo Grau, and Navin Ramankutty. 2015. "Cropland/Pastureland Dynamics and the Slowdown of Deforestation in Latin America." *Environmental Research Letters* 10 (3): 034017.

Hairong, Yan, Chen Yiyuan, and Ku Hok Bun. 2016. "China's Soybean Crisis: The Logic of Modernization and Its Discontents." *The Journal of Peasant Studies* 43 (2): 373–395. https://doi.org/10.1080/03066150.2015.1132205.

He, Y., X. Yang, J. Xia, L. Zhao, and Y. Yang. 2016. "Consumption of Meat and Dairy Products in China: A Review." *Proceedings of the Nutrition Society* 75 (3): 385–391. https://doi.org/10.1017/s0029665116000641.

High Quest Partners and Soyatech. 2008. *How the Global Oilseeds and Grain Trade Works.* South West Harbour: Soyatech.

Howard, Philip. 2016. *Concentration and Power in the Food System: Who Controls What We Eat?* Vol. 3. Edited by D. Goodman and M. K. Goodman. Contemporary Food Studies: Economy. Culture and Politics. London: Bloomsbury.

Hymowitz, Theodore. 1990. "Soybeans: The Success Story." In *Advances in New Crops*, edited by J. Janick and J. E. Simson, 159–163. Portland: Timber Press.

Interagency Agricultural Projections Committee, USDA. 2011. "USDA Agricultural Projections to 2020." In *Outlook Report*. Office of the Chief Economist, World Agricultural Outlook Board, U.S. Department of Agriculture.

ISAAA. 2017. *Global Status of Commercialized Biotech/GM Crops: 2016*. Ithaca, NY: ISAAA.

ISGA. 2017. *Biotech Soy Production in the Americas and the EU—Today and Tomorrow*. The International Soybean Growers Alliance.

Krausmann, Fridolin. 2004. "Milk, Manure, and Muscle Power: Livestock and the Transformation of Preindustrial Agriculture in Central Europe." *Human Ecology* 32 (6): 735–772. https://doi.org/10.1007/s10745-004-6834-y.

Lang, Tim, and Michael Heasman. 2015. *Food Wars: The Global Battle for Mouths, Minds and Markets*, 2nd ed. New York: Routledge.

Lapitz, Rocío, Gerardo Evia, and Eduardo Gudynas. 2004. *Soja y carne en el Mercosur: comercio, ambiente y desarrollo agropecuario*. Montevideo: Coscoroba Ediciones.

MacDonald, Graham K., Kate A. Brauman, Shipeng Sun, Kimberly M. Carlson, Emily S. Cassidy, James S. Gerber, and Paul C. West. 2015. "Rethinking Agricultural Trade Relationships in an Era of Globalization." *BioScience* 65 (3): 14.

Martínez Alvarez, Diego Leonardo. 2012. Historia de la soja en la Argentina: Introducción y adopción del cultivo. In *El Cultivo de Soja en Argentina*, edited by Baigorri and Salado Navarro, 1. Buenos Aires: Agroeditorial.

McKay, Ben, and Gonzalo Colque. 2016. "Bolivia's Soy Complex: The Development of 'Productive Exclusion'." *The Journal of Peasant Studies* 43 (2): 583–610. https://doi.org/10.1080/03066150.2015.1053875.

McMichael, Philip. 2009. "A Food Regime Genealogy." *The Journal of Peasant Studies* 36 (1): 139–169. https://doi.org/10.1080/03066150902820354.

McMichael, Philip. 2013. *Food Regimes and Agrarian Questions*. Edited by Kate Kennedy. Agrarian Change & Peasant Studies. Winnipeg, MB: Fernwood Publishing.

Mintz, S. W., C.-B. Tan, and C. M. Du Bois. 2008. "Introduction: The Significance of Soy." In *The World of Soy*, edited by C. M. Du Bois, C.-B. Tan, and S. W. Mintz. Singapore: NUS Press.

Morgan, Kevin, Terry Marsden, and Jonathan Murdoch. 2006. *Worlds of Food—Place, Power and Provenance in the Food Chain, Oxford Geographical and Environmental Studies Series*. Oxford, UK: Oxford University Press.

Moss, Diana L. 2009. *Transgenic Seed Platforms: Competition Between a Rock and a Hard Place?* Washington, DC: The American Antitrust Institute.

Mottet, Anne, Cees de Haan, Alessandra Falcucci, Giuseppe Tempio, Carolyn Opio, and Pierre Gerber. 2017. "Livestock: On Our Plates or Eating at Our Table? A New Analysis of the Feed/Food Debate." *Global Food Security* 14: 1–8. https://doi.org/10.1016/j.gfs.2017.01.001.

OECD/FAO. 2016a. *OECD-FAO Agricultural Outlook 2016–2025*. Paris: OECD Publishing.

OECD/FAO. 2016b. "Oilseeds and Oilseed Products." In *OECD-FAO Agricultural Outlook 2016–2025*, 13. Paris: OECD Publishing.

Oosterveer, Peter, and David A. Sonnenfeld. 2012. *Globalization and Food Production and Consumption*. New York: Routledge.

Otero, G., M. Poitras, and Pechlander G. 2008. "Political Economy of Agricultural Biotechnology in North America: The Case of rBST in La Laguna, Mexico." In *Food for the few: Neoliberal Globalism and Biotechnology in Latin America*, edited by G. Otero, 159–188. Austin, TX: University of Texas Press.

Patel, Raj. 2009. "Food Sovereignty." *The Journal of Peasant Studies* 36 (3): 663–706. https://doi.org/10.1080/03066150903143079.

Ray, Rebecca, and Kevin P. Gallagher. 2016. "China in Latin America: Environment and Development Dimensions." *Revista Tempo do Mundo (RTM)* 2 (2): 131–154.

Roser, Max, and Esteban Ortiz-Ospina. 2018. "World Population Growth." *Our World in Data*. Accessed 14 July 2018. https://ourworldindata.org/world-population-growth.

Sandoval, Lazaro. 2016. "Oilseeds and Products Annual: Uruguay." In *GAIN Report*, edited by Global Agricultural Information Network. USDA.

Service/USDA, Foreign Agricultural. 2018. "World Agricultural Production." In *Circular Series WAP*. Foreign Agricultural Service/USDA, Office of Global Analysis.

Shiva, Vandana. 2009. Soil Not Oil. *Alternatives Journal* 35 (3): 19.

Shurtleff, William, and Akiko Aoyagi. 2009. *History of Soybeans and Soyfoods in South America (1882–2009): Extensively Annotated Bibliography and Sourcebook*. Lafayette: Soyinfo Center.

Souto, Gonzalo. 2013. "Oleaginosos y derivados: situación y perspectivas." In *Anuario 2013, Opypa: Cadenas productivas; temas de política; Proyectos, estudios y documentos*, edited by OPYPA—MGAP, 17. Montevideo: Editorial Agropecuaria Hemisferio Sur.

Souto, Gonzalo, and Alejandra Ferenczi. 2010. Organismos genéticamente modificados: avances en la instrumentación del nuevo marco regulatorio. In *Anuario Opypa 2010*. Montevideo: MGAP.

Stahler-Sholk, Richard, and Harry Vanden. 2011. "A Second Look at Latin American Social Movements Globalizing Resistance to the Neoliberal Paradigm Introduction." *Latin American Perspectives* 38 (1): 5–13. https://doi.org/10.1177/0094582x10384204.

Stiglitz, Joseph. 2008. "Is There a Post-Washington Consensus Consensus?" In *The Washington Consensus Reconsidered: Towards a New Global Governance*, edited by Serra and Stiglitz, 3. New York: Oxford University Press.

USDA. 2016. *Overview of the United States Cattle Industry*, edited by National Agricultural Statistics Service (NASS), Agricultural Statistics Board.

USDA. 2018a. "Comparison of EU Tariff Rate Quotas for High Quality Bovine Meat." In *GAIN Report*, edited by Foreign Agricultural Service. Global Agricultural Information Network.

USDA. 2018b. *Livestock and Poultry: World Markets and Trade.* United States Department of Agriculture, Foreign Agricultural Service.

USDA. 2018c. "Oilseeds: World Market and Trade." In *World Agricultural Reports*. United States Department of Agriculture, Foreign Agricultural Services.

USDA, Foreign Agricultural Service. 2011. "Oilseeds: World Market and Trade, March 2011." In *Circular Series*. United States Department of Agriculture.

Waltz, Emily. 2015. "First Stress-Tolerant Soybean Gets Go-Ahead in Argentina." *Nature Biotechnology* 33: 682. https://doi.org/10.1038/nbt0715-682.

Ward, Michael. 2018. "Grain and Feed Update, Peoples Republic of China, Drought, Floods, and Storms Buffet China's Grain Market." In *GAIN Report*: USDA—Foreign Agricultural Service.

Wolford, W., S. M. Borras Jr., R. Hall, and I. Scoones. 2013. "Governing Global Land Deals: The Role of the State in the Rush for Land." In *Governing Global Land Deals: The Role of the State in the Rush for Land*, edited by W. Wolford, S. M. Borras Jr., R. Hall, and I. Scoones, 1–22. Malaysia: Wiley Blackwell.

Zhang, Qing, and Lifeng Liu. 2014. "Perspective on the Trend of Soybean Production and Trade in China." In *Proceedings of Selected Articles of 2013 World Agricultural Outlook Conference*, edited by Shiwei Xu, 19–28. Berlin and Heidelberg: Springer.

# CHAPTER 4

# Regulative Shifts Paving the Way for Agrarian Change

Agrarian change in Latin America is driven by rising global demand for animal products, which requires both more pasturelands for livestock grazing and more croplands for high-protein vegetable feed grains, particularly soybeans. This is in turn driven by rising global demand on animal products, coupled with agrofood liberalization, economic growth in China, and a stronger global intellectual property rights regime. As shown in Chapter 3, the overall organization of the production and distribution systems of soybeans and meat have become increasingly complex with diverse constellations of economic actors linked together in long and distant supply chains. These chains are increasingly dominated by a concentrated group of mega agribusiness firms engaged in strategies of vertical integration as well as widespread outsourcing of productive functions. This is particularly true for the soybean chain since it is the most integrated into world trade (Craviotti 2016). In this context, Latin America had the geographic conditions—availability of relatively cheap fertile land, ports, and existing export-infrastructure—to emerge as mainland provider (Clapp 2016, 61; ECLAC et al. 2017; Morgan et al. 2006, 10; OECD/FAO 2016; Oosterveer and Sonnenfeld 2012, 61, 73–77). In this way, the Mercosur bloc (Argentina, Brazil, Uruguay, and Paraguay) has become the largest soybean and beef exporter in the world (Bureau et al. 2005; Lapitz et al. 2004, 162; USDA 2018).

© The Author(s) 2020                                                      165
M. Baraibar Norberg, *The Political Economy of Agrarian Change in Latin America*, Governance, Development, and Social Inclusion in Latin America, https://doi.org/10.1007/978-3-030-24586-3_4

However, exogenous drivers and land availability cannot exclusively explain recent dramatic agrarian change. Historically formed institutions, including political-economic arrangements and regulations, also matter. As explored in Chapter 2, the region has a long history as specialized agro-food provider in the capitalist world system. There are important historical legacies from earlier periods that influence the ways agrarian change are articulated. For example, land concentration, well-organized landed inter-est groups, foreign dominance of processing and trading stages, and spe-cialization in agrofood exports appear to constitute important continuities in Latin American history. Besides institutional legacies of *longue durée*, more recent national regulative shifts also shape agrarian change. In par-ticular, the regulative shifts and neoliberal reforms of the 1980s and 1990s hold important explanatory value for the later articulation of the soybean and beef complexes, and their "boom" during the twenty-first century in Argentina, Paraguay, and Uruguay. These liberalizing reforms in the region reflect wider trends in the international agrofood system, which many *food regime* scholars refer to as the third, corporate or neoliberal, food regime (Burch and Lawrence 2009; McMichael 2009, 2013).

This chapter explores agrarian change and regulative shifts in Argentina, Paraguay, and Uruguay since re-democratization and until the fall of neoliberalism and the rise of a turn to the left—the *"Pink Tide"*—in the early twenty-first century. While many *food regime* scholars argue the corporate regime to continue still today, there was in Latin America a rather abrupt break with neoliberalism after the economic meltdown in Argentina in 2002, which is why this chapter exclusively deals with the late twentieth century, while Chapter 5 in depth deals with regulative shifts of the twenty-first century.

Focus is on the role of national regulations for agrarian change, but as mentioned many times throughout this book, administrative and political boundaries are not the only determinants of change. The chapter thus also addresses the role of new technologies (not least RR soybeans) and transnational capital, of the creation of a common market (Mercosur) and leakage effects between the countries (not least Argentine farm-ers spreading throughout the region). It also considers how different social-ecological eco-regions—Pampas, the Gran Chaco, and the Atlantic forest region—also co-shape articulation of agrarian change and form dif-ferent pathways. The remaining parts of this chapter are organized as fol-lows: The first section introduces public governance trends in the region that preceded and partly made possible the accelerated pace of agrarian change during the past decades (Sect. 4.1). Three sections then explore the main regulative shifts and their role for agrarian change in Argentina, Paraguay, and Uruguay, respectively (Sects. 4.2, 4.3, and 4.4). The chapter ends with a comparative analysis (Sect. 4.5).

## 4.1    BEFORE THE "BOOM": REGULATIVE SHIFTS AND AGRARIAN CHANGE IN THE 1980s AND 1990s

The soybean "boom" since the turn of the millennium in the Southern Cone is often summarized as the result of private business actors (foreign and domestic) building up a cost-efficient productive complex exclusively responding to rapidly increasing external (Chinese) demand that has nothing to do with public policies and regulations. This narrative forgets and ignores some important facts. First, it forgets that commercial soybean production in Argentina and Paraguay started to take off already during the 1970s (Chapter 2 of this book). It also tends to forget all the prior public (and private) investigation and support programs. In Argentina, a high-yielding and cost-effective soybean complex emerged gradually after public and private investments in research and development on soybean seeds for local adaption, combined with credit and extension services prior to the 1960s (Shurtleff and Aoyagi 2009, 88, 173–175, 242, 318, 449).[1] In Paraguay, the soybean also started to expand several decades before the recent "boom" due to public research and promotion under the "wheat program", forming part of the agrarian modernization impetus of Stroessner. In addition, production in Paraguay leapfrogged on technological advances from decades of Brazilian public research on soybeans and adaption to the geographic conditions of the Atlantic forest. In Uruguay, there was also public research on conventional soybeans for many decades, but yields remained low in Uruguayan soils.

Besides the importance of public funded research since well before the post new millennium "take-off", equally important is to explore the role of the public regulative shifts of the 1980s and 1990s. While macroeconomic stability, privatization, and liberalization had been high on the policy agenda since the 1970s, it was not until after democratic governments were reinstalled (1983 in Argentina, 1985 in Uruguay, and 1989 in Paraguay), that the recipes from the so-called *Washington Consensus* were firmly adopted (Bulmer-Thomas 2003, 393–395; Grugel and Riggirozzi 2007, 89). Although the restructuring reforms varied in intensity, timing, and forms, all countries in Latin America took significant steps toward trade liberalization: dismantling of nontariff barriers, dispersing of tariffs, cutting public spending deregulation, and privatization of public enterprises (Baraibar 2014; Clapp and Rowlands 2014; de Castro et al. 2016; Ocampo 2004). The international lending organizations spurred the reorientation by imposing loan conditionality for the rescheduling of the external debt (Bárcena Ibarra and Prado 2015, 19; Katz 2015, 243; Margulis and Porter 2013, 71).[2] The *World Bank*, the *Inter-American Development Bank* (IADB) and the *International Monetary Fund* (IMF) argued that the policies adopted in the region had been strongly biased against agriculture and advised the elimination of export taxes. The structural adjustment programs also sought to strengthen private property rights to land and improve the functioning of the land markets—by supporting land-titling programs and make governments abandon or toned down policies in line with "the social function of land" doctrine. In this way, restrictions on foreign or anonymous shareholders to own land were removed, rules for land renting were relaxed, and many other "obstacles" to foreign investments inland markets taken away (Arezki et al. 2015; Hernández 2015; McMichael 2013, 45; Ocampo 2015; Sassen 2010; Vergara-Camus and Kay 2017; Wolford et al. 2013, 2). Many bilateral and multilateral investment agreements further diminished the ability of the state to use and allocate land according to other principles than the market (Ankersen 2006, 113–119; Romson 2012, 25–26, 359).

The role of the state for planning and development of the economy, including the agrarian sector, was reduced, reformed, and replaced by an emphasis in attracting foreign direct investments, ensuring predictable rules, strengthening private property rights (including intellectual property), and facilitating exports (Lapitz et al. 2004, 11). This was also combined with targeted programs for poverty alleviation, market integration, decentralization, and transparence. This process was also reinforced by a strong decentralization impetus, combined with budget cuts, which further undermined the capacity of the states for long-term planning and development of the economy. All three countries decentralized many functions and entities of the state apparatus, a measure in line with policy recommendations from powerful international organizations (Willis et al. 1999). In Argentina, already a federal country since the nineteenth century, Menem further decentralized both tax policies and expenditure, empowering the provinces (and making them responsible for the budget cuts). Paraguay and Uruguay remained much more centralized systems than Argentina (and closer in size to some of the Argentine Provinces), but also implemented decentralization of the state apparatus and public offices.

In addition, Argentina, Brazil, Paraguay, and Uruguay agreed to eliminate customs duties and promote "free movement" between the countries, as well as to adopt a common trade policy toward outside countries, including a common external tariff, through the creation of the Southern Common Market; *Mercosur.*[3] The common market was established by the treaty of Asunción in 1991. In 1994, the Protocol of *Ouro Preto*, formalized the status of Mercosur as a customs union.[4] On 1 January 1995, approximately 80% of all products traded between its members were no longer subject to tariffs, while their external tariffs were also substantially reduced in a general trend toward increased multilateral openness (García et al. 2013).[5] In contrast to previous regional trade agreements in Latin America, Mercosur radically departed from protectionist inward-looking policies and was instead in line with the general trend toward economic reform and more open trade regimes (García et al. 2013). This represented an important step toward increased internal free trade and multilateral openness.[6]

Besides increased regional integration, agrofood market integration worldwide in 1990s brought important regulative shifts in the international arena, such as the *World Trade Organization* (WTO) and its *Agreement on Agriculture* (AoA) and *Agreement on Trade-Related Aspects of Intellectual Property Right* (TRIPS). International investment started to flow back to the region, particularly into land deals and agriculture (Morgan et al. 2006, 10; Oosterveer and Sonnenfeld 2012, 61, 73–77; Turzi 2016, 12). In this way, national liberalization policies were reinforced by the consolidation of Mercosur and by the general trade liberalization impetus in the international agrofood system.

## 4.2   Neoliberalism in Argentina and *Sojization*

A discourse of deregulation and free trade had been dominant in Argentina since the 1970s, and yet Argentina had remained one of the most protected economies. This changed as Carlos Menem (1989–1999) took office and turned the country into a shop window for neoliberal reforms. The IMF had a clear guiding role and provided funding from international credit sources. The reforms included the dismantlement of a myriad of import protection measurements, a reorientation of the agricultural sector toward export markets, severe cuts in public investments, and social spending combined with decentralization (Director of CARI, and Special Adviser to the General Director of FAO, and former sub-secretary of Agriculture of the Government, Buenos Aires, 1 March 2017).[7]

### 4.2.1    Privatization, Deregulation, and Liberalization

The government closed down the various agriculture boards linked to the *Junta Nacional de Granos* and the *Junta Nacional de Carnes* (Teubal 2008b, 6). These had regulated supply, quality, and prices since 1933—first mainly to protect national producers from falling international prices, and then under Perón to protect domestic consumers from rising international prices instead (Chapter 2 in this book). As market prices on agricultural commodities were rising, the segments of commercial farming producers benefitted from elimination of the boards (Teubal 2008a, 203; Turzi 2016, 12). Menem also dismantled in 1991 the *Argentine National Forestry Institute* (IFONA), and turned it into weaker units, which made it easier to transform forests into more profitable land uses (Burns and Giessen 2016).

Basic infrastructure for agricultural export, such as silos, elevators, and ports were privatized (Teubal 2008b, 6). Private transnational firms, mainly traders, came instead to fill the vacuum and invest heavily in developing the infrastructure for the export-oriented soybean complex (Lapitz et al. 2004, 34–35). *The National Agricultural Technology Institute* (INTA) had severe cuts in the budget and there was even an attempt to privatize it (Pellegrini and Balatti 2017, 113). While INTA escaped privatization and remained as a government research agency, the Argentine agrarian sector in general terms had all of a sudden become one of the world's most deregulated and privatized (Teubal 2008b, 6).

The traditional landed ranchers' organization, the *Argentine Rural Association* (SRA) outspokenly supported the privatization, deregulation, free market, and state reducing reforms, at the same time as it pushed for still lowering the taxes for the agrarian sector. Menem listened carefully to the claims of producers' organizations, institutionalized arenas for negotiation, and eliminated agricultural export taxes (Manzetti 1992, 615). The pro-market reforms also included the removal of quantitative export restrictions and the reduction of tariffs on imported agricultural inputs (FAO 2017, 6). In turn, SRA and the *Argentine Rural*

*Confederations* (CRA) urged their members to improve compliance with other taxes (VAT, property and income taxes) and called for increase in production. Menem also had close relations with the agribusiness firms, not least symbolically illustrated in his decision to hand over the Ministry of the Economy to former executives of the mega trader and agrofood company *Bunge and Born* (Manzetti 1992, 614–615). While aligned with the liberal trade policies of Menem, the producers' organizations still asked for a more active role of the state in agrarian planning and coordination (Tkachuk and Dossi 2014). According to the agrarian consultant, researcher, and former National Director of Agricultural Production of SAGPyA, as well as former President of INTA, the overall situation for the agrarian sector was positive due to overall sound policies:

> When I entered the ministry in 1991, we took away the export taxes that had been there for 20 years, because export taxes take away competiveness from the sector. This gave good results. For example, the production of wool almost doubled between 1991 and 1996, and this was exclusively because the export taxes were removed and inflation controlled. Who dares to invest if you do not know if inflation will all of a sudden rise? (Consultant and former President of INTA, Buenos Aires, 28 February 2017)

Argentina actually became a more important exporter of whole soybeans than Brazil, since the giant neighbor at this time used differential export taxes to support domestic crushing and ultimately domestic chicken and pork production (Shurtleff and Aoyagi 2009, 318). While removal of export taxes benefitted the agrarian exports, the introduction of the convertibility regime, or "one to one", which pegged the Argentine new peso to the US dollar by law from the same year, did not (Abuelafia et al. 2010; Toledo Lopez 2017).[8] This macroeconomic reform managed to come to terms with the waves of inflation and stabilize prices, but the exchange rate was highly unfavorable to national industry, particularly for the export sector. However, the agricultural organizations were willing to accept the convertibility regime, as long as it led to long-term stability and recovered international investors' trust in Argentine economy (Manzetti 1992, 615). In addition, Argentina got unprecedented access to global finance and a rapid rise of foreign investment (Grugel and Riggirozzi 2007, 90–91).[9] In spite of the overvalued Argentine Peso, the soybean exports remained competitive in the international markets and expanded in the fertile Pampas region.

### 4.2.2    Approval of RR Soybeans

The policy regulation with the most dramatic impact on the soybean expansion was the approval of RR soybeans (GTS-40-3-2) for food, feed, and planting. When Argentina approved Monsanto's RR soybean in 1996, it became the first country in the region to allow genetically modified crops, almost at the same time as in the US (Chapter 3 in this book).[10] The rate of adoption of RR soybeans in Argentina was extremely fast; the share of genetically modified (GM) soybean production went from 0 to well over 90% in only six years, and by the year 2010 practically all soybeans were GM (Leguizamon 2014). This is by far the most rapid adoption of any seed variety in Argentina, including those introduced in the *Green Revolution* (Qaim and Traxler 2005). Soybeans RR are designed to be combined with no-tillage farming—sowing without disturbing the soil through tillage. One contributing factor to the rapid RR diffusion was that no-tillage had already become standard production technology among most crop farmers of the Pampas in order to reduce soil degradation and erosion risks (Peiretti and Dumanski 2014; Trigo and Cap 2003). Another contributing factor was that Monsanto had licensed the technology to a firm in Argentina that was bought up by another seed company, and Monsanto did not manage to get the Argentine farmers to pay any price premium for the technology. Monsanto tried together with other biotech and seed companies, to make Argentina change its legislation on intellectual property of plant varieties and ratify the much more restrictive UPOV '91 version of plant breeders' rights. Argentina had only ratified UPOV '78 version, which involves "farmer's privilege", giving the right to farmers to use save seeds stipulated as an exception to the plant breeders' rights, also in line with the national 1973 seed law. The government did not give into the demands of the transnational biotech companies, which spurred a years-long dispute. The fact that only a minority share of soybeans seeds were purchased at the formal seed market put a downward price pressure, so that even the minority of farmers who bought authorized seeds still paid a considerably lower markup over the price of conventional seeds than farmers in the US (Peiretti and Dumanski 2014; Trigo and Cap 2003). Soybean planted area showed year-over-year record levels, with almost a doubling between 1993—5.8 Mha and 2000—11 Mha (FAO 2017, 5–9). The soybean by far replaced the country's former export stars; beef and wheat.

While soybeans had been expanding in the very heartland of the fertile humid Pampas in the provinces of Buenos Aires and Entre Ríos since the 1970s, the increase in area and productivity snowballed after the adoption of RR soybeans after 1996. The rate of expansion even exceeded the pace of the agricultural revolution of the "wheat boom" for more than a hundred years ago in the same area (Chapter 2 in this book), and created a likewise rapid transformation. Under the new technological package of RR soybeans, the production model of the Pampas region became extremely cost-effective and internationally competitive, even in relation to the subsidized soybean production of the US (Lapitz et al. 2004, 53). Soybean cultivation became the most profitable land-use, at least in croppable land not too far away from storage and export facilities (Acosta Reveles 2008, 9; Bisang et al. 2009). In this way, the RR soybean expanded from the Pampas heartland provinces into the wider Pampas region, including the provinces of Santa Fe, La Pampa, and Cordoba (Turzi 2016, 12). Its fertile land and the new technological package fueled the cost-effectiveness of the soybean production model in the Pampas, but previous public research and development in soybean seeds had played a role. Research efforts since the 1960s had contributed to the development of locally adapted soybean seed varieties and allowed for dual cropping—wheat during winter and soybeans during summer (Martínez Alvarez 2012, 17–23). The combination of these nationally developed seed varieties with the new RR technological package with no-tillage allowed for a break with the traditional mixed pastures-crop systems of the Pampas. Continuous cropping became the dominant model, spurred by changed price relations in favor of the soybean. As presented in Chapter 1 in this book, this resulted in the displacement of pastures for beef and dairy production from the area. As land and leasing prices soared in the wake of increased competition for land, beef production was intensified—producing more on less land. The herd of cattle actually also increased slightly between 1988 and 2002 due to intensification (Gras and Hernández 2013, 103). Another contributing factor to the soybean "success" was that the significant economies of scale involved in soybean production made a good match with the concentrated land structure of the Pampas, where only rather big capitalized farmers remained active, while the poorest and smallest farmers already had lost access to land before the soybean "boom" (see Chapter 2 in this book).

However, the new mega-big drills and harvesters, airplane fumigation, the increasing access to "cheap" capital seeking new markets, and the previous liberalization of land leasing regulation favored even bigger farming units than in the past. The bigger the scale, the higher the profit margins, seemed to be the new economic rationale. Moreover, new commercialization tools, such as the possibility to sell the soybean harvest already at seeding time, or after sprouting (but well before harvest) increased access to capital to cover all production costs without savings, which further spurred increments in scale. In response to this new scenario, a new type of organization of the agrarian activity and business emerged on the Pampas in the 1990s. Many medium size farmers got together to be able to lease larger amounts of land and by "pooling" resources to be able to invest in new technology and share the net benefits after harvesting (Teubal 2008b, 7). They created a new type of firm called *pools de siembra*, who often managed to attract capital from investors from outside the sector that financed production in return for a share of the profits after harvest.

The *pools the siembra*, often referred to as "network firms", went from nonexistent to managing hundreds of thousands of hectares of land within a couple of years (Bell and Scott 2011; Caligaris 2017; Gras and Hernández 2013, 86; Varrotti and Frederico 2018). This remarkable expansion was driven by the high economic margins offered by increasing international soybean prices, and decreasing costs of fuel and labor due to the new technological package and the new form of organization. The mega firms would not have been able to grow so fast, however, without the previous deregulation and liberalization of land and leasing markets, allowing for short-term contracts on leased land (Previous advisor for Productive Processes at the Secretary of Family Agriculture, Ministry of Agroindustry, 2016–2017, in Stockholm and Bella Vista, of September and December 2018). The new firms had almost no fixed assets (the companies typically own neither land, nor labor, nor machinery), but access it all through short-term contracts. The model is thus centered in the coordination of a dense network of subcontracted and outsourced services and activities (Acosta Reveles 2008, 10). Hence, the new model brought upon a separation between the place where the production takes place and the territorial origin of the people working in the land. Moreover, the big new crop firms vertically integrated into many other stages in the chain, engaging as input providers, as intermediaries between producers and traders, as providers of credits and insurance to other producers, as processors, as producers, and sellers of feed to livestock and as biodiesel producers (Gras and Hernández 2013, 223). Transnational traders and investors also created proper firms following a similar model.

In spite of soybeans expanded over pastureland, the relation between the new agribusiness oriented crop firms and the traditional landed rancher elite and organized ranchers, was in many respects characterized by cooperation rather than by competition. The traditional organizations SRA and the *Confederaciones Rurales de Argentina* (CRA) never argued for any policy restrictions on soybeans. This was probably partly because the rising leasing and land values ultimately benefitted many of their members, who rent out their land to the "network" firms, and thus gained from the soybean business through land rent.[11] In addition, they have throughout history in general been faithful to a liberal discourse centered in advocating for free trade, strong private property rights and stability, and against state interventionism, while also from time to time asking for specific policy interventions in support of the agrarian sectors, such as debt clearance, favorable credits, and public investments in infrastructure. At the same time, with leasing prices hitting new records, the crop farmers who leased the land could often not afford to use the land in any other way than for the crop with the highest economic margins, that is, soybeans. Many sharecroppers and small farmers also left farming. The *Argentine Agrarian Federation* (FAA) representing mainly tenants and family farmers, expressed some early preoccupations about displacement of tenants, family farmers, and rural laborers (Gras and Hernández 2013, 19). However, as a conflict regarding export taxes broke out between the government and producers' organizations in 2008 (see Chapter 5), the critique from FAA on the process of *sojization* of the Pampas silenced. At the same time, several targeted programs—often through initiatives from international organizations such as the World Bank, IFAD, IDB, and IICA, in support of rural development and family farming and smallholders—were launched to "soften" the impact of neoliberalization (Lapegna 2016). Besides the programs of the IOs, targeted public policies for rural development also start to emerge (Senior consultant for various international organizations and former high official of INTA and SAGPyA, in Buenos Aires, 28 March 2017). Several new credit lines were also made available to small producers in order to stimulate the "modernization" of agriculture and encourage investments in machinery and seeds, but agrarian rents for small producers did not follow the pace of indebtedness and many small farmers ended up forced to sell their land at auctions to clear debts (Giarracca 2008, 19). As a whole, Argentina lost around 87,000 farmers, or one-fourth of all producers, between the agrarian census of 1988 and 2002, of which 86% had less than 200 ha. The amount of farmers managing more than 1000 ha, however, increased (Teubal 2008b). It is thus clear that commercial farming was doing quite well, while many family producers abandoned agricultural activities.

The adoption of RR soybeans also allowed for expansion of cultivation into areas that were previously considered unsuitable for cultivation, such as most of the Argentine area of the *Gran Chaco* region (Acosta Reveles 2008; Bisang et al. 2009; Garcia 2015; Jobbágy et al. 2015; Prieto and Ernst 2010). Accordingly, soybeans moved into the provinces of Salta, Tucumán, Jujuy, Santiago del Estero, San Luís, Chaco, and Formosa. In these areas, the soybean expanded over land previously used for cotton, grazing fields for livestock farming, and food crops (including self-subsistence production), but most of all it expanded over native forests (Graesser et al. 2015, 4–6; Gras and Hernández 2013, 61; Teubal 2008b, 7). Between the Agrarian census of 1988 and 2002, in Salta, Chaco, and Santiago del Estero, the crop area (mainly soybeans) increased with 4% (Giarracca 2008, 20). Land clearing was not only done for crop production, but also ranching, as capitalized livestock farmers from the Pampas sold or leased out their land to the advancing crop production (Tkachuk and Dossi 2014; Volante and Seghezzo 2018). Thus, both soybeans and livestock expanded the agrarian frontier into areas of native forests and savannah. However, the livestock sector suffered the outbreak of foot and mouth disease leading to the slaughtering of many cows in 2001.

### 4.2.3    Economic Recession

Argentina's agrofood industry was a key growth driver in the economy by the late 1990s. While this business was doing well and rising, the economy as a whole was in recession. The over-valued peso was also increasingly problematic for the export sectors, even though the agrarian exports at the same time were benefitted by the absence of export taxes, low inflation, and rising international prices. After the 1995 *tequila crisis*, Argentina faced difficulties to get new loans, and thus for the government to obtain sufficient foreign currency reserves to meet both its payment obligations to foreign lenders and its conversion obligations to local savers. In 1998, this situation got worse as Brazil, Argentina's main trading partner, devalued the Brazilian Real. A new center-left government was elected in 1999. It had promised reform, but largely maintained the neoliberal agenda until it was not possible anymore (Grugel and Riggirozzi 2007, 92–93). Around Christmas 2001, a full-blown economic crisis broke out. Argentina defaulted on its financial obligations (USD 141B in public debt—the largest sovereign default in history at the time), after IMF withdrawn its support (Fridman 2010). The country suffered the most severe economic crisis in its history. On 6 January 2002, under the *Public Emergency and Exchange Regime Reform Act* (Law 25.561),[12] Argentina broke the decade-long fixed peg of one-to-one parity. As the convertibility was dropped, the peso shrunk as a stone; a 75% devaluation (Grugel and Riggirozzi 2007, 92–93).

Faith in neoliberalism fell in parallel with the value of the Argentine peso and full-blown economic and political turmoil broke out. Poverty rose alarmingly. Social rebellion soared, not least under the newly organized *piquetero* movements (Grugel and Riggirozzi 2007, 92–93). While the immediate effects of the crisis were devastating, the end of the convertibility also brought an abrupt transformation of the structure of relative prices in benefit of the agrarian export sector. Moreover, private debts were *pesified*, benefiting farmers and firms who could repay debts taken in dollars in devaluated pesos (Former Secretary of SAGPyA, 2013–2015, Buenos Aires, 2 March 2017). The export-oriented soybean complex benefitted greatly (Gras and Hernández 2013, 337). However, the government soon reintroduced export taxes as an immediate, but temporary, response to the desperate need to increase fiscal revenue (Richardson 2009).[13] This marked the beginning of the end of neoliberalism and with the election of Néstor Kirschner in 2003 state interventionism was back.

## 4.3   THE END OF THE STROESSNER ERA AND CONTINUOUS COLORADO RULE IN PARAGUAY

The neoliberal trend that swept the region and the world after the 1982 debt crisis, also reached Paraguay. The change was nevertheless not very dramatic in Paraguay, since it had never adopted any strong industrialization policies and had remained rather open to trade and with low taxes throughout the long and authoritarian Stroessner rule (1954–1989).

Moreover, Paraguay was less affected than its neighbors were by the debt crisis, since it had never attracted any significant amount of foreign loans (Former Minister of MAG, 1989–2001, Asunción 21 February 2017).[14] However, after the fall of Stroessner, neoliberal reforms advanced, and long-term state planning and programs were dismantled, while decentralization and deregulation increased.

### 4.3.1   Agrarian Modernization and the Fall of Stroessner

Soybeans and beef started to expand under the Stroessner regime promoted agrarian "modernization", including agrarian frontier expansion (into the Atlantic forest region) and commitment to export promotion, through long-term planning and big public programs, as presented in detail in Chapter 2 of this book.[15] There were also important research and extension programs linked to cotton, which was the most important export crop of Paraguay up until the late 1980s. By 1987, cotton and soybeans accounted for over 1.1 Mha or over 40% of all land in crops and contributed over 60% of exports (Seyler 1988, 113–115). Commercial beef farming likewise expanded under the *livestock fund* and other public programs (Plant and Animal Health Specialist IICA Paraguay, Asunción 22 February 2017). Livestock, cotton, and soy caused massive deforestation, fragmentation, and habitat destruction of the Atlantic forest region, alongside demand for fuelwood, and small-scale food producing agricultural activity. The vast forests of Paraguay had been mainly perceived by the authorities as "unproductive"; an untapped resource with policies designed to incentivize to put it under "productive use" (Walcott 2014, 4–5), but as explored in detail in Chapter 5, policies to protect the forests have gradually been adopted.

After half a century of so-called "agrarian reform", Paraguay in the late 1980s still had among the most unequal land structures in the world. Stroessner had expanded the agrarian frontier with 12.23 Mha of public land (mainly in the Atlantic forest region), but in spite of land distribution being an explicit aim, the great majority of the population remained poor and engaged in more or less self-subsistence farming on very small parcels of land. The vast majority of the Paraguayan population was poor peasants, *campesinos*, engaged in subsistence farming combined with a small surplus sold at local markets. They had very small parcels of land, they made very low use of modern technology, and they were largely invisible in political terms (Turzi 2016, 102). The lion's share of the "new" land added up to big landholders and capitalist farming was rapidly expanding. The development of commercial farming in Paraguay, particularly of soybeans, was also closely tied to the important influx of farmers from the South of Brazil, so-called *Brasiguayos*, attracted by Paraguay's low land prices and fertile land.[16] Brazilian investments competed with traditional Argentine investment in land and extractive industries, and Brazilians soon came to dominate the soybean production and a large part of the commercial farming in the Atlantic forest region. Foreign investments were supported by a preferential exchange rate between dollars and Paraguayan pesos for investors (Innovation Specialist IICA Paraguay, Asunción 22 February 2017).

By the late 1980s, discontent with Stroessner had grown. Several peasant organizations that had been loyal to the regime claimed that Stroessner had failed to carry out his promised agrarian reform (Hetherington 2009). In addition, Stroessner's close ally, the US, had started to criticize the violence, corruption, and authoritarianism of the regime, and opposition had grown within his own Colorado party. Organized rural dissent eventually contributed to the downfall of Stroessner's 35-year rule, deposed by dissidents within his own party through a *coup d'état* in 1989, led by General Andrés Rodriguez. In 1992, a new constitution paved the way for free elections. In May 1993, Juan Carlos Wasmosy, supported by the Army and a part of the Colorado party, was elected President in the first fully open election in more than two generations. The 1992 Constitution also included proposals from the traditional ranchers' organization the *Rural Association of Paraguay* (ARP), which excluded the traditional usufruct right to land. It also removed the idea of "social function of land" (Chapter 2 in this book) and established that expropriation of unproductive lands must be accompanied by full compensation, paid in advance to the landholder (Ezquerro-Cañete and Fogel 2017).

While the Paraguayan economy already was open to trade and the lowest tax rates in the region, the post-Stroessner administrations further liberalized the economy. The financial market was deregulated and foreign exchange market liberalized, the interest rate ceilings removed, several credits removed, and direct price interventions in agricultural markets removed (Turzi 2016, 51). There was nevertheless a strong internal split within the still ruling Colorado party. Rather than a coherent neoliberal path, political turmoil and instability characterized Paraguay for at least a decade after the end of the Stroessner rule. In this context, the illegal market of land titles and contraband trade exploded. Paraguay already had a long history of smuggling and corruption, but these phenomena became worse under the immediate post-Stroessner era. Moreover, cultivation and processing of illegal drugs were "booming" activities in the Paraguayan forests and further smuggled into other countries.[17] The turbulence created loss of control, according to several observers and here illustrated in a quote form by one of the technical specialists and consultants working for IICA Paraguay:

> During this period, everything became fragmented and coordination between different public entities was lost. A lot of changes, politicians entering and leaving, no stability and the technical work started to become politicized. Moreover, the international organizations, such as the Inter-American Development Bank started spreading the idea that everything needed to be decentralized and that public offices should function autonomously with their own budget and personnel. They sold us this idea that the country would improve through decentralization, and they took good money for it, offering huge loans. But, what happened was a total loss in state capacity; public policies became fragmented and isolated, loss of human resources, of know-how and coordination. (Innovation Specialist IICA Paraguay, Asunción 22 February 2017)

Government power alternated between factions leaning on neoliberal policies and factions favoring employment creation through public works, forgiveness of agricultural debt, and free distribution of seed (The World Bank 2001). In this way, policies lacked coherence and capacity for long-term planning. In addition, the country's legal and institutional framework was weak, leading to evasion and selective enforcement of laws and regulations (Former Minister of MAG, 1989–2001, Asunción 21 February 2017). Moreover, the widespread corruption further eroded efficacy of all policies. Nostalgic policy officials argue that after the fall of Stroessner, policies lost coordination, long-term planning, order, knowledge-based decisions, efficiency, and discipline (Plant and animal health specialist IICA Paraguay; Innovation specialist IICA Paraguay; Senior Technical Advisor UNDP Paraguay, Asunción 20 and 22 February 2017).

Not only the state was weak, but also domestic civil society was weak after more than 35-years of repressive authoritarian rule. In this post-Stroessner vacuum, many international organizations strengthened their influence in the country. Thus, the majority of the state reforms and development programs of the 1990s were designed by the World

Bank, IADB, UNDP or other organizations of international coopera-tion. The World Bank was particularly active in reforming the land mar-kets away from state-initiated land reforms to market land reforms. While land titling was high on the agenda for decades, the trend was negative; over 60% of farms held title to land in 1991, but only 37% did in 2001 (Masterson 2007).

### 4.3.2   Soybeans Expanding, but Cotton Retracting

In the midst of the political instability of the 1990s, the cotton sector—the traditional number one export item—started to face severe problems as black-market cotton seeds of inferior quality started to spread rapidly among farmers. This resulted in a harvest of low quality, which yielded low prices paid for Paraguayan cotton. In addition, the productivity was not improving, while international cotton prices were falling, not least due to US surpluses (Turzi 2016, 10–12). The state responded with programs subsidizing cotton production and offering free seeds. The programs failed to fulfill their purpose. Paraguay's deputy Minister of Agriculture described the problem in the following way:

If you look back, and follow the agricultural performance of this country since the 80's, you will see that Paraguay had a good position; quality and productivity of cotton were high and small producers living well from it produced it. But, the policies of MAG during the 1990s did not help to maintain what we had. There was a lot of state failure. The Paraguayan cotton production stagnated; yields were low, the quality deteriorated and the soils were badly managed. The state could have worked to solve the emerging problems early, but missed the opportunity. Instead, the state assisted in paternalistic ways giving away inputs for free, giving price guar-antees in the middle of falling market prices. Another problem was the bad management of the soils. Tilling, tilling, tilling… Under the hot sun and without anything to cover and protect the soil in between the harvests. There you had many errors in public policies… When the state saw that the cotton market started to fall, it desperately tried to maintain the same path as before, to keep the castle from falling… Instead of helping the poor small producer to shift away from the ox, the plow and the rattle, and away from the continuous degradation of the soil, the state locked in itself in doing charity and assistance-based policies. Besides the failures of the state, we have the failures of the market, with lack of transparence in price formation. So the cotton sector fell like a falling building of 15 floors;

first one floor, then the following, then the following, then the following and so on. (Paraguay's Deputy Minister of Agriculture, Asunción 27 February 2017)

As vividly expressed in above quote, the state tried, but did not manage to hinder the fall of the cotton complex. According to the Minister, the main "state failure" was a historical legacy from Stroessner, in which agrarian policies were characterized by populism, paternalism, and *asistencialismo* (a policy of social handouts, charity, and short-term aid). It was a pattern hard to break since the citizens expected the state to give them small gifts, rather than the tools to become more productive in sustainable ways and stand on proper legs (Paraguay's Deputy Minister of Agriculture, Asunción 27 February 2017). In this way, the cotton producers remained trapped in a vicious cycle of low-yielding and low-paying farming, surviving through punctual assistance from the state that did not provide any long-term solution. At the same time, commercial soybean farming continued to advance throughout the 1990s, leading to a rapid substitution process:

The white ocean of 500,000 ha disappeared... And of course, in this context it was easy for the soybean to advance. Now, we have a green ocean of 3.3 Mha of land instead... That transition, from cotton to soy in the 1990s, occurred in the context of deteriorating and corrupt policies towards family agriculture, in which the IBR[18] was corrupt. It had no problem in giving away land to whoever; to themselves, to you, to foreigners, to anyone who passed by and wanted a piece of land. At the same time, a chain of stimulus for soybean productions emerges, seed adaption, mechanization, silos, etc. But, this is not coming from the state, no, it is all private. This is why I tell you, Paraguay, you do not know if it is free market or if it is libertinism as far as it goes. Obviously, this leads to a change in land tenure. Those who suffered the fall of cotton lost their property and other actors advanced. (Paraguay's Deputy Minister of Agriculture, Asunción 27 February 2017)[19]

As clearly described in above quote, the rapid advancement of soybeans in Paraguay occurred in a context of crisis of the cotton sector, but it was also spurred by the widespread corruption of the IBR,[20] as well as by investments of the private sector. At the same time, the percentage of farms receiving some kind of credit assistance dropped from over 34% of farm households in 1991 to less than 8.4% in

2001 (Masterson 2007). In this context, transnational traders invested in infrastructure to facilitate further expansion of the soybean complex. At farmers' level, capitalized Brazilian soy-farmers gradually expanded from the areas close to the Brazilian border (the departments of Itapúa, Alto Paraná, and Canindeyú) further west (the departments of Caagazú, Caazapá, Amambay, and San Pedro). In 2002, when the Agrarian Statute from 1963 was redrafted, IBR was reformed and renamed to the *Rural Development and Land Institute* (INDERT). Together with this reform it was stated that increased productivity, agroindustrial growth, and overall reduction of market interventions were the means to achieve rural development and poverty alleviation (Ezquerro-Cañete and Fogel 2017). The social consequences of soybeans replacing cotton became dramatic as former cotton producers lost their connection to agriculture and migrated to the cities (Mansourian et al. 2014). The great majority of the cotton farms were smaller than 20 ha, while the soybean farmers typically managed more than 100 ha. According to the director of the Paraguayan office of agrarian censuses and statistics, however, not all cotton producers that lost land to soybeans abandoned agriculture, but left to area that is more marginal:

> The trend from 1991 is that in areas where soybeans are produced, as in Alto Paraná and Itapúa, the big are getting bigger and the small—often previously involved in cotton—migrate. There, a wave of migration of smallholders are moving into land that is more marginal. So the small producers that sold their hectares in the areas of soybean production go and migrate to poorer land areas. (Director of DCEA-MAG, Asunción 21 February 2017)

While the state did not explicitly plan, facilitate through investment or monitor the rapidly expanding soybeans, it provided a beneficial free-trade framework, a low-cost setting, and most importantly it enabled the acquisition of land at a low price. Paraguayan exports also benefitted from several devaluations of the *Paraguayan Guaraní* in relation to the US dollar. In addition, it closed a bilateral agreement with Taiwan, in which Taiwan showed tribute to Paraguay's political support and stance against China (another Stroessner legacy) through buying soybeans for above market price (Turzi 2016, 12; Former Minister of MAG, 1989–2001, Asunción 21 February 2017). The lack of active participation of the state in the rapidly evolving soybean complex is also illustrated by the illegal spread of RR soybeans from Argentina in the late 1990s (Trigo 2002, 52). This process was recalled by the Minister of the time in the following way:

> When I was Minister during the end of the 1990's, the transgenic soybeans started to arrive. The state was busy trying to save the cotton sector and attend the surplus of rural population that emerged from the cotton retraction. So it did not pay attention to the soybean complex at all. It was completely informal and not institutionalized. Not until 2004, proper biosafety norms and an institutional structure for approval was created. (Former Minister of MAG, 1989–2001, Asunción 21 February 2017)

As illustrated by the ex-Minister of Agriculture and Livestock, the diffusion of RR soybeans by far preceded the regulatory approval. Accordingly, the Paraguayan soybean farmers did not pay any technology fee at this moment. The RR technology made pace and scope of soybean expansion accelerated further. From 1996 to 2003, the soybean area expanded from 0.8 to 2 Mha (Turzi 2016, 12). Yields also increased rapidly, which increased the economic returns of the already profitable soybean production.

The rapid advancement of foreign agribusiness firms doing large-scale soybean production, and the retraction of the generally small-scale and more labor-intensive cotton production created dramatic social consequences. This provoked increased social tensions in the eastern border departments of the Atlantic forest region (Nickson 2015). Peasant mobilization against the Brazilian penetration increased markedly, not least through an increased amounts of land occupations, resulting in several violent confrontations between *campesinos* and the army, as well as between *campesinos* and Brazilian farmers, followed (Zoomers and Kleinpenning 1990, 143). Moreover, the continuous soybean expansion did not only displace cotton, but also spurred continuous deforestation of the remnants of the Atlantic rainforest. By the end of 1990s, the forest cover had been reduced into a fraction.[21]

### 4.3.3   Livestock Expansion

In a parallel way, without much state attention or support, the commercial livestock sector in Paraguay also started to increase productivity in the late 1990s and to attract foreign investors. The increasing land values in the Eastern region (Atlantic forest) both led to additional pressures on remaining forests and to land-use intensification, which in turn caused considerable environmental challenges, such as biodiversity loss, land erosion, loss of productive agricultural land, and degradation of water quality (Public Policy Officer at WWF Paraguay; Sustainable Development Specialist UNDP Paraguay; Natural Resource Management Specialist IICA Paraguay, Asunción February, 2017).

Pastures also started timidly to expand from the Atlantic forest region into the vast and "unexploited" Chaco forests in the western region. This area was too far away from the export markets and with too little service and infrastructure to previously been considered attractive for agricultural firms and investors. Due to these harsh conditions, it had remained without much anthropogenic modification, dominated by immense forests. The exception was the Mennonite communities and their dairy farmers and cooperatives, which had developed proper techniques of water storage and turned the region into a main producer of dairy products (Cartes et al. 2015; Mereles and Rodas 2014). The sparsely populated Chaco region was also the home of several indigenous peoples, both nomadic hunters and gatherers and sedentary communities, involved in different types of farming. The Mennonites often hired the indigenous inhabitants as laborers (Human Rights Council 2015). The farming activities of the Mennonites grew in size from modest meat and dairy enterprises into large cattle agribusiness, and became the most important economic activity in the region by the late 1990s (Caldas et al. 2015). Soon large-scale ranchers from other South American regions started to establish in the area. This development would soon come to "explode", as thoroughly presented in Chapter 5.

## 4.4   Uruguayan Regulative Shifts and Late Insertion in the Soybean Complex

Compared with Argentina and Paraguay, Uruguay is the country with the longest and strongest tradition of state intervention in the economy and welfare policies (Chapter 2 in this book). However, during the military dictatorship (1973–1985), there was a policy shift toward liberalization—albeit full of exceptions—and trade remained highly regulated. The governments following re-democratization (1985 and onwards) took more steps toward the opening up of the economy, lowering substantially tariffs to trade, and restructuring the state. However, compared to other countries in Latin America, the Uruguayan pro-market liberal reforms were significantly more gradual and moderate, and the old welfare-state remained rather intact (Canzani and Midaglia 2011; Lanzaro 2013, 252).

### 4.4.1    Liberalization à la Uruguaya

The boldest regulative steps toward privatization, deregulation, freer trade, and state restructuring (austerity) reforms in the modern Uruguayan political-economic history, were taken after the debt crisis and after re-democratization. The Colorado party with *Julio María Sanguinetti* ruled two terms (1985–1990 and 1995–2000). The National Party, *Blancos* with *Luis Alberto Lacalle* (1990–1995) ruled in-between. Finally, yet another Colorado government, led by *Jorge Batlle* (2000–2005), followed the same path. Reduction of fiscal deficit, liberalization, and privatization were top priorities. These governments largely removed industrial support and protection. The explicit development strategy was that Uruguay should specialize in line with its comparative advantage, not least in agrarian-based exports (export taxes were reduced or eliminated, while import tariffs on agricultural products remained with 30%). Import tariffs on capital goods and intermediate inputs were eliminated.

In combination with the gradual reduction of custom duties and other trade restrictions within the *Asunción treaty* (from 1991) and the later consolidation of Mercosur, the manufacturing sector of Uruguay followed a path of relative deindustrialization, and the previous trend toward economic equality was drastically reversed (Paolera et al. 2018). It was tacitly agreed that the Uruguayan role within Mercosur was restricted to development in land-intensive agriculture, financial services, and tourism. However, the agricultural sector also faced many problems, and it was in general difficult for tiny Uruguay to compete successfully with its neighbors due to a relatively higher cost structure. The livestock sector remained the most important, but international meat prices were low and protectionism high in the international meat markets (Baraibar 2014, 146). Moreover, a lot of academic knowledge and technical experience within veterinary and agricultural sciences, as well as within the state apparatus, had exiled during the military regime, and it took a long time to rebuild these tacit assets. The overall balance of trade for Uruguay remained negative. Export performance was bad, and even declining in absolute terms, by the end of the 1990s.[22]

Moreover, the degree of indebtedness among farmers had steadily increased during the 1990s and it represented 70% of annual sectorial GDP in 1997 (Piñeiro 2004, 27–33).

As support programs to promote and support cultivations and family farming, including extension services and subsidies on fertilizers had been largely dismantled and eliminated, the amount of producers and cultivation area retracted very fast.[23] Small crop farmers managing less than 100 ha could not remain competitive and almost completely disappeared from agricultural activities (Figari et al. 2007, 80; Rossi 2010, 69). The general crop productivity was low and the quality of the products was not up to international standards. The economic margins of crops were in general low, since cultivations were costly and low yielding compared to its neighbors, and most crop producers had to pay in order to access land, since most of them did not own the land, but were sharecroppers (Chapter 2 of this book). The Uruguayan land structure had remained rather intact between the late nineteenth and the late twentieth century, with most of the land, even the most fertile area suitable for crops, in the hands of big ranchers (Piñeiro 1998, 2011). However, the ranchers often wanted their pastures to rotate every three or four years with crops, in order to improve soil fertility, protect soils from erosion, and give themselves access cheap fodder (the mixed crop-pasture rotation, *agrícola-ganadero*). Accordingly, sharecroppers typically cultivated the land for two-three years, after every three years of pastures, paying a rather "subsidized" price for land (often a share of harvest income minus production costs) as they contributed with more fertility to the soils (Errea et al. 2011, 12). Particularly soybeans were popular to enter in the rotation schemes due to their nitrogen fixating capacity. Thus, while the soybean yields (under conventional tilling) were too low to be profitable on their own right, they still sometimes entered as a rotation crop in the mixed systems, improving the soils for the pastures (Political advisor of the National Director of MVOTMA, in Montevideo, 6 March 2017). Malt, wheat, barley, and soybeans created a small exportable surplus, together accounting for around 5% of total exports during the 1990s.[24]

While intensive, mixed systems were common on the most fertile land of the Lítoral, most ranching activities in the rest of the country relied almost exclusively on extensive plains of natural pastures with very little technology incorporation and labor generation (Shardul Agrawala 2004). Moreover, the process of regional integration (Mercosur) required the elimination of government-owned stocks of frozen beef (used for price regulation) and deregulation of the processing industry, which increased concentration and transnationalization of meatpackers (Bervejillo et al. 2012). However, some initiatives and policies to improve technology incorporation and sanitary control during the 1990s were successful. The livestock sector that had been rather stagnated since 1915 entered a period of increasing productivity and "late" modernization after 1993. In addition, the benefits of purely grass-fed cattle for the local ecosystems, as well as for the quality and taste of the beef (and for animal well-being) started to be recognized in some markets. Uruguay was increasingly able to position itself as a high-quality beef provider. Last but not least, in 1994 Uruguay managed to access the status of free of foot and mouth disease (FMD) without vaccination, which provided access to the most demanding high-paid markets in the world, such as the Japanese market (Mederos Porto 2014, 62–64).[25]

### 4.4.2    Export Promotion and Policies to Attract Foreign Investments

The government facilitated export promotion and sought diversification by providing institutional support and tax incentives, especially for barley, citrus, rice, dairy, and forestry (Bervejillo et al. 2012). One specific, and controversial, initiative to promote exports was the 1987 *Free Trade Zone Law* (15.921) and the 1988 Regulatory Decree (454/988), exonerating businesses established in specific areas (the free zones) from several taxes (Corporate Income Tax, Income Tax, and Value-Added Tax [VAT]) and from customs duties.[26] Goods and raw materials may be imported into the free zones without paying customs duties. Exclusively social security taxes (BPS) need to be paid. The same year, the parliament passed the *Afforestation Promotion Law* (15.939) to provide an alternative use for marginal agricultural lands in an effort to promote export diversification and attract foreign investment, while protecting native forests (Morales Olmos and Siry 2009). Forestry development was promoted by large subsidies for the development of forest plantations (large-scale eucalyptus and pine plantations) and wood manufacturing industries (pulp mill projects and plywood production), including exemptions from a number of taxes for 15 years (Land tax, Rural Property Tax, and VAT on several imported goods). They were also given huge loans with generous repayment conditions and they were allowed to set up operations in free trade zones, with additional tax benefits (Redo et al. 2012). The generous economic incentives and the geographical conditions allowing for high tree growth rates, led to fast arrival of several foreign mega-companies doing large investments in forest plantations. In addition, the state upkeep of the necessary highway facilities to transport timber to the mill at no expense to the companies (Redo et al. 2012, 128). Forest plantations cover grew rapidly, from 97,000 ha in 1990 to 751,000 ha in 2004. The foreign companies also made large investments in pulp-processing mills, and pulpwood exports increased from 46,251 $m^3$ in 1988 to 1.6 million $m^3$ in 2004 (Morales Olmos and Siry 2009, 63). The forestry sector became characterized by the presence of large vertically integrated firms (Redo et al. 2012, 128).

Besides rapid development of the forest sector (albeit not contributing much fiscal revenue and according to some observers at the price of high ecological costs), dairy and rice had growth in export output. Another tool used to make the agricultural sector more "flexible" and reduce production costs was the 1991 *Leasing Law* (16.223), allowing for shorter-term (12 months) leasing contacts and reducing the rights of the tenants for longer-term land access. Moreover, the 1999 new *Land Title Law* (17.124), allowed for corporations of limited responsibility (corporation with bearer shares) to own and lease land (Baraibar 2014, 149–197; Paolino et al. 2013; Paolino 2015). The government was also preparing for more exports by the 1992 *Port Law*, in order to increase private investments in the ports to rapidly enhance trade capacity and reduce transport costs.[27] Diversification of agrarian production and increased productivity were also prioritized through the 1990 creation of the *National Institute of Agrarian Research*, INIA, revitalizing funding for agricultural R&D in Uruguay (Bervejillo et al. 2012).[28] While initiatives such as INIA spurred long-term growth in agricultural productivity (Bervejillo et al. 2012), overall production and export performance remained weak and dominated by the "classical" items of meat products, wool, and leather.[29]

Besides these laws and export-promoting initiatives, the Lacalle/ Sanguinetti administrations—in a similar way to Menem in Argentina— committed to create an overall attractive climate for investments and exports, including the sound macro-economic policies, strong property rights to land and intellectual property, transparent rules, and guarantees to investors (Baraibar 2014). The government changed regulations and taxes to promote foreign investment, and to encourage international economic integration. The 1996 creation of the investment and export promoting institute *Uruguay XXI* had the explicit aim to attract foreign investments and internationalize the Uruguayan economy. The 1998 *Investment Law* (16.906), aimed at stimulating national and foreign direct investment by offering important tax incentives for investments, as well as allowing foreign investors to freely remit profits and transfer capital abroad (Redo et al. 2012, 128). In 1999, FDI as proportion of GDP was among the lowest in the region with proportion of GPD only 0.7% (Barrios et al. 2010), but rose significantly a couple of years later, particularly with investments in the agro-exporting sectors (Durán and Salgado 2013; Uruguay XXI 2015).

In 1996, RR soybeans (GTS-40-3-2), was authorized for production in Uruguay, only a few months after Argentina, in spite of the fact that at the time there was almost no soybean production in the country, and in spite of the fact that Uruguay had no regulatory biosafety framework in place (Bianco-Bozzo et al. 2010; UNEP-DINAMA 2007). In 1997, the intellectual property rights for seeds were strengthened in the *Law on the Development, Production, Distribution and Internal and External Marketing of Seeds and Phytogenetic Creations* (16.811) and the creation of the *National Seed Institute* (INASE).

All these policy shifts were vital in opening the way for the later advancement of transnational soybean, beef, and forestry-related exports and in general the advancement of foreign investments in agriculture and agribusiness.

### 4.4.3   Discontent with Neoliberalism and the Arrival of RR Soybeans

While Uruguay was the country in the region that privatized the least during the neoliberal period 1985–1999 (for a value of less than 0.1% of GNI, compared to the privatizations in Argentina that represented 9% of GNI during the same period), public discontent with neoliberalism had still grown stronger throughout the 1990s (Lanzaro 2013, 252). Several privatization initiatives met such popular resistance that they were voted down by the citizens in referendums and could not be implemented.[30] As the Uruguayan economy stagnated, the critique amplified. The loudest political articulation against neoliberalism came from an alliance integrated by the union movement (particularly PIT-CNT), the cooperative movement, the newly formed "ecologist" movement, and the left-center coalition *Frente Amplio* (FA) (Berrón and Freire 2004, 297;

Moreira 2010, 290). FA criticized the reforms of free trade zones, the heavy subsidies for deforestation, and the elimination of industrial policies, among other specific reforms. The electoral support for FA had been growing for every election held since it was founded in 1971 and represented 40% of the votes in 1999. As FA grew stronger, cooperation and collaboration between the traditional parties, *Blancos* and *Colorados*, also increased in order to hinder FA getting political influence and power (Lanzaro 2013, 238).[31]

Uruguay is a small country, squeezed in between Argentina and Brazil, and hence its political and economic life is never exclusively responding to domestic affairs, but always hugely affected by the macroeconomic and political shifts in its giant neighbors. The gradual devaluation of the Real in Brazil in 1998 weakened the competitiveness of Uruguayan exports. In 2001, outbreak of foot and mouth disease in Argentina spread to Uruguay with devastating effects for the livestock sector, leading both to the slaughtering of many cows, the loss of the recognized international status of being free from FMD, and the exclusion from export markets for Uruguayan beef (Mederos Porto 2014, 140).[32] Last and worst was the political and economic collapse of Argentina in 2002, with a 75% devaluation of the Argentine peso against the US dollar. Uruguay also entered a full-blown economic crisis. Foreign debt and fiscal debt soared (Baraibar 2014, 229). This resulted in a depositor run on banks and a massive default on foreign debt and the national acquisition of private banks (Redo et al. 2012). The Uruguayan peso de facto devaluated in 2002. However, almost without any notice, in the middle of the economic crisis, Uruguay started to receive an inflow of investment and business actors from Argentina. In particular, big Argentine crop firms and farmers crossed the river and started to plant soybeans in Uruguay.

The devaluation of the Uruguayan Peso, the low land values (around two-thirds of the price in the Argentine Pampas), the absence of export taxes, and the political stability attracted many Argentine crop producers—often *pools de siembra* or network firms. Several of these Argentine crop firms were also expanding in Bolivia, Brazil, and Paraguay (Guibert et al. 2011). Traditionally, cultivations in Uruguay had faced constraints of low yields and erosion of the soils caused by tilling (Chapter 2 in this book), but RR soybeans combined with glyphosate and no-tillage farming allowed for increased yields in less perfect soils. On average, yields increased from 1000 kg/ha in 1990 (conventional soybeans) to around

2000 kg/ha in 2002 (IICA 2009; Souto 2013). All of a sudden, soybeans became the most profitable land-use. The big Argentine crop firms
took their ways of doing business with them; subcontracts, direct selling
to multinational traders, future contracts, geographical diversification,
and economies of scale, channeling investments from outside the sector,
integrating vertically with other stages of the chain (commercialization,
seeds, processing), and leasing land. As the Argentine farmers started to
grow soybeans in Uruguay, the international traders started to invest in
infrastructure (silos, stockpiles, and port facilities) to meet the growing
domestic production, as well as transport and commercialization close to
the farm (ex ante the port of Nueva Palmira).[33] Besides setting up trade
in the physical market, these traders take positions in the futures market. They started to offer grain cooperatives and big farmers in Uruguay
contracts to commit their future harvest at a fixed price to reduce their
exposure to future movements (price falls) and to be able to cover production costs before harvest (while the traders gain from the price dis-
parity between the spot and future prices). Within one year, Uruguay
already produced an important exportable surplus of soybeans.[34] While
the Argentine crop firms expanded extremely fast in the Uruguayan
Litoral, "traditional" crop producers who owned land—many who had
accumulated high debts—sold a piece, or all, of their land, in order to
clear debts when the Argentines arrived and land prices started to rise
(Baraibar 2014, 158–159). The degree of indebtedness among farmers
had increased substantially during the 1990s, representing 70% of annual
sectorial GDP in 1997 (Piñeiro 2004, 27–33).

All previous experiences of cropland expansion in Uruguayan agrarian history had been the result of proactive public policy (Chapter 2 in
this book), but soybeans expanded rapidly over pastures for the first time
in history based on market values. While the soybean expansion was not
the result of intentional public policy—policymakers were even said to be
"taken by surprise"—it is nevertheless clear that many prior public liberalizing regulations were decisive for allowing the crop "boom" to occur
in Uruguay. The rapid expansion of foreign agribusiness corporations
leasing on short-term contracts would for example not have been possible without the prior shifts in leasing law. Furthermore, big soybean producing firms made extensive use of the new investment law to receive tax
exonerations, as well as the *Free Trade Zone* law as the lion's share of the
soybeans are exported from free trade zones. These regulations, together
with a strong private property right regime (including both land and

intellectual property), low land prices (compared to the Argentine Pampas), the introduction of a free-floating peso, and no export taxes on soybeans, turned soybean cultivation into a highly profitable land-use in Uruguay.

During the second half of 2004, Uruguay was busy preparing for national elections combined with a national referendum about privatization of water. *Frente Amplio* (FA), the labor union, and other social movements led the campaign against the government's privatization plans and declared that neoliberalism was over. At this time, most Uruguayans were unaware of the advancement of big Argentine agribusiness firms, linked to financial capital, doing RR soybeans in the countryside. Still, in the electoral platform of FA, it was argued that if winning the election, a new agrarian development model based on land distribution and support to family farmers would be implemented and against land concentration and foreignization (Frente Amplio, 22 December 2003). In the elections of October 2004, the Uruguayan citizens rejected the water privatization plans and the left-center party coalition (FA) won the elections, marking an end for the first time in history to the bipartite system and power alternation between *Blancos* and *Colorados*.

## 4.5    National Regulations and Their Intersection with Other Drivers

The previous sections presented some of the most important shifts in public policies during the 1980s and 1990s in Argentina, Paraguay, and Uruguay. They showed how the narrative of agrarian change as an almost direct and spontaneous response to increasing international demand for agrarian commodities, particularly soybeans, is too simple and needs to be complemented with the regulative shifts that created the necessary preconditions for the agribusiness-led export "booms" of the twenty-first century. In the case of Argentina and Uruguay, where inward-oriented development strategies and state interventionism had dominated public governance at least since 1930, the gradual dismantlement of industrial policies and export taxes, private investment promotion, and the deregulation of financial activities radically changed the conditions for the financial, agricultural, and land markets. For example, the removal of direct price interventions in agricultural markets implied a strengthening of the agribusiness sectors (Teubal 2008a, 203; Turzi 2016, 12).

Agrarian productivity, export performance, and inflow of financial capital increased during the 1990s, particularly in Argentina, but also to some extent in Uruguay.[35] In particular, the cattle sector managed to become more dynamic and increase productivity after many decades of relative stagnation.[36]

Moreover, shifts in legislation on leasing from the 1990s allowed for short-term contracts in both Argentina and Uruguay. Coupled with the traditional pattern of ranchers owning the land and sharecroppers doing the cultivation, this reinforced the logic of seeking the highest returns in the shortest time as the overarching principle for land-use decisions. With rising international soybean prices and a cost-effective technological package this meant soybeans became the dominant choice.[37] This logic privileges the annual soybean over more long-term productive orientations, such as cattle or sheep, subordinating all long-term risks and costs (for example erosion and salinization of the soils, pollution of waterways or health hazards) to short-term economic profits. An immediate effect of the high soybean prices was skyrocketing land and leasing prices, which in turn further exacerbated *sojization*, since no other land-uses could cover the high rents (Bisang et al. 2008). In Argentina, soybean related exports were already important, but they were increasing at a rather stable pace until 1996 (when the adoption of soybean RR started), when pace increased rapidly, and production and exports exploded after 2002.[38] In Uruguay, there was a considerable time lag between policies designed to promote agricultural exports (abolition of export taxes, authorization of genetic soybeans, investment promotion, relaxation of rules surrounding land and leasing markets, strengthening of intellectual property rights, free trade zones, and so forth), and the actual arrival of agribusiness firms and foreign investments. However, in Argentina and Paraguay, where conventional soybeans were already important, the response was more immediate. This indicates that favorable policies alone are not enough, but when these coincide with increased "cheap" capital searching for new markets, higher commodity prices, a new technological package, coupled with "push" factors (instability and export taxes in Argentina), then agrarian change can happen extremely fast.

At the same time, the decrease in protection, credits, and extension to family farmers and small farmers, combined with a further industrialization of agriculture, made it harder for small producers and sharecroppers to stay in activity. The economies of scale involved in both soy and beef production "fit perfectly" with the historical legacies of a concentrated land structure (Chapter 2 in this book), which was allowed to become even more concentrated through the liberalization of land markets and the relaxation of leasing regulation. The adverse social effects of concentration were particularly severe in Paraguay, characterized by a large peasant, *campesino* population who in addition suffered the collapse of the labor-intensive and generally small-scale cotton production. In Paraguay, soybeans and beef had started to expand already under the export trade friendly and low taxing regime (the lowest in the region) of Stroessner. After the fall of Stroessner, during the 1990s, the country further liberalized the economy, deregulated financial markets, removed direct price interventions in agricultural markets, and restructured the state. These reforms, combined with political turbulence, corruption, and decreasing amount of credits for small farmers, coinciding with the quality and price crisis in cotton sector, led to a rapid substitution process. Soybean products became extremely important in the export basket, accounting for more than 40% of total export value by the end of the past millennium.[39] Soybeans, often in the hands of foreigners, expanded over cotton, food crops, and forestland, and displaced small-scale peasant farmers.

In short, neoliberal policies, with their emphasis on a smaller, less interventionist state, in combination with export promotion in line with comparative advantage (agriculture) created the necessary preconditions that allowed for the agribusiness-led export "booms" of the twenty-first century. Domestic public policies are important, but they hardly determine agrarian change on their own. The Washington-consensus inspired regulative shifts in Argentina, Paraguay, and Uruguay, and interacted with shifts in the international agrofood system (technological, regulative, demographic, and geopolitical) that increased global demand for soybeans/beef. These in turn provoked shifts in the business strategies of lead firms along commodity chains at both transnational and national levels (Chapter 3 in this book). Besides the changes in the international agrofood system and the intentional liberalizing and export promoting policy shifts in the region, a very influential shift behind the post-2002 commodity boom was the dramatic devaluations of national currencies at the turn of the millennium in all three countries. This was a result of necessity rather than choice, but it implied a sudden increase in competitiveness for all export sectors in the region, and particularly for the agrarian sector, as it is typically less dependent on intermediary imports, and thus less affected by the higher costs of imported goods. Last, but not least, there are beneficial agro-ecological conditions for pastures and crops in the Southern Cone. Through a combination of these factors, the pace and scope of agropastoral expansion and intensification from 2002 and onward took historically unprecedented forms (Chapter 1 in this book). The agrarian systems in the region thus started to transform in an accelerated pace into being increasingly export-oriented, large-scale, specialized, cost-effective (if not taking "externalities" into consideration), and agribusiness-led.

However, while the year 2002 marked the beginning of the "super-cycle" of high commodity prices (Vivares 2017), Latin America citizens increasingly criticized the neoliberal policy reforms and the export-led growth model. On average, GDP per capita rates declined 1998–2002; the region entered a new "lost half-decade" (Ocampo 2004, 68). The severe economic crisis in Argentina that broke out around Christmas 2001 and spread the region was seen as a direct consequence of neoliberal policies. Market economy alone had shown incapable of securing full employment and good quality jobs. As poverty rates soared, faith in "trickle-down" vanished, while faith in the interventionist state returned. The articulation of agrarian change in Argentina, Paraguay, and Uruguay 2002–2019, with focus on the main regulative responses to shifts in forms of production and in social relations brought by the expansion and intensification of soybean and beef, is thoroughly explored next, in Chapter 5. Besides scrutinizing public policies, it also addresses the ways social groups and organizations have strived to put their interests and concerns on the agenda by influencing the regulatory framework and sometimes resisting governmental policies.

## Notes

1. Particularly research from the National Institute of Agrarian Research (INTA), with its advanced domestic seed science, managed to develop locally adapted seed varieties with improved yields (Martínez Alvarez 2012, 17–23).
2. Rates of inflation remained high, real wages were falling and external debt (in Argentina and Uruguay) was still high throughout the 1980s (Bulmer-Thomas 2003, 399). During this "lost decade", the GDP/c in Latin America declined on average 0.9% annually (Ocampo 2014). The steepest decline was in Argentina: GDP/c (in 1998 U.S. dollars), was 3228 in 1981 and 2672 in 1990. In Paraguay GDP/c was 1641 in 1981 and 1557 in 1990. In Uruguay, GDP/c was 2883 in 1981 and 2755 in 1990 (Bulmer-Thomas 2003, 399).

3. See webpage: http://www.mercosur.int/. Accessed 21 February 2019.
4. Mercosur was first launched in 1991 by the *Treaty of Asuncion*, but the scope was later deepened and expanded. Venezuela joined Mercosur as a full member in 2012, but was suspended in late 2016.
5. The external tariff spans between 35% on certain imports from outside the bloc.
6. While it faced several setbacks—internal trade disputes have been common—the estimated overall effect of Mercosur within its member is understood as economically positive, while differentiated; it is larger for Brazil and Argentina and more moderate for Paraguay and Uruguay (García et al. 2013).
7. Argentina is a federal system with 23 Provinces. Menem increased significantly the authority of the Provinces over fiscal transfers, education, health, housing, social security, and public works, making them responsible to make the sever budget cuts in national social welfare policies.
8. To make the Argentine peso equivalent to a dollar was an attempt to restore a viable monetary and financial system in a country that had lost confidence in its authorities' ability to manage the currency.
9. Since Argentina had no control over monetary policy because of the peso-dollar convertibility—it could not devalue nor expand the monetary base to encourage economic growth, but covered fiscal expenditures by taking on foreign loans.
10. The regulative process started already in October 1991, the *National Advisory Committee on Agricultural Biotechnology* (CONABIA), within *the Secretariat of Agriculture, Livestock, Fisheries, and Food* (SAGPyA), as one of the first GMO regulatory institutions in the world (Pellegrini 2013).
11. An important part of the soybean area expanded on leased land—facilitated by the relaxation of land regulation under Menem. The crop firms typically paid in advance for a fixed price, allowing the landowners to benefit from the soybean expansion without having to take any risks. Some big landowners sold their land in the Pampas region for a high price and bought cheaper land in the Gran Chaco region or retired with a lot of money in the pocket.
12. Ministerio de Justicia y Derechos Humanos, Información Legislativa, "Emergencia Publica y Reforma del Regimen Cambiario, Ley 25.561", http://servicios.infoleg.gob.ar/infolegInternet/anexos/70000-74999/71477/texact.htm. Accessed 23 February 2019.
13. Export taxes have existed for the most part of Argentine history, and they played a particularly important role under the ISI-period (Chapter 2 in this book), but Menem removed them during the 1990s. The real value of the Argentine peso relative to the USD fell by more than 60% from

1.04 pesos per dollar in 2001 to 2.68 pesos per dollar in 2002 (FAO 2017, 20). Export taxes were set at 10%.

14. At the time for the interview, the respondent was the President of the Paraguayan Chamber Pro Agro (CAPROA) and President of the Seed Chamber and the Vice-President of the Chamber China-Paraguay.

15. Soybeans had emerged as a new crop within the framework of one of these programs, the *National Wheat Plan*, promoted as an ideal supplementary crop to wheat (winter crop), since the soybean is a summer crop, and in combination they could yield two harvests a year.

16. Infrastructure improvements in Brazil allowed the soybean frontier to expand into the vast farmland of the Atlantic forest (Shurtleff and Aoyagi 2009, 449, 501). The binational construction of the mega hydroelectric power plant, *Itaipu*, close to the Brazilian border (during the 1970s and 80s) also strengthened the Brazilian presence and increased the stream of arriving Brazilian capital, firms, and farmers.

17. The drug trade was one of the most lucrative and pervasive sources of illegal income, and party officials and military leaders were principal beneficiaries. In 1999, the value added from different types of smuggling was equivalent to 25% of GDP (The World Bank 2001).

18. IBR was the Rural Welfare Institute in charge of land reform, later replaced by INDERT.

19. The respondent is also a lecturer in rural development and family farming, at the faculty of Agronomy, University of Asunción, and he is Paraguay's representative at The Specialized Meeting on Family Farming of Mercosur (REAF).

20. Since 1963, IBR had been responsible of the provision of land titles, settlement, and agrarian reform (see Chapter 2 of this book).

21. Global Forest Watch. "FAO reforestation in Paraguay". https://www.globalforestwatch.org/dashboards/country/PRY. Accessed 19 July 2018.

22. Country Profile: Uruguay. Atlas media MIT education. (2018). Creative Commons Attribution-Sharealike 3.0 Unported License. Available at https://atlas.media.mit.edu/en/profile/country/ury/#Trade_Balance. Accessed 21 September 2018.

23. Crop area under the protected ISI-period had reached 1½ Mha, but retracted to around one third of that area during the 1990s.

24. Exports Uruguay: 1994. Atlas media MIT education. (2018). Creative Commons Attribution-Sharealike 3.0 Unported License. Available at https://atlas.media.mit.edu/en/visualize/tree_map/sitc/export/ury/all/show/1994/. Accessed 5 December 2018.

25. In 1989, Law 16.082 declares of national interest the control and elimination of FMD, and that MGAP is responsible to control, facilitate, and monitor this process and to develop new organizational and technological solutions (Mederos Porto 2014, 134–135).

26. Official Record (Diario Oficial), 26 January 1988. Montevideo: IMPO.
27. All laws and resolutions can be accessed at the official website of the Uruguayan Parliament: https://parlamento.gub.uy/documentosyleyes/ leyes. Accessed 1 December 2018.
28. INIA was/is partly funded from a farm sales tax of 0.4%. The private sector holds two seats on the board of four directors, the other two being appointed by MGAP, of which one is designated president of the board.
29. Atlas media MIT education. (2018). Creative Commons Attribution-Sharealike 3.0 Unported License. Available at https://atlas.media. mit.edu/en/visualize/tree_map/sitc/export/ury/all/show/1994/. Accessed 5 December 2018.
30. In the midst of the regional privatization wave, the Uruguayan citizens opposed through referendum the law for privatization of public utilities in 1992, and later the privatization of the state petroleum company in 2003 and the privatization of the public water company in 2004 (Canzani and Midaglia 2011, 118–119).
31. FA was forbidden during the military dictatorship, but already participated in the first re-democratizing elections in 1984.
32. The immediate effect was that all export markets closed, and well-paid market outlets for Uruguayan beef remained closed until the country got back the status of being free from FMD with vaccination in 2003.
33. ADM was already installed in the country, reloading the Paraguayan soybeans for re-export from the port of Nueva Palmira. LDC was installed for offshore reasons, but responded rapidly with investments in export-infrastructure to "take care of" the emerging domestic soybean surplus. As soybeans continued to expand in Uruguay, Cargill also settled in the country in 2005, using offensive strategies to expand market share, and in 2008 Bunge started to operate in the country (Country manager of Cargill, Paysandú and at FAGRO-Udelar).
34. Already in 2004, Uruguay exported soybeans for a value of USD 180M, representing 5.2% of total export value. Atlas media MIT education. (2018). Creative Commons Attribution-Sharealike 3.0 Unported License. Available at https://atlas.media.mit.edu/en/visualize/tree_ map/sitc/export/ury/all/show/1994/. Accessed 21 October 2018.
35. Access the export profile per year and country at the Observatory of Economic Complexity website—an open source platform at the MIT Media, built with data from Comtrade FAO statistics: http://atlas.media. mit.edu.
36. The dynamism of the 90s, however, ended abruptly with the outbreak of foot and mouth disease in 2001 that spread the region.
37. In Uruguay, 70% of soybean production was cultivated on leased land in 2013 (Carriquiry 2015).

38. In Argentina, the value of whole soybean exports oscillated around USD 500 million (M) throughout the 1990s, but rose to over I billion (B) at the turn of the millennium, to more than 2B in 2004, and more than 3.5B in 2007. Soybean oil exports oscillated around 400 M up until 1995, but more than doubled, to over 1B at the end of the 1990s, and was exported to a value of more than 4B in 2007. Soybean cake moved from 1B in the beginning of the 1990s to around 2B at the turn of the millennium, to more than 4B in 2005 and more than 6.5B in 2007. Source: Observatory of Economic Complexity, atlas.media.mit.edu.

39. Source: Observatory of Economic Complexity, data from Comtrade FAO statistics, http://atlas.media.mit.edu.

## REFERENCES

Abuelafia, Emmanuel, Sergio Berensztein, and Miguel Braun y Luciano di Gresi. 2010. "¿Quién toma las decisiones sobre el gasto público? La importancia del proceso informal de determinación del presupuesto en la Argentina." In ¿Quiénes deciden el presupuesto? La economía política del proceso presupuestario en América Latina, edited by Mark Hellerberg, Carlos Scartascini, and Ernesto Stein, 17–47. Colombia: Banco Interamericano de Desarrollo, BID, Mayol Ediciones S.A.

Acosta Reveles, Irma L. 2008. "Capitalismo agrario y sojización en la pampa Argentina. Las razones del desalojo laboral." Lavboratorio, Cambio Estructural y Desigualdad Social (CEyDS) / Facultad de Ciencias Soicales, UBA 10 (22): 8–12.

Agrawala, Shardul, Annett Moehner, Frédéric Gagnon-Lebrun, Walter E. Baethgen, Daniel L. Martino, Eugenio Lorenzo, Marca Hagenstad, Joel Smith, and Maarten van Aalst. 2004. "Development and Climate Change in Uruguay: Focus on Coastal Zones, Agriculture and Forestry." In Working Party on Global and Structural Policies, Working Party on Development Co-operation and Environment, edited by Environment Policy Committee Environment Directorate. Paris: OECD.

Ankersen, Tom, and Thomas Ruppert. 2006. "Tierra y Libertad: The Social Function Doctrine and Land Reform in Latin America." Tulane Environmental Law Journal; University of Florida Levin College of Law Research Paper 19 (69): 52.

Arezki, R., K. Deininger, and H. Selod. 2015. "What Drives the Global 'Land Rush'?" The World Bank Economic Review 29 (2): 207–233.

Baraibar, Matilda. 2014. "Green Deserts or New Opportunities? Competing and Complementary Views on the Soybean Expansion in Uruguay, 2002–2013." Doctoral thesis Monograph, Economic History, Stockholm University. Stockholm Studies in Economic History 64.

Bárcena Ibarra, Alicia, and Antonio Prado. 2015. "Introducción." In *Neoestructuralismo y corrientes heterodoxas en América Latina y el Caribe a inicios del siglo XXI*, edited by Alicia Bárcena Ibarra and Antonio Prado. Santiago de Chile: CEPAL.

Barrios, Juan José, Nestor Gandelman, and Gustavo Michelin. 2010. "Analysis of Several Productive Development Policies in Uruguay." IDB Working Paper Series. Montevideo: Universidad ORT Uruguay.

Bell, Daveid E., and Cintra Scott. 2011. "Los Grobo: Farming's Future?" *Harvard Business School General Management Unit* (Case 511-088): 23.

Berrón, Gonzalo, and Rafael Freire. 2004. "Los movimientos sociales del Cono Sur contra el mal llamado 'libre comercio'." *OSAL* V (13): 297–307.

Bervejillo, J. E., J. M. Alston, and K. P. Tumber. 2012. "The Benefits from Public Agricultural Research in Uruguay." *Australian Journal of Agricultural and Resource Economics* 56 (4): 475–497.

Bianco-Bozzo, M., M. Chiappe, C. Hernández, and M. Carámbula Pareja. 2010. "Agrobiotecnologías en Uruguay: Posicionamiento de actores en torno a un debate incierto." *Agricultura, sociedad y desarrollo* 7 (3): 18.

Bisang, R., G. Anlló, and M. Campi. 2008. "Una revolución (no tan) silenciosa. Claves para repensar el agro en Argentina." *Desarrollo Economico* 48 (190): 165–207.

Bisang, R., M. Campi, and V. Cesa. 2009. "Biotecnología y desarrollo." In *Documento de proyecto*. Santiago de Chile: Comisión Económica para América Latina y el Caribe (CEPAL).

Bulmer-Thomas, Victor. 2003. *The Economic History of Latin America Since Independence*. Cambridge: Cambridge University Press.

Burch, David, and Geoffrey Lawrence. 2009. "Towards a Third Food Regime: Behind the Transformation." *Agriculture and Human Values* 26 (4): 267.

Bureau, Jean-christophe, Maria Priscila Ramos, and Luca Salvatici. 2005. "Tariffs, TRQs and Import Composition: The Case of Beef Trade Between the EU and Mercosur." European Trade Study Group, Seventh Annual Conference.

Burns, Sarah L., and Lukas Giessen. 2016. "Dismantling Comprehensive Forest Bureaucracies: Direct Access, the World Bank, Agricultural Interests, and Neoliberal Administrative Reform of Forest Policy in Argentina." *Society & Natural Resources* 29 (4): 493–508. https://doi.org/10.1080/08941920.2015.1089608.

Caldas, Marcellus M., Douglas Goodin, Steven Sherwood, Juan M. Campos Krauer, and Samantha M. Wisely. 2015. Land-Cover Change in the Paraguayan Chaco: 2000–2011. *Journal of Land Use Science* 10 (1): 1–18.

Caligaris, Gastón. 2017. "Las grandes empresas agropecuarias en Argentina: los casos de Cresud y El Tejar." *Cuadernos de Economía* 36 (71): 469–488.

Canzani, Agustín, and Carmen Midaglia. 2011. "Entre el riesgo de la segment-ación y el desafío de construir ciudadanía. El futuro posible del Estado de bie-nestar." In *La aventura uruguaya. Las Memorias del Futuro. Debates sobre el Uruguay que viene*, edited by Gerardo Caetano and Rodrigo Arocena, 111–132. Montevideo: Random House Mondadori.

Carriquiry, Florencia. 2015. "El papel del agro en el desarrollo económ-ico nacional." In *El desarrollo agropecuario y agroindustrial de Uruguay: Reflexiones en el 50 aniversario de la Oficina de Programación y Política Agropecuaria (OPYPA-MGAP)*, edited by Unidad de Comunicación Organizacional y Difusión, 26–35. Montevideo: OPYPA-MGAP.

Cartes, José L., J. J. Thompson, and A. Yanosky. 2015. "El Chaco paraguayo como uno de los últimos refugios para los mamíferos amenazados del Cono Sur." *Paraquaria Natural* 3 (2): 37–47.

Clapp, Jennifer. 2016. *Food*, 2nd ed. Cambridge: Polity Press.

Clapp, Jennifer, and Ian H. Rowlands. 2014. "Corporate Social Responsibility." In *Essential Concepts of Global Environmental Governance*, edited by Jean-Frédéric Morin and Amandine Orsini. Earthscan: Routledge.

Craviotti, Clara. 2016. Which Territorial Embeddedness? Territorial Relationships of Recently Internationalized Firms of the Soybean Chain. *The Journal of Peasant Studies* 43 (2): 331–347.

de Castro, Fabio, Barbara Hogenboom, and Michiel Baud. 2016. *Environmental Governance in Latin America*. Hampshire: Palgrave Macmillan.

Durán, Verónica, and Lucía Salgado. 2013. "Avances en el régimen de promo-ción de inversionen (COMAP)." In *Anuario Opypa 2013*, edited by MGAP-OPYPA. Montevideo: Agropecuaria Hemisferio Sur.

ECLAC, FAO, IICA. 2017. *The Outlook for Agriculture and Rural Development in the Americas: A Perspective on Latin America and the Caribbean 2017–2018*. San José, Costa Rica: Economic Commission for Latin America and the Caribbean (ECLAC), Food and Agriculture Organization of the United Nations (FAO), Inter-American Institute for Cooperation on Agriculture (IICA).

Errea, Eduardo, Juan Peyrou, Joaquín Secco, and Gonzalo Souto. 2011. *Transformaciones en el agro uruguayo - Nuevas instituciones y modelos de organización empresarial*. Edited by Dámaso Antonio Larrañaga Universidad Católica del Uruguay, Facutad de Ciencias Empresariales, Programa de Agronegocios. Mastergraf ed, Universidad Católica. Montevideo: Facutad de Ciencias Empresariales, Programa de Agronegocios.

Ezquerro-Cañete, Arturo, and Ramón Fogel. 2017. "A Coup Foretold: Fernando Lugo and the Lost Promise of Agrarian Reform in Paraguay." *Journal of Agrarian Change* 17 (2): 279–295.

FAO. 2017. "Soybean Prices, Economic Growth and Poverty in Argentina and Brazil." In *Background Paper to the UNCTAD-FAO Commodities and Development Report 2017 Commodity Markets, Economic Growth and*

*Development*. Rome: Food and Agriculture Organization of the United Nations.

Figari, Mercedes, Virginia Rossi, and Rosario González. 2007. "Los Productores Familiares." In *El sector agropecuario en el Uruguay: Una mirada desde la sociología rural*, edited by Marta Chiappe, Matías Carámbula, and Emilio Fernández, 7393. Montevideo: Departamento de ciencias sociales, FAGRO, Udelar.

Frente Amplio. 2003, December 22. *Grandes lineamientos programáticos para el gobierno 2005–2009 - Porque entre todos otro Urugauy es posible*. Edited by Frente Amplio. IV Congreso extraordinario del Frente Amplio.

Fridman, Daniel. 2010. "A New Mentality for a New Economy: Performing the Homo Economicus in Argentina (1976–83)." *Economy and Society* 39 (2): 271–302.

García, Eduardo Cuenca, Margarita Navarro Pabsdorf, and Estrella Gómez Herrera. 2013. "The Gravity Model Analysis: An Application on MERCOSUR Trade Flows." *Journal of Economic Policy Reform* 16 (4): 336–348. https://doi.org/10.1080/17487870.2013.846857.

Garcia, Fernando. 2015. "Agricultura en el Cono Sur ¿Qué se conoce, qué falta por conocer?" *Siembra, Repositorio Digital UCE* 2: 103–115.

Giarracca, Norma. 2008. "La Argentina y la democratización de la tierra." *Lavboratorio, Cambio Estructural y Desigualdad Social (CEyDS) / Facultad de Ciencias Soicales, UBA* 10 (22): 18–21.

Graesser, Jordan, T. Mitchell Aide, H. Ricardo Grau, and Navin Ramankutty. 2015. "Cropland/Pastureland Dynamics and the Slowdown of Deforestation in Latin America." *Environmental Research Letters* 10 (3): 034017.

Gras, Carla, and Valera Hernández. 2013. *El agro como negocio: producción, sociedad y territorios en la globalización*. Buenos Aires: Editorial Biblos.

Grugel, Jean, and Maria Pia Riggirozzi. 2007. "The Return of the State in Argentina." *International Affairs* 83 (1): 87–107.

Guibert, Martine, Susana Grosso, María Eva Bellini, Pedro Arbeletche. 2011. "De Argentina a Uruguay: espacios y actores en una nueva lógica de producción agrícola." *PAMPA, Revista Interuniversitaria de Estudios Territoriales* 7 (Suplemento especial tematica: Impactos territoriales asociados a la reconfiguración del sistema productivo primario): 13–38.

Hernández, René. 2015. "Transformación del Estado y paradigmas de desarrollo en América Latina." In *Neoestructuralismo y corrientes heterodoxas en América Latina y el Caribe a inicios del siglo XXI*, edited by Alicia Bárcena Ibarra and Antonio Prado. Santiago: CEPAL.

Hetherington, Kregg. 2009. "Privatizing the Private in Rural Paraguay: Precarious Lots and the Materiality of Rights." *American Ethnologist* 36 (2): 224–241.

Human Rights Council. 2015. "Report of the Special Rapporteur on the Rights of Indigenous Peoples, Victoria Tauli-Corpuz, Addendum the Situation of Indigenous Peoples in Paraguay." In *Human Rights Council, Thirtieth Session, Agenda Item 3*. United Nations General Assembly.

IICA. 2009. "Evolución y situación de la cadena agroalimentaria Sojera." In *Uruguay Agroalimentario en cifras*. Montevideo: Inter-American Institute of Cooperation on Agriculture.

Jobbágy, E. G., H. R. Grau, J. M. Paruelo, and E. F. Viglizzo. 2015. "Farming the Chaco: Tales from Both Sides of the Fence." *Journal of Arid Environments* 123: 1–2. https://doi.org/10.1016/j.jaridenv.2015.07.011.

Katz, Jorge. 2015. "La macro- y la microeconomía del crecimiento basado en los recursos naturales." In *Neoestructuralismo y corrientes heterodoxas en América Latina y el Caribe a inicios del siglo XXI*, edited by Alicia Bárcena Ibarra and Antonio Prado, 243–259. Santiago: CEPAL.

Lanzaro, Jorge. 2013. "Continuidad y cambios en una vieja democracia de partidos: Uruguay (1910–2010)." *Opinião Pública* 19 (2): 235–269. https://doi.org/10.1590/s0104-62762013000200001.

Lapegna, Pablo. 2016. "Genetically Modified Soybeans, Agrochemical Exposure, and Everyday Forms of Peasant Collaboration in Argentina." *The Journal of Peasant Studies* 43 (2): 517–536. https://doi.org/10.1080/03066150.2015.1041519.

Lapitz, Rocío, Gerardo Evia, and Eduardo Gudynas. 2004. *Soja y carne en el Mercosur: comercio, ambiente y desarrollo agropecuario*. Montevideo: Coscoroba Ediciones.

Leguizamon, Amalia. 2014. "Roundup Ready Nation: The Political Ecology of Genetically Modified Soy in Argentina." PhD dissertation, Sociology, City University of New York.

Mansourian, Stephanie, Lucy Aquino, Thomas Erdmann, and Francisco Pereira. 2014. "A Comparison of Governance Challenges in Forest Restoration in Paraguay's Privately-Owned Forests and Madagascar's Co-managed State Forests." *Forests* 5 (4): 763–783.

Manzetti, Luigi. 1992. "The Evolution of Agricultural Interest Groups in Argentina." *Journal of Latin American Studies* 24 (3): 585–616.

Margulis, Matias E., and Tony Porter. 2013. "Governing the Global Land Grab: Multipolarity, Ideas, and Complexity in Transnational Governance." *Globalizations* 10 (1): 65–86.

Martínez Alvarez, Diego Leonardo. 2012. Historia de la soja en la Argentina: Introducción y adopción del cultivo. In *El cultivo de soja en Argentina*, edited by Baigorri and Salado Navarro, 1. Buenos Aires: Agroeditorial.

Masterson, Thomas. 2007. "Land Rental and Sales Markets in Paraguay." Economics Working Paper Archive No. 491, Levy Economics Institute.

McMichael, Philip. 2009. "A Food Regime Genealogy." *The Journal of Peasant Studies* 36 (1): 139–169. https://doi.org/10.1080/03066150902820354.

McMichael, Philip. 2013. *Food Regimes and Agrarian Questions*. Edited by Kate Kennedy. Agrarian Change & Peasant Studies. Winnipeg, Canada: Fernwood Publishing.

Mederos Porto, Leticia. 2014. "La fiebre aftosa como problema para la producción ganadera en Uruguay y la demanda de ciencia, tecnología e innovación endógenas 1870–2001." Master of Science in Economic History, Facultad de Ciencias Sociales. Unidad Multidisciplinaria, Universidad de la Montevideo: Udelar.

Mereles, María Fátima, and Oscar Rodas. 2014. "Assessment of Rates of Deforestation Classes in the Paraguayan Chaco (Great South American Chaco) with Comments on the Vulnerability of Forests Fragments to Climate Change." *Climatic Change* 127 (1): 55–71. https://doi.org/10.1007/s10584-014-1256-3.

Morales Olmos, Virginia, and Jacek P. Siry. 2009. "Economic Impact Evaluation of Uruguay Forest Sector Development Policy." *Journal of Forestry* 107 (2): 63–68.

Moreira, Carlos. 2010. "Movimientos populares y luchas sociales en Uruguay." *Interseções: Revista de Estudos Interdisciplinares* 12 (2): 283–300.

Morgan, Kevin, Terry Marsden, and Jonathan Murdoch. 2006. *Worlds of Food— Place, Power and Provenance in the Food Chain, Oxford Geographical and Environmental Studies Series*. Oxford, UK: Oxford University Pres.

Nickson, A. 2015. *Historical Dictionary of Paraguay*. Edited by J. Woronoff, 3rd ed. Historical Dictionary of the Americas. Lanham, MD: Rowman & Littlefield.

Ocampo, José Antonio. 2004. "Latin America's Growth and Equity Frustrations During Structural Reforms." *Journal of Economic Perspectives* 18 (2): 67–88.

Ocampo, José Antonio. 2014. The Latin American Debt Crisis in Historical Perspective. In *Life After Debt*, edited by J. E. Stiglitz, and D. Heymann. International Economic Association Series. London: Palgrave Macmillan.

Ocampo, José Antonio. 2015. "América Latina frente a la turbulencia económica mundial." In *Neoestructuralismo y corrientes heterodoxas en América Latina y el Caribe a inicios del siglo XXI*, edited by Alicia Bárcena and Antonio Prado. Santiago de Chile: CEPAL.

OECD/FAO. 2016. *OECD-FAO Agricultural Outlook 2016–2025*. Paris: OECD Publishing.

Oosterveer, Peter, and David A. Sonnenfeld. 2012. *Food, Globalization and Sustainability*. London: Routledge.

Paolera, Gerardo della, Xavier H. Duran Amorocho, and Aldo Musacchio. 2018. "The Industrialization of South America Revisited: Evidence from Argentina, Brazil, Chile and Colombia, 1890–2010." NBER Working Papers 24345, National Bureau of Economic Research Inc.

Paolino, Carlos. 2015. "La política públca y el apoyo al sector agropecuario." In *El desarrollo agropecuario y agroindustrial de Uruguay: Reflexiones en el 50 aniversario de la Oficina de Programación y Política Agropecuaria (OPYPA-MGAP)*, edited by MGAP Unidad de Comunicación Organizacional y Difusión, 37–45. Montevideo: OPYPA-MGAP.

Paolino, Carlos, Lucía Pittaluga, and Mario Mondelli. 2013. "Cambios en la dinámica agropecuaria y agroindustrial del Uruguay y las políticas públicas." In *Anuario Opypa 2013*. Montevideo: OPYPA-MGAP.

Peiretti, R., and J. Dumanski. 2014. "The Transformation of Agriculture in Argentina Through Soil Conservation." *International Soil and Water Conservation Research* 2 (1): 14–20. https://doi.org/10.1016/S2095-6339(15)30010-1.

Pellegrini, Pablo A. 2013. "What Risks and for Whom? Argentina's Regulatory Policies and Global Commercial Interests in GMOs." *Technology in Society* 35 (2): 129–138.

Pellegrini, Pablo A., and Galo Ezequiel Balatti. 2017. "Los bancos de semillas: entre la preservación y la apropiación de recursos naturales. El acceso a los recursos fitogenéticos en la Argentina." *Desenvolvimento e Meio Ambiente* 41: 105–123. https://doi.org/10.5380/dma.v41i0.46802.

Piñeiro, Diego. 1998. "Cambios y permanencias en el agro uruguayo. Tendencias y coyuntura." In *Las agriculturas del Mercosur: El papel de los actores sociales*, edited by Norma Garraca and Silvia Cloquell, 248. Buenos Aires: La Colmena/FLACSO.

Piñeiro, Diego. 2004. "Rentabilidad o muerte: La propuesta rural en el Uruguay." In *En busca de la identidad : la acción colectiva en los conflictos agrarios de América Latina*, edited by Diego Piñeiro, 344 s. Buenos Aires: Consejo Latinoamericano de Ciencias Sociales, CLACSO.

Piñeiro, Diego. 2011. "Dinámicas en el mercado de la tierra en América Latina: el caso de Uruguay." In *Procesos de concentración y extranjerización de tierras en América Latina y el Caribe*, edited by FAO. Santiago de Chile: Oficina Regional de la FAO.

Prieto, G. S., and O. Ernst. 2010. "Manejo del suelo y rotación con pasturas: Efecto sobre la calidad del suelo, el rendimiento de los cultivos y el uso de insumos." *Informaciones Agronómicas del Cono Sur* 45: 22–26.

Qaim, Matin, and Greg Traxler. 2005. "Roundup Ready Soybeans in Argentina: Farm Level and Aggregate Welfare Effects." *Agricultural Economics* 32 (1): 73–86. https://doi.org/10.1111/j.0169-5150.2005.00006.x.

Redo, Daniel J., T. Mitchell Aide, Matthew L. Clark, and María José Andrade-Núñez. 2012. "Impacts of Internal and External Policies on Land Change in Uruguay, 2001–2009." *Environmental Conservation* 39 (2): 122–131. https://doi.org/10.1017/s0376892911000658.

Richardson, Neal P. 2009. "Export-Oriented Populism: Commodities and Coalitions in Argentina." *Studies in Comparative International Development* 44 (3): 228.

Romson, Åsa. 2012. "Environmental Policy Space and International Investment Law." Doctoral dissertation, Environmental Law, Acta Universitatis Stockholmiensis, Stockholm University. Studia Juridica Stockholmiensia 88.

Rossi, Virginia. 2010. "La producción familiar en la cuestión agraria uruguaya." *Revista NERA* 13 (16): 63–80.

Sassen, Saskia. 2010. "A Savage Sorting of Winners and Losers: Contemporary Versions of Primitive Accumulation." *Globalizations* 7 (1–2): 23–50.

Seyler, Daniel. 1988. "The Economy." In *Paraguay: A Country Study*. Area Handbook Series, edited by Dennis M. Hanratty and Sandra W. Meditz, 97–156. Washington, DC: Library of Congress, Federal Research Div.

Shurtleff, William, and Akiko Aoyagi. 2009. *History of Soybeans and Souyfoods in South America (1882–2009): Extensively Annotated Bibliography and Sourcebook*. Lafayette: Soyinfo Center.

Souto, Gonzalo. 2013. "Oleaginosos y derivados: situación y perspectivas." In *Anuario 2013, Opypa: Cadenas productivas; temas de política; Proyectos, estudios y documentos*, edited by OPYPA-MGAP, 17. Montevideo: Editorial Agropecuaria Hemisferio Sur.

Teubal, Miguel. 2008a. "Genetically Modified Soybeans and the Crisis of Argentina's Agriculture Model." In *Food for the Few—Neoliberal Flobalism and Biotechnology in Latin America*, edited by G. Otero, 189–216. Austin: University of Texas Press.

Teubal, Miguel. 2008b. "Soja y agronegocios en la Argentina: la crisis del modelo." *Lavboratorio, Cambio Estructural y Desigualdad Social (CEyDS) / Facultad de Ciencias Soicales, UBA*. 10 (22).

The World Bank. 2001. "Paraguay Country Assistance Evaluation." Edited by Operations Evaluation Department, Washington.

Tkachuk, Maximiliano, and Marina Dossi. 2014. "Dinámica de la producción ganadera Argentina: Análisis de variables intervinientes y de escenarios futuros." FAUBA, Apuntes Agroeconomicos.

Toledo Lopez, Virginia. 2017. "La política agraria del kirchnerismo. Entre el espejismo de la coexistencia y el predominio del agronegocio." *Mundo Agrario* 18 (37): 25.

Trigo, E., D. Chudnovsky, E. Cap, and A. Lopez. 2002. *Genetically Modified Crops in Argentine Agriculture: An Open Ended Story*. Buenos Aires: Libros del Zorzal.

Trigo, E. J., and E. J. Cap. 2003. "The Impact of the Introduction of Transgenic Crops in Argentinean Agriculture." *AgBioForum* 6 (3): 87–94.

Turzi, Mariano. 2016. *The Political Economy of Agricultural Booms: Managing Soybean Production in Argentina, Brazil, and Paraguay*. Cham, Switzerland: Springer.

UNEP-DINAMA. 2007. "Propuesta de Marco Nacional de Bioseguridad para Uruguay." In *Proyecto de Desarrollo del Marco Nacional de Bioseguridad*, edited by UNEP-GEF. Montevideo: Dirección nacional de Medio Ambiente (DINAMA).

Uruguay XXI. 2015. "Informe de comercio exterior - exportaciones e importaciones de Uruguay." In *Informe de comercio exterior*. Montevideo: Uruguay XXI Instituto de Promoción de Inversiones y Exportaciones.

USDA. 2018. "Livestock and Poultry: World Markets and Trade." United States Department of Agriculture, Foreign Agricultural Service.

Varrotti, Andrea Patricia Sosa, and Samuel Frederico. 2018. "Las estrategias empresariales del agronegocio en la era de la financiarización. El caso de El Tejar." *Mundo Agrario* 19 (41): e086–e086.

Vergara-Camus, Leandro, and Cristobal Kay. 2017. "Agribusiness, Peasants, Left-Wing Governments, and the State in Latin America: An Overview and Theoretical Reflections." *Journal of Agrarian Change* 17: 239–257.

Vivares, Ernesto. 2017. *Regionalism, Development and the Post-commodities Boom in South America*. Cham: Springer.

Volante, José Norberto, and Lucas Seghezzo. 2018. "Can't See the Forest for the Trees: Can Declining Deforestation Trends in the Argentinian Chaco Region Be Ascribed to Efficient Law Enforcement?" *Ecological Economics* 146: 408–413. https://doi.org/10.1016/j.ecolecon.2017.12.007.

Walcott, J., J. Thorley, G. Casco, L. Coronel, V. Kapos, L. Miles, R. Blaney, and S. Woroniecki. 2014. *Mapeo de los beneficios múltiples de REDD + en Paraguay: el uso de la información espacial para apoyar la planificación del uso de la tierra*. Cambridge: Programa de las Naciones Unidas para el Medio Ambiente.

Willis, Eliza, Christopher da C. B. Garman, and Stephan Haggard. 1999. "The Politics of Decentralization in Latin America." *Latin American Research Review* 34: 7–56.

Wolford, W., S. M. Borras Jr., R. Hall, and I. Scoones. 2013. "Governing Global Land Deals: The Role of the State in the Rush for Land." In *Governing Global Land Deals: The Role of the State in the Rush for Land*, edited by W. Wolford, S. M. Borras Jr., R. Hall, and I. Scoones, 1–22. Malaysia: Wiley-Blackwell.

Zoomers, E. B., and J. M. G. Kleinpenning. 1990. Farm Size, Land Tenure and Rural Prosperity: Effects of Rural Development Policy on the Struggle for Land in Central Paraguay. *Revista Paraguaya de Sociologia* 27 (79): 7–29.

# CHAPTER 5

# Regulative Shifts and Agrarian Change of the Twenty-First Century

This chapter explores how the main protagonists of agrarian change of the twenty-first century; soy and beef in Argentina, Paraguay, and Uruguay, are leading the way for land-use change and intensification, and provoking important shifts in *forms of production* and *social relations*.[1] These shifts brought new pressing social-ecological concerns over, for example, rapid biodiversity loss, contamination of water, extreme market concentration, and loss of small farmers, rural depopulation, and degradation of land throughout the territory, albeit divergently.[2] Specific local social-ecological conditions of different places have created different pathways of agrarian change between different regions, with important differences between the Pampas, the Gran Chaco, and the Atlantic forest region (Chapter 1 of this book). The chapter explores the different ways Argentina, Paraguay, and Uruguay acted through public policies and regulations co-shaping the accelerating agrarian change. While focus is on the role of national regulations on agrarian change, it also addresses the differentiated patterns of change that emerge in the different ecoregions (Pampas, Gran Chaco, and the Atlantic forest).

Public policies and regulations may facilitate, accept, renegotiate, ignore or resist the new social relations and forms of production brought by exogenous pressures. Notwithstanding continuities, a significant policy reorientation away from the neoliberal reforms of the 1980s and 1990s (Chapter 4 in this book), influenced public policies

© The Author(s) 2020                                                      209
M. Baraibar Norberg, *The Political Economy of Agrarian Change in Latin America*, Governance, Development, and Social Inclusion in Latin America, https://doi.org/10.1007/978-3-030-24586-3_5

throughout the region in the early twenty-first century. This was centered in bringing the state back in as a central unit for strategic development planning, to make export-oriented commodity chains more "developmental" through redistribution and upgrading. In addition, an important part of agrarian public policies in Argentina, Paraguay, and Uruguay is funded and designed by International Organizations (IOs), and their new post-Washington Consensus agenda also included "state reform" and "participatory decision-making processes" for strengthened (market) "competitiveness", (social) "inclusion" and (environmental) "sustainability", as well as higher acceptance and compliance among the population.

Hundreds of new laws and resolutions and other policy shifts have appeared during the past decades with some relevance for agrarian change. Instead of providing an extensive list over all these, I have chosen to present both the relevant overall policy framework in each country and in more depth some flagship regulations, including the conflicts, negotiations, and effects of the same. This chapter does thus not only describe what is said in legislation and public policies, but also explores the ways social groups and organizations (including producers' organizations, environmental organizations, social movements, IOs as well as agribusiness actors) strive to put their interests and concerns on the policy agenda by influencing the regulatory framework and sometimes resisting governmental policies. By following the trajectory, or genealogy, of specific policies—how they emerged on the agenda, were negotiated, institutionalized, translated into practice, and interpreted divergently across actors—a deeper and fuller understanding of their actual role for agrarian change, including their acceptance and enforcement, is here duly addressed. Thus, this chapter brings the state into action as a field of dispute between a variety of interests and actors, domestic and foreign.

The remaining parts of this chapter are organized in the following way: The first section introduces the public governance trends in the region after the turn of the millennium, when the increasing discontent with neoliberalism provoked the rise of Latin American *neostructuralism*, including the so-called *Pink Tide*. This is followed by one section for each country, exploring the interaction between agrarian change and national policies and regulations. The chapter ends with a short concluding section.

## 5.1  The Twenty-First-Century Commodity Boom—And Policy Trends in the Region

After 2002, agropastoral area expansion as well as land-use intensification in Argentina, Paraguay, and Uruguay increased faster than in any other time in recorded history. It is safe to say that the remarkable growth of agricultural activities was primarily privately driven and governed in all three countries. This contrasts to previous waves of crop expansion, particularly in Paraguay under Stroessner and in Uruguay under Batllismo y neo-Batllismo, when crops received significant amount of public support, ranging from R&D, free seeds, subsidies, fixed prices, free silos, credits, and so forth.[3] While the state was not a direct driver, its long tradition of R&D programs, coupled with the liberalization reforms of the late twentieth century, clearly facilitated the twenty-first century-"boom". At the same time, neoliberalism was increasingly discredited throughout the region. Instead, new policy ideas announced that truedevelopmentrequiredaninterveningstatethatcould(re)balancethesocial-environmental trade-offs involved in purely market-driven development (i.e., taxing the "winners"—compensating the "losers"). A heterodox alliance of scholars, political movements, social organizations, national political forces, and IOs pushed policies in a direction that (again) emphasized the importance of the state and public planning for agrarian change and development of the twenty-first century.

*Neodesarrollismo*, or *neostructuralism*, emerged as an updated version of Latin American structuralism. This new reform agenda argued that the state should intervene more actively in the economy, in order to make markets more redistributive and universalist in social matters and more conservationist and sustainable in environmental matters (Bárcena Ibarra and Prado 2015, 21). As during the inward-oriented period of Latin American structuralism (Chapter 2 in this book), the *United Nations Economic Commission for Latin America and the Caribbean* (CEPAL) played a prominent role in articulating the intellectual underpinnings for the reform agenda and in spreading it across the region. In contrast to its predecessor, however, the neostructuralist approach sees international trade and foreign investments as necessary ingredients for economic development in the region (Gonzaga Belluzo 2015). However, not all trade is equally "developmental". The re-primarization of the export structure—with high dependence on a handful of agrarian-based exports—risks to create "Dutch disease" related problems and hinder access to more dynamic and well-paid market segments, according to this vein of thought (Ocampo 2004, 2015, 103).[4] The role of the developmental state is then to transform the exports structure away from high reliance on natural resources, to more diversification and more value-added activities. Thus, to make agrarian change "developmental" the state should use long-term development planning for the gradual change of the productive structure. By making specific value chains incorporate more technology and knowledge, they will "upgrade", compete more with quality than quantity, in order to access better-paid and less volatile market opportunities, and offer good well-paid jobs. In the end, more inclusive and more sustainable national development is achieved (Ffrench-Davis 2005; Hernández 2015). The commodity booms can be used as springboards in this respect, by making value chains upgrade.

The policy turn from neoliberalism to a renewed faith in the "developmental" state was also materialized in the national political forces elected to govern in most countries in the region during the early years of the 2000s, often referred to as the *"Pink Tide"*. In Mercosur, the *"Pink Tide"* started with the triumph of the workers' party PT and *Luiz da Silva* in Brazil (2003–2011), followed by *Dilma Rousseff* (2011–2016). In Argentina, anti-neoliberalism was materialized in the election of *Néstor Kirchner* (2003–2007), from the center-left Peronist faction *Front of Victory* (FPV) of the *Justicialist Party* in Argentina. The election of Kirchner marked the beginning of a 12-year-old era, often referred to as *Kirchnerismo*, which also included two government terms of *Cristina Fernández de Kirchner* (2007–2011 and 2011–2015), of the same political force. In Uruguay, the triumph of *Tabaré Vázquez* (2005–2010), from the center-left coalition *Frente Amplio* (FA) was followed by a new FA government with *José "Pepe" Mujica* (2010–2015), and once again with *Tabaré Vázquez* (2015–2020). Even in Paraguay, the 61-year consecutive Colorado rule was broken with the electoral triumph of former Catholic bishop *Fernando Lugo* (2008–2012) and his *Patriotic Alliance for Change*. The majority of the countries of the continent had periods ruled by different variations of "leftist" governments.[5]

The new political forces in the region criticized the previous reforms of trade liberalization, structural adjustments, and strengthening of property rights, and their polarizing and anti-developmental results. All these governments expressed instead faith in allowing the state take an active part in foster institutions to correct market failures and plan for long-term inclusion and sustainability. In this way, in clear contrast to the policy trends of the 1990s, there was a general move toward a development model, which favors a more intervening and strategically planning state. They also had a focus on redistributive public policies and explicitly expressed that the state should rebalance the relations of force between different social actors, in order to protect family farmers from the advancement of powerful agribusiness corporations.

Policies and programs in support of smallholders and family producers expanded, and labor legislation for rural workers advanced. In addition, South–South cooperation activities and programs strengthening family farming increased, not least within the realm of Mercosur, with the 2004 creation of *The Specialized Meeting on Family Farming* (REAF)—a meeting place for governments and family farmers and rural organizations in the region. REAF was created on initiative from Brazil in order to strengthen public policies in support of family farming in the Mercosur countries and to develop a joint framework of regional public policies for family farming (CLAEH 2015). REAF stated that family farming is a way of living with vital "multifunctional" positive consequences for rural development, social cohesion, national food security and sovereignty, sound localized ecosystem stewardship, rootedness, and local knowledge production (FONAF 2008, 7). It also expressed that the states have a responsibility to create conditions for family farmers to be able to continue playing these roles without being out-competed by agribusiness farming.

Post-neoliberal policies were not an exclusive trend to Latin America, but a kind of "post-Washington consensus" became increasingly mainstream in the international policy community and cherished by IOs (Ocampo 2015, 94). This post-Washington consensus involved the promotion of a state that engages in active productive policies, development planning, universal social policies (not targeted), and progressive tax systems (no flat tax). It also included "state reform" for strengthened (market) "competitiveness", (social) "inclusion", and (environmental) "sustainability". The state should foster "good institutions" and correct markets, as well as set strategies for poverty reduction and tackle corruption. One particularly dominant paradigm for agrarian policy the past decades has been the so-called *Rural Territorial Development* (RTD) strategy. While RTD shares many elements with the previously dominant

"Washington Consensus"-paradigm—such as poverty alleviation, market integration, decentralization, and transparence—it focus more on public policy planning, which ideally should involve broad "stakeholder participation" and consensus-based decisions, territorial and multi-sectorial planning (Officials at IICA Argentina and Paraguay, UNDP Paraguay, OPYPA-MGAP and MVOTMA Uruguay, Asunción, Buenos Aires, Montevideo, February–March, 2017).[6] In line with broader trends within the international governance for sustainable development, there has also been a considerable increase in projects dealing with environmental sustainability particularly linked to climate change adaption and mitigation. A great deal of projects also works to align national legal frameworks with international declarations, not least the UN 2030 Agenda for Sustainable Development and the *Sustainable Development Goals* (SDGs), succeeding the *Millennium Development Goals*.

However, in spite of new social concerns on the agenda and targeted programs, family farmers have been increasing leaving agriculture in all three countries at a historically unprecedented pace, while agribusiness farming has been advancing throughout the period. All countries continued to be open to trade (albeit with some restrictions and ISI-inspired taxes in the Argentine case) and to foreign investments, allowing for export-oriented commodity activities and agribusiness to grow stronger (Gonzaga Belluzo 2015). Thus, when the favorable world economic context, 2002–2008, translated into ever-higher agro-commodity prices, the overarching result in the three countries was dramatic increase in production and exports (FAO 2017, 20). Moreover, the region got a boost in liquidity and access to credit at low interest rates during the period, as international financial capital searched for new markets, investing heavily in Latin American land, agriculture, and exports infrastructure. Terms of trade (TOT) in the region also improved between 2003 and 2008.[7] Some public policies further actually benefitted the agribusiness firms, such as enhanced protection of private property rights to land and intellectual innovations, fiscal incentives for investments, devaluated exchange rates, publicly funded export-oriented infrastructure, tax exonerations for purchase of machinery, and other input (Vergara-Camus and Kay 2017b, 225–226). This is sometimes referred to as "the commodity consensus" (Svampa 2013). The remaining parts of this chapter explores in greater detail the new policies and regulations in each country, and discuss the relation between the same and agrarian change.

## 5.2    Argentina—State Interventionism
## and Back Again

The newly installed government of Néstor Kirchner (2003–2007) marked a rather abrupt change in policy orientation from the neoliberal 1990s of Carlos Menem; *laissez-faire* was over, state interventionism was back. The election of Kirchner marked the beginning of 12-year of *Kirchnerismo*, which also included two government terms of Cristina Fernández de Kirchner (2007–2011 and 2011–2015), the political force backed up by a multi-class alliance between business and labor similar to the historic populist coalition under Perón (Chapter 2 in this book). The government was nevertheless also inspired in the renewed attention on family farming, not least in the policy proposals of the previously mentioned newly installed *permanent meeting on family farming of Mercosur*, REAF. It was reasserted that after a long period of cuts in credits, extension services, and price controls, the state should proactively rebalance the relations between social actors, counteracting the polarizing effects of "pure" market relations and create inclusive development.[8] *Kirchnerismo* was also highly influenced by neostructuralism and particularly the idea that it was good to engage in international markets, but that the country should try to escape from the historical negative trade balance by exporting more value-added products and by increasing domestic purchasing capacity, to make domestic demand complement international market demand as engine to growth (Richardson 2009). In this way, the government leaned on Keynesian macroeconomic policy centered in support to domestic manufacturing industry, redistributive programs for family farmers, and a lot of money put on public research and education (Lapegna 2016). Post-crisis Argentina managed to recover lost ground and expand the size of the middle class. Along with proactive social protection policies, economic growth enabled the country to reduce poverty and unemployment. Until 2010, export growth and improved tax revenue led to fiscal and external surpluses, allowing Argentina to accumulate international reserves, reduce debt, and build fiscal space as buffers against shocks (IBRD 2014). Argentina's GDP growth rates averaged 6% annually from 2003 to 2013, well above the 1.6% average annual growth rate Argentina achieved during 1980–2000 (IBRD 2014). With social redistribution high on the agenda, the trend of increasing economic inequality of the 1990s was reverted; the Gini coefficient increased 8% from 1992 to 2002, but decreased 11% from 2002 to 2012 (FAO 2017, 27). However, land concentration continued and an increasing number of family farmers left the activity.

While the soybean became a symbolically laden discursive battle-field under *Kirchnerismo*, the actual soybean exports uninterruptedly increased throughout the period. The value of Argentine soy products was in 2002 USD 5.57B and USD 13.78B in 2017. In this way, agrarian change the past two decades in Argentina can almost be reduced into one word: soybeans. Soy represents the most important export product and the main driver of land-use change the past two decades (CAS 2018).[9] At the same time, Argentina is traditionally a livestock society, and while the combination of increased competition of land from soybeans and export restrictions during *Kirchnerismo* resulted in a drop of beef production and exports, the livestock sector remained important, not least as a political actors through its powerful organizations. With the election of Macri in 2015, the "*Pink Tide*" in Argentina was over and liberalism back. This policy reorientation has coincided with economic slowdown, depreciation of the peso and financial turmoil. After a period of strong economic growth, with a nearly doubling of the economy from 2002 to 2011, growth rates declined and became highly fluctuating.[10] The government responded with austerity policies. During 2018, Argentina entered economic recession, with extremely high fiscal and external deficits and inflation.[11] However, there are also lines of policy continuity. This section explores deeper the main regulations—their changes and continuities; their wider acceptance and resistance among important social groups—in relation to the changed social relations and forms of production brought by the twenty-first-century agrarian change.

### 5.2.1    Agrarian Interventionism

When Kirchner took office, the soybean complex represented the most dynamic sector in an overall context of crisis. The devaluation benefitted all agricultural export commodities since they are priced in dollars and all of a sudden their value in peso increased by a factor of three. Moreover, international soybean prices were rising and Argentina was producing soybeans at a very cost-efficient way (Former Secretary of MAGyP, 2013–2015, Buenos Aires, 2 March 2017). This scenario rendered soybean activities high economic margins in spite of the reintroduced export taxes on grains and oilseeds. In the beginning of the Kirchner administration export taxes oscillated around 13.5% on unprocessed agricultural products and around 5% for processed products (Richardson 2009). The government rose levies several times and by 2007 taxes on unprocessed grains (whole soybeans) were 27.5%, and on processed products (oil, cake, meal) 24%. The government argued it to be fair for the sector "payback" some of the profits to society, not least because it had benefitted greatly from the devaluation (Consultant and Former President of INTA, Buenos Aires, 28 February 2017). Moreover, the export tax is in contrast to most other fiscal sources (such as the value-added tax and the income tax) not regulated by federal sharing rules, and thus possible for the central government to freely decide on how to use them (Former Secretary of MAGyP, 2013–2015, Buenos Aires, 2 March 2017).[12] The revenues were used to payback rents on the public debt, as well as for social reforms (particularly unemployment assistance) and improvement of transport infrastructure (Nogués 2011; Richardson 2009; Tkachuk and Dossi 2014). After a couple of years, the increasing soy exports and a successful debt restructuring created a surplus in the balance of payment (Leguizamon 2014, 60).[13] The lion's share of revenue came from soybean products, both because they were by far the most exported products (see Appendix B), but also because they were levied with higher taxes than other agricultural products. This was justified as a means to put a break on the rapid *sojization* by reducing the profit gap between planting soybeans and other land-uses, not least food staples to the domestic market (Gomez 2008, 23).[14]

Moreover, the differentiated rates (higher taxes on whole beans than on processed products) aimed to encourage the development of domestic processing and value-adding activities and to counterbalance the differentiated import taxes in China (Chapter 3 in this book). Argentina became the world's biggest exporter of soybean cakes (meal) and soy-based biodiesel. The neostructuralist-inspired government wished to "upgrade" the soybean chain even further. The ultimate goal was to transform soybean production into higher value-added products such as soybean meal-fed chicken and pork, soymilk, soybean-based oleo-chemical polyols for the polyurethane industry (for example foams, coatings, and floorings), and margarine (Castellano and Goizueta 2011; FAO 2017; Shurtleff and Aoyagi 2009). While the chicken industry grew slightly and started to have an exportable surplus, the great majority of the soybeans were shipped away in the form of meal to become feed input for broiler, pork, and beef production elsewhere.[15] This could still be seen as a relative success, as expressed by a senior consultant with a long trajectory in agrarian policy from the state, private firms, research institutes, IOs, and national organizations:

> The result of these policies were very good. The value captured selling processed soybeans is more than double than the value of selling whole beans. Of course the valued added is even higher if you sell soybeans in the form of meat, but processed soybeans still have a much more added-value than whole beans. It is much better than nothing. (Director of Agricultural Affairs CARI, Buenos Aires, 1 March 2017)

The central government wanted to avoid its classical problem of negative balance of trade. One benefit of soybeans in this respect is that they can be sold massively at high prices to the international market without inflating domestic food prices too much, since they do not play an important part in households' consumption basket. This in stark contrast to beef. Beef is a central part of the traditional Argentine diet and of national identity. As international meat prices were rising, they provoked persistent price inflation with regressive effects (Former Secretary of MAGyP, 2013–2015, Buenos Aires, 2 March 2017). High beef prices in Argentina comes with the risk of social protests. The Kirchner administration tried in different ways to keep down domestic prices on beef, including a ban on beef exports for 180 days during 2006 (with the exception of prearranged shipments and the *Hilton quota*, see Chapter 3). The ban was later replaced by a quantitative restriction policy in order to make meat-packers prioritize the domestic market. Another measurement to curb the rises was the implementation of consumer price controls (FAO 2017, 26). One experienced agrarian consultant commented these policies in the following way:

> As the consumers were losing, the state tried to fix it. It is problematic, but the temptation is big... The political cost of high meat prices can be very high. This is always an additional problem coming with rising meat exports, the inflationary pressure from increasing food prices. The temptation to tax and restrict become enormous.... (Director of Agricultural Affairs CARI, Buenos Aires, 1 March 2017)

The traditional ranchers of the *Argentine Rural Association* (SRA) and the meat processing industry loudly opposed these policies (Gras and Hernández 2013, 342). They argued that the policies destroyed the livestock sector, reduced rural income, and created trade imbalance. At the same time, the livestock sector had by 1996 lost more than 5 Mha of pastureland to soybeans (Gras and Hernández 2013, 61). The government, however, was loyal to the demands of its main supporter groups among urban workers and middle classes, while also trying to compensate ranchers with economic support and credits (Tkachuk and Dossi 2014).[16]

When Cristina Fernández de Kirchner, also known as "CFK" (2007–2015) took over office from her husband in 2007, she continued applying export restricting policies on wheat, maize, and beef to keep domestic food prices down, while no restriction on soybeans but high export taxes to contribute with income to national treasury (Former Secretary of MAGyP, 2013–2015, Buenos Aires, 2 March 2017). In order to compensate the loss of land of the livestock sector, some of the revenues from the soy taxes were used to create a compensatory fund, which included granting subsidies to feedlots to buy corn to spur intensification of the livestock sector, so that it could maintain production volume with less land. The *Office for Control of Agricultural Business* (ONCCA) administrated the fund, which also gave subsidies for a wide range of other food-producing activities for domestic consumption. ONCCA was also in charge of setting quantitative export restrictions through quotas on several food products to hold down domestic prices (Davila et al. 2017; Gras and Hernández 2013, 338). As the value of the peso appreciated and trade surplus started to decline, the government also implemented hundreds of nonautomatic import licenses to avoid a negative trade balance. ONCCA was nevertheless surrounded by strong suspicion of corruption and there were no clear rules for the allocation of quotas (Consultant and Former President of INTA, Buenos Aires, 28 February 2017). In 2011, the government dissolved ONCCA and replaced by the *Unit of Coordination and Assessment of Domestic Consumption Subsidies* (UCESCI).[17]

Besides difficulties to compete with soybeans and adversary policies, the cattle industry suffered a new outbreak of foot and mouth disease in 2006, which, beyond the immediate necessary slaughter of many sick animals, damaged the reputation of Argentine beef.[18] Argentine beef exports fell significantly. In 2005, it still ranked fourth biggest exporter in the world of bovine meat with 0.77 mmt, while in 2012 it only ranked eleven, with 0.18 mmt. However, not only exports were falling; there was a significant decrease in the size of the national cattle herd (FAO 2017, 26). While the cattle herd retracted, soybean area, production, and exports continuously increased in an accelerating pace at the same time as soy exports generated an increasingly high amount of fiscal revenue (Richardson 2009).[19] Soybean exports had no restriction and the government did not influence the prices downward beyond the tax (FAO 2017).

In March 2008, the National Ministry of Economy declared a system of sliding export taxes on wheat, corn, sunflower, and soybean, following the international commodity price (Lindenboim 2008, 13). Since international soybean prices had been rising fast (from USD 305/t in June 2007 to USD 590/t in March 2008) levies rose from 35 to 44% (FAO 2017, 20; Leguizamon 2014).[20] This was the trigger for one of the most important agrarian conflict in Argentine history; *el conflicto del campo*.

The big umbrella producers' organizations, SRA, FAA, CRA and the grain cooperatives CONINAGRO organized a nationwide lockout in response to the sliding taxes. Other organizations also joined the call against the taxes.[21] National highways were blocked with trucks and agricultural machinery throughout Argentina, leading to major shortages in cities' supermarkets (Caligaris 2017, 125; Leguizamon 2014, 61). Young people living in the countryside with no previous experience of political involvement became engaged in the protests (Gomez 2008, 25). Historically, these organizations have had different ideological views and political alliances (Manzetti 1992, 610–611), but now for the first time in history, the countryside, *el campo*, seemed to talk with one united voice, with no expressed difference in interest between tenants, big agribusiness firms, and transnational traders. *El campo* demanded the complete elimination of export taxes (Lapegna 2016). The big media conglomerates supported the rural protests, and many urban citizens sided with the countryside. A researcher in agronomy, engaged in several extension service programs with many different agricultural producers' groups in the Pampas provinces, expressed that while the conflict tapped into many discussions, it could ultimately be boiled down to one basic fault line:

The discontent with the export taxes on soybeans really managed to engage people living in the countryside, even the ones with no previous engagement in politics or in any type of organization. All of a sudden, a rather heterogeneous group of people started to see themselves as sharing one united interest against what was perceived as a vacuum cleaner sucking all the money from the countryside into the state apparatus [...] Really, it is the same story as always: Who gets the wealth? Who can appropriate it? Who produces? This tapped into the long history of a love-hate relationship between the city and the countryside, the rural-urban divide. But a lot of this is really about one question; who gets the profit? (Researcher FAUBA and Agricultural Extensionist, Buenos Aires, 24 February 2017)

As illustrated in above quote, the conflict of the export taxes can ultimately be seen as a conflict over the distribution of surplus value. Both *el campo* and the government used a discourse about meritocracy and fairness to justify their right to the revenue. The *campo* claimed that the wealth from the soybean production was the fruit of the producers' risk-taking and hard work. The export taxes were accordingly deemed as an unjust "appropriation" by the central government of the wealth of the countryside (to finance the fiscal deficit created by an inefficient but expensive central state bureaucracy and the urban industrial sector). This revoked historical distributional fights between the national government and the provinces. In the narrative of the producers, the countryside represents "the people" (*el pueblo*), republican and federal values (against the central state). Landowners and producers have the right to do whatever they want with their land (in the name of private property), without the bureaucratic, incompetent, and paternalist central government; *the political class* intervening through redistributive programs in benefit of people who do not like to work in any way (Giarracca 2008, 21; Gomez 2008, 28–313). Thus, the historical suspicion toward the central government and the city was combined with classical conservative tropes about the poor receiving too much benefit and parasitizing on the true wealth generating sectors (the agrarian sector).

The government responded in equally polemic ways, talking about *gorillas*,[22] coups, and oligarchs (Lindenboim 2008, 13).[23] Soybeans were discursively constructed as equivalent with Monsanto and other ultra-capitalistic firms acting in line with short-term profits, while destroying the soils, intoxicating flora and fauna, and displacing small farmers in miscopy with the rural, archaic oligarchs (Giarracca 2008, 2; Leguizamon 2014, 139–142). Moreover, it was underlined that the lucrative soybean business has benefitted from public investments in infrastructure and research, a low value of the peso, as well as of a favorable temporary international conjuncture, and was thus not exclusive "merit" of private producers and investors. In this way, the government argued that it was fair and just to make the highly profitable soybean business to "payback" to society. Government supporters, including the central unions and human rights' organizations in the cities and local rural organizations organized countermarches. The rural strikes placed soy production on every TV channel and almost overnight GM soy mono-cropping and its consequences became a public issue. New critical voices about the soybean model emerged (Director of Research Promotion CLACSO, Buenos Aires, 3 March 2017). The discursive field thus was reduced into two camps. One urban, pro-government, pro-taxes, and anti-continuous cropping of RR glyphosate soybeans. Another rural, anti-government, anti-taxes, and pro continuous cropping of RR glyphosate soybeans.

However, the CFK-administration also tried to break the "power-block" of the *campo* by stressing other lines of conflict than the rural–urban divide. One tool in this respect was to use the export taxes for increased public funding and support for agricultural development and upgrading. The government created for example the *Fondo Solidario de la Soja*, a fund based on revenue from soy export taxes for local governments to use for development projects, such as for infrastructure, improved public service, housing projects, and support programs for vulnerable parts of the population. Many indebted local governments appreciated the fund and became more positive toward the export taxes (Leguizamon 2014, 63–66).

After several months of conflict, the government decided on 17 June 2008, to let the Congress decide on the controversial sliding export tax in order to get to an end of the conflict. While the lower house, the Chamber of Deputies, approved the mobile tax, it was voted down by one vote in the senate. Accordingly, export taxes became fixed again and returned to the export duties established in November 2007, which were 35% on whole soybeans, 32% on sunflower; 23% on wheat; 20% on maize; and 15% on bovine meat (FAO 2017). These levels remained in force and unchanged until the change of government in December 2015. The most violent confrontations, strikes, and blockades disappeared as soon as the mobile tax was off the table, but the antagonistic positions between the government and the countryside remained. The former (2013–2015) Secretary of Ministry of Agriculture, Livestock and Fisheries (MAGyP) self-critically analyzed that the unity of the "*campo*" was created by the aggressive discourse of the CFK administration:

> Really, it was we [the government] who united them [the agrarian organizations] by fighting against them all, indiscriminately; saying to them that 'you are all the same; you are all sons of bitches'. We made no difference between small or big. In that sense, it was not them - it was we! They only reacted to that. Therefore, one of the biggest legacies of the 'Kirchnerismo' in the agrarian sector was the new rural unity, which had never existed before, and we managed to create it… Hahaha…. But it is the truth; we have to recognize that. Probably if the government had not attacked the sector so hard, the sector would not have organized itself so well, because the government also demonized the sector. It said that the soybean is a weed; that it eats the soil; that it poisons the people; kills the children… So that narrative remains in the social imaginary of the people even as the government has changed. (Former Secretary of MAGyP, 2013–2015, Buenos Aires, 2 March 2017)

While the self-named "*campo*" criticized the government for too much interventionism, many local small farmer, peasant, indigenous, and ecological organizations instead criticized the government for being too permissive and supportive to the rapid *sojization*. The government was particularly criticized for lack of action and support as large-scale soybean production spread into the poor Northern provinces and displaced traditional small-scale *campesinos* systems as well as threatened the traditional uses of the forests of indigenous communities (Jobbágy et al. 2015; Leguizamon 2014, 73; Volante and Seghezzo 2018).[24]

The government nevertheless tried to have good relations with many of these organizations. One particularly important actor was the *National Forum for Family Farming* (FONAF), a network of local family, peasant, community, and indigenous organizations was assigned the task to develop sectorial policies and distribute funds.[25] The organizations of FONAF and the newly formed *National Campesino Front* (FNC) received a significant amount of funds from the government and supported it strongly in the *conflicto del campo* (Gisclard et al. 2015). In this way, the legitimacy in which SRA, FAA, CRA, and CONINAGRO claimed to represent one united countryside was challenged, and in particular the claims of FAA to represent all family farmers (Former Secretary of MAGPyA, 2013–2015, Buenos Aires, 2 March 2017).[26]

### 5.2.2    Shifts in Social Relations and Regulations

While family farming was already on the agenda under the first Kirchner government period (2003–2007), the government increased support in the wake of the 2008 rural conflict. Extension services, credits, and different types of funds had increased significantly in support to small farmers (Lapegna 2016; Nogueira et al. 2017).[27] A national register of family agriculture (RENAF) was launched in order to make possible differentiated policies with specific support for the farmers that fulfilled criteria for family farming. The government also built up strong direct ties to several rural movements and incorporated some of their leaders into the government. Many of the new programs and projects received financial received support from IOs, such as IFAD and IICA, but belonged institutionally to the *Secretary of Agriculture, Livestock and Fisheries* (SAGPyA)—which in 2009 had been converted into the *Ministry of Livestock, Agriculture and Fishery*—MAGyP (Nogueira et al. 2017). A new sub-secretariat of *Rural Development and Family Farming* (SSDRAF) to MAGyP was launched and staffed with more than 1200 employees providing support and extension services to family farmers (Gisclard et al. 2015; Nogueira et al. 2017, 45).[28] All resources from SSDRAF were channeled in a decentralized way through provincial and regional round-tables organized by FONAF. The government also supported many specific programs directed to smallholders, food sovereignty projects, and poverty alleviation in the countryside (Researcher and former Public Official, Stockholm, 10 September 2018). Moreover, new legislation for improved working conditions for rural laborers and the prohibition of "foreignization" of land to curb "land grabbing" was passed (Papotto 2013). A 2014 law *Historical Reparation of Family Agriculture* (No. 27.118), underlined land as a social good (in line with policy recommendations from FONAF), and created new mechanisms for improved rural infrastructure and service (Toledo Lopez 2017).[29] However, within the CFK-administration, different strands had different views on how much public support should be directed toward family farming and rural development, and the law did not receive sufficient funds for implementation (Nogueira et al. 2017, 42).

*Kirchnerismo* policies toward family farmers have been criticized from both left and right. The right, including the traditional producers' organizations argued that *Kirchnerismo* skyrocketed public spending for a variety of ineffective initiatives for social inclusion, transferring money from the hard working and productive parts of the country to lazy people who do not like to work. Left-wing scholars argued that policies were of typical *Peronista-style-welfare*; clientelistic, paternalistic, coopting specific leaders, demobilizing people, and making them dependent on the state, while not changing the structural inequality and extractive character involved in the model (Lapegna 2016; Vergara-Camus and Kay 2017a). Many social movements and NGOs (for example MOCASE, Via Campesina, Friends of the Earth, WWF, and Rural Reflection Group) also argued that many of the government's programs were not sufficiently funded and that the systems of allocation lacked efficiency and transparency (Director of Research Promotion CLACSO, Buenos Aires, 3 March 2017). Moreover, the approval of new genetically modified events, the lack of regulative protection against displacement of small farmers, the lack of land-use planning for both social and environmental protection, as well as the cases of intoxication and high pesticide residue in food were critically addressed. This often formed part of a broader criticism against neoliberal globalization, land grabbing, Monsanto control, and "extractivism", and in favor of land reform and food sovereignty (Dominguez and Sabatino 2008, 38–39; Giarracca 2008, 20; Leguizamon 2014, 129–135; Muñoz 2016). Some peasant and landless organizations also engaged in radical actions, such as land occupations. These sometimes ended up with paramilitary-like forces evicting the occupants and threatening their leaders (Dominguez and Sabatino 2008, 38–39; Leguizamon 2014, 130). While the net effects of the government's policies for family farming are contested, it is clear that the general trend of agribusiness advancement and gradual disappearance of small farmers, native forests, and indigenous groups were not reverted (Colla 2017; Gomez 2008, 26; Gras and Hernández 2013, 19; Muñoz 2016).

At the same time, it is estimated that the whole soybean value chain in Argentina, from seeds to crushing, involves around 395,000 workers (FAO 2017, 12–13). This by far exceeds the number of family farmers that abandoned agriculture between the two last censuses in Argentina. This majority of new employments are linked to the sales of input and agrarian services, but also to processing plants and port services (the soybean processing industry in Rosario is the largest in the world). However, in the poorer regions, far away from the logistic hubs, the soybean business took more the form of isolated extractive activities with little backward and forward linkages (Gras and Hernández 2013, 64). Even in the areas where some rural small towns became revitalized, the countryside as a whole has been increasingly depopulated and urbanization has intensified throughout the region. This should nevertheless be analyzed against the historical backdrop (Chapter 2) of a countryside that for most parts never became fully developed, and where infrastructure and service were always lacking.

### 5.2.3    Shifts in Forms of Production and Regulations

Besides the changes in social relations brought by agribusiness expansion in the Northern provinces, the rapid deforestation of millions of hectares of native forest in the Argentine Chaco, driven by agropastoral expansion, also caused new environmental concerns (Torrella et al. 2018).[30] The withdrawal of the state in the 1990s, particularly the dismantling of the *National Forestry Institute* (IFONA), had weakened the capacity of the federal state, as well as the province governments in the Chaco region, to monitor and plan land-use change and the snowballing deforestation (Volante and Seghezzo 2018). The research community, peasant organizations of the Chaco, and environmental NGOs increasingly stressed the various negative impacts and risks involved in the rapid forest loss in the early 2000s. Fragmentation, habitat destruction, and dramatic biodiversity loss emerged as forestland was replaced by annual crops, such as soybeans. Moreover, this LULCC provoked shifts in the soil physical properties that regulate water circulation and storage, which in turn incited deeply stored soil salts to come to the surface by raising water tables (Amdan et al. 2013; Jobbágy et al. 2015). As subterranean water-flow increased in speed, problems of flooding emerged and in some provinces a sudden appearance of a network of new rivers (Magliano et al. 2017). The role of soybeans and beef as the central drivers behind forest loss led to a strong societal debate. Under pressure from civil society organizations, several provincial governments of the Gran Chaco region declared temporary moratoria on the clearing of native forests, between 2003 and 2005 (de Waroux et al. 2017).

One policy response to the rapid deforestation was the establishment of a *Federal System of Protected Areas* (SIFAP) with 452 protected areas for biodiversity conservation, as of 2019, representing 8% of total land area.[31] While important, these areas still only represent a small proportion of the dry Chaco, not near the size required to compensate for the rapid forest loss. A stronger regulative response to the rapid deforestation eventually emerged in the 2006 law proposal *Native Forest Protection Law* (26.331). The proposal involved obligatory territorial planning, in which each Province had to map its forests and reorganize its provincial laws in order to meet minimum federal requirements for forest protection (Vallejos et al. 2015). The World Bank formed an active part in the law formulation process, including an extensive participatory and multi-stakeholder decision-making processes for increased legitimacy and acceptance for the outcome (MAyDS 2017). However, the law proposal caused strong tensions within and across the government party, FPV, and allied parties to the government (Saylor 2014). It also received a lot of critique from many leaders and legislators representing the Northern Provinces, who argued that it posed a threat to Argentine federalism, that it limited the possibilities for agrarian development, and that it threatened local autonomy (Researcher FAUBA and Agricultural Extensionist; Former President INTA; Rural Development Specialist IICA; Director of Research Promotion CLACSO, Buenos Aires, February–March 2017). At the same time, environmental NGOs were active campaigning in favor of the law, and managed to collect one million signatures in support of the proposal (Saylor 2014). A fund for monetary compensation to provinces with a high degree of forestry conservation was added to the proposal, to meet the complaints of the Northern Provinces.

After many debates, the law regulating the protection, enrichment, restoration, utilization, and management of native forests and the environmental services they provide through land-use planning (26.331) was approved in the senate and parliament in November 2007 (de Waroux et al. 2017).[32] Accordingly, each Province had to map all its native forests into three conservation categories. Category I (red) includes areas of high conservation value—not allowed to deforest. Category II (yellow) means medium conservation value—deforestation is accepted exclusively for "sustainable uses" and requires an approved sustainable forest management plan. Category III (green) includes areas of low conservation value— allowed to be converted into other land-uses after a previous assessment of environmental impact (Burns and Giessen 2016; Torrella et al. 2018). As the Province decides which areas are to be considered green, yellow or red, there is considerable space of local maneuver. The law also established a moratorium on deforestation until completed categorization (Saylor 2014).

According to a report from the *Ministry of Environment and Sustainable Development* (MAyDS), 10.47 Mha of native forests nationwide has been categorized as red (representing 19% of all native forestland), 32.646 Mha labeled yellow (61%) and 10.54 Mha labeled green (20%) between 2010 and 2016 (MAyDS 2017). The decentralized character of the territorial planning resulted in a high degree of variance in the categorization. Some provinces marked a high share of its native forests red, others did not (MAyDS 2017; Torrella et al. 2018; Vallejos et al. 2015). This made some producers to move into Provinces that were known to proportionally categorize more native forest into green (category III) area, such as Formosa (Volante and Seghezzo 2018). Moreover, the meaning of the categories was also ambiguous. Particularly the interpretation of what could count as "sustainable use" for the yellow area varied from almost nothing, to selective logging and ranching (Torrella et al. 2018). Accordingly, forest loss still occurred in both yellow and red-labeled area, particularly in the poorest provinces after the implementation of the law (Volante and Seghezzo 2018). An agronomist from FAUBA illustratively explained the strengths and weaknesses of the regulation:

> Finally, after many discussions and difficulties, the forest law became a reality, and it marks an important step. It is a huge leap forward, but it is not perfect. In some places, it is implemented based on rigorous definitions and good control, but in other provinces it does not work very well. The pressures are very strong. It is an attempt to protect the forest, but of course, the law is like throwing you out to a war of machine guns with a fork. The interests are very strong. The persons working with the forest inspections have to deal with powerful bigshots and they risk losing their jobs. I was on the countryside with a guy who spoke very freely and with a high voice. After two months, they had got rid of him. It is not easy. The landowners have a lot of power. In provinces like Salta, you see that the persons engaged in politics have some 50,000 ha of land, or they have leasing-rights free for 20-30 years. (Researcher FAUBA and Agricultural Extensionist, Buenos Aires, 24 February 2017)

As illustrated in above quote, many provinces suffer strong pressures from actors engaged in the agrarian sector, sometimes installed in the proper political Provincial structures. Besides the difficulties to resist pressures from strong interest groups, many provinces lack administrative and technical capacity for rigorous planning and implementation, which require more time and resources than the permanent funding from the national budget granted by the law. In this way, the decentralized design of the legislation increased the difficulties of strict implementation (Researcher and former Public Official, Stockholm, 10 September 2018). Nonetheless, deforestation rates dropped significantly and the law in protection of native forest was accordingly an important step forward (de Waroux et al. 2017). Many researchers and NGOs argue, however, that Argentina should take implement land-use planning over the entire territory and consider all land-uses, in order to address the whole territory, standardize categorization, and consider other values than only native forest. In spite of all the weaknesses, the forest legislation is often mentioned as the most important environmental regulative response to the soybean boom (and the "displacement" of livestock) in Argentina.

With the exception of the law for protection of native forests, *Kirchnerismo* showed a rather *laissez-faire* approach to environmental concerns. Different types of neighbor associations in the Pampas region criticized the spraying from a health perspective, claiming increased cases of illness—from cancer to intoxication—in nearby communities due to the intensive agrochemical spraying on the fields (Toledo Lopez 2017). The Argentine association of beekeepers, *Sociedad Argentina de Apicultores* (SADA), also organized against the glyphosate, both because the residues in honey and because of the loss of fauna and therefore habitat for bees.[33] Some local politicians, however, have responded to the critique by prohibiting the use of glyphosate in their municipalities for public health reasons.

Glyphosate in Argentina is nevertheless equally contested as in most parts of the rest of the world (Chapter 3 in this book). Several Argentine researchers have published extensively on glyphosate and some argue that it poses several health hazards on animal and people, ranging from direct intoxication, to increased long-term risk for cancer, embryos producing malformations, neural, intestinal, and heart disorders (Arancibia 2016; Avila-Vazquez et al. 2018). At the same time, other actors claim that glyphosate, used as prescribed, poses absolutely no threat to living beings or the environment. This position is for example taken by the *Argentine Association of the Soybean Chain* (ACSOJA) and the *Argentine Association of Direct Seeding Producers* (AAPRESID), but also by the public entity in charge for agrochemical control use and regulation; the Argentine *National Service of Sanitation and Food Quality* (SENASA).[34] Many scholars also stress that while all agrochemicals imply some type of alteration and have some degree of toxicity, glyphosate is well-known and less toxic than most alternative herbicides (Consultant and former President of INTA; Director of Agricultural Affairs CARI; Buenos Aires, 28 February and 1 March 2017).

Another problematized aspect of the "soybean model" of the Pampas region in research papers was soil depletion. For example, a study conducted by the *Argentine National Institute of Agrarian Technology*, INTA, on the nutrient extraction of the main monocultivars showed a very negative nutrient balance for soybean production (−60%), resulting in millions of tons of soil nutrients mined without replenishment.[35] As mentioned in Chapter 1, the traditional crop-grazing system of the Pampas region, integrating row crop cultivation and pasture grazing for beef and dairy cattle production, provided several ecosystem services which both held erosion and weed under control. With the shift in price relations and the new soybean technological package, pastures were displaced by continuous cropping under no-tillage soil management with glyphosate as a total weed killer. Besides the loss of pastures, diversification of crops was also reduced. Many producers started planting soybeans over soybeans with no winter crop in-between:

> We made a study showing that almost 80% of the agricultural area in Argentina during winter has no crops on it; only around 20% of soybean area is rotated with wheat as a winter crop. One reason is the low economic returns on winter crops and another is that producers want to save water for the dry summer months, and fear that the winter crop will reduce available water for the high-price soybean. We have shown that they can store water anyway, depending on the time of sowing. (Researcher FAUBA and Agricultural Extensionist, Buenos Aires, 24 February 2017)

As illustrated in above quote, an extreme specialization in soybeans emerged leaving the soils naked during the winter with high exposure for erosion. Besides soil degradation, the land-cover change replacing permanent grasslands with continuous soybeans resulted in specialized weeds and plagues, rising groundwater levels, limiting soil water storage, and may explain increased flooding events on the Pampas (Kroes et al. 2019). The rapid process of soil degradation and erosion in the Pampas region—in the wake of continuous cropping and high agrochemical use was thus by some actors argued in need of stricter environmental regulation—obligatory crop rotations and/or more restricted agrochemical use (Lende 2017).

While there is agreement among researchers and experts that it would be desirable with more diverse farming systems and that continuous soybean production without rotations are not sustainable, many still argue that forceful regulation is seldom effective.[36] Besides that regulation often require a costly meta-structure of auditing and monitoring, many stressed that they are not reconcilable with the Argentine strong private property rights to land—understood as the right of the landowner to do whatever he wishes with the land.

An important part of the critique from the more "ecologist" and urban organizations was centered in the discussion about genetically modified crops in general, and the role of Monsanto in particular, organizing protests including blockading routes and rallies (Blois 2016; Pagnussatt 2018). However, they also asked for stricter environmental regulation, territorial planning, and land reform. With the exception of the new regulations to protect native forests, however, *Kirchnerismo* expressed a rather *laissez-faire* approach to environmental concerns, in spite of the increasingly close ties between the government and rural civil society groups, and their common critical narrative about the soybean model. Environmental problems have not had a central position in the policy agenda of the government, or in any other political force in Argentina (Saylor 2014).

The policy priority was clearly economic growth and social redistribution, not environmental sustainability. Accordingly, in spite a rising radicalization against the soybean model among an important part of the urban middle classes and working classes supporting *Kirchnerismo*, the government did not pose any restrictions to the model, besides taxes. On the contrary, several public policies facilitated agribusiness expansion. For example, the value of the Argentine peso was kept down; export-oriented infrastructure was assisted, public support to agrarian innovation and R&D (including a public–private partnership for GM seed innovation), increased and specific trade deals with China and Russia were made; CFK struggled hard in negotiations with China in 2010 to make China reduce import tariffs on Argentine soybean oil—but she did not succeed (Vergara-Camus and Kay 2017b, Toledo Lopez 2017). One reason for the permissive and sometimes even supportive attitude toward soybean exports, in spite of the massive critique expressed against the soybean model, may be the high fiscal revenue provided by the controversial export taxes, which improved fiscal space and allowed to use cash transfers for poverty alleviation programs.[37] High soybean prices also contributed to more favorable TOT, which is estimated to have contributed with more than 10% of the gross national income growth between 2003 and 2011 (FAO 2017, 32–33). Another reason was that the government hoped to be able to use the soybean complex as a springboard for development by the incorporation of more technology and knowledge, in order to make it upgrade into more value-added products, as already mentioned. Specific strategic plans for upgrading of the soybean chain and other agrarian-based commodity chains were adopted (Anichini et al. 2013; Castellano and Goizueta 2011; Dabezies 2009).

In general, *Kirchnerismo* believed in long-term planning of the agrarian sector. This was not least reflected in the ambitious nationwide (federal) creation of a *Strategic Agrofood and Agroindustry Plan*, 2010–2020 (PEA). Together with a broad range of different agrarian actors, the government set quantitative productive targets and qualitative "upgrading" targets for specific commodity chains, and a battery of public policies to fulfill them (MAyDS 2010, 35–39).[38] The livestock sector seems to have benefitted from some of the policy reorientations established in the PEA, and the cattle herd has slowly recovered since 2011 (when cattle numbers bottomed at 48.1 million head, the lowest since 1964), and has since registered a consistent yearly increase. Besides neostructuralist development ideas, the plan reflects that the overarching priority is to increase productivity, intensify land-use, and spur economic growth. The role of the state is to smoothen this process, whereas also "tweak" rules and create market incentives to make agriculture "bear its own costs" and internalize negative social-ecological externalities (MAyDS 2010).

### 5.2.4   Macri—A New Liberalization Impetus

In the elections 2015, Mauricio Macri and *Cambiemos* (Let's Change) won a narrow victory on a right-liberal political platform, which received the explicit support of *El campo*. Macri promised liberalization reforms including the elimination of export taxes and quotas. After taking office in December 2015, Macri led a significant change in Argentine economic policy. The new administration implemented core reforms such as the unification of the exchange rate, the agreement with international creditors, the modernization of the import regime, reduction of inflation, and reform of national statistics system (IBRD 2014). Moreover, Macri removed foreign exchange restrictions, devalued the peso by 45%

to restore export competitively, eliminated export permits for grains and oilseeds, and removed export restrictions on beef exports. Export taxes on agricultural goods were removed, including meat, but excepted soy products. Taxes on soybeans were reduced by from 35 to 30% and Macri announced that they would be further reduced with 0.5% per month until completely eliminated (FAO 2017). While the plan was delayed, Macri gradually reduced export taxes on soybeans. In August 2018, they were down on their lowest level—26%. However, in the wake of the later IMF emergency deal export taxes on soybeans were risen again (see below).[39]

MAGyP was renamed to the Ministry of Agroindustry, and it was filled with private agribusiness sector leaders as well as with profiles from the producers' organizations, some already well-known in the media since the "conflict of the campo" (Nogueira et al. 2017).[40] In general, the relation between Macri and the organized actors of the agrarian sector was very close. Bovine stocks and exports started to rise almost immediately. While "El campo" had demanded complete elimination of the export taxes, there was an expressed acceptance of a gradual reduction of tax on soybeans. In a similar way, many scholars and agricultural consultants who in theory argued against export taxes, expressed comprehension over their persistence.

> I know that it is a very bad tax, very distortive, and yet I agreed when *Duhalde*[41] reinstalled it at the 2002 crises. There was no alternative at that point. Even Macri, who is totally against it, still has not been able to remove it away on soybeans. For Macri it is a necessary bad because the state always needs money. (Director of Agricultural Affairs CARI, Buenos Aires, 1 March 2017)[42]

As illustrated in above quote, actors that consider that it is a bad tax still acknowledge that sometimes the state needs to collect fiscal revenue from where it can get it. Besides revenue, the tax is functioning as incentive away from soybeans. Some observers fear that if export taxes were to be removed on soybeans, continuous monocultural soybeans would further expand to the detriment of soils, water and national food security (Former President INTA; Rural Development Specialist IICA; Director of Agricultural Affairs CARI; Director of Research Promotion CLACSO, Buenos Aires, February–March 2017). The combined effects of declined international soybean prices and export tax reduce the relative higher economic margins for soybeans vis-à-vis other land-uses. It makes it economically viable also in the shorter term to rotate soybeans with other crops, which is beneficial for the long-term sustainability of the soils. In the absence of other regulations to protect the soils, to reduce the profits from soybeans may be the most efficient policy tool. At the same time, soybeans still offer the best margins in the Pampas region and area has not (yet) retracted. There are, in addition, some doubts whether the lowering of export taxes would actually provide any benefits for the producers, as explained by an agronomist working with producers in the Pampas region:

> When you take away the export taxes, what happens? It is not going to be translated into higher profits for the soybean producer. No. What is going to happen is that the costs of leasing will rise. It is obvious. This is exactly what happened when the government lowered the taxes. The margins for producers remained the same. Why? Now that he gets paid more for the soybeans, he has to pay more for the land. If he leased land for USD 300/ha, now he has to rent it for 350/ha. So, who won with this reform? Well, the guy sitting in *Recoleta*[43] now has USD 50 more in his pocket. The landowners gained with this reform. But, not so much. The taxes on soybeans were only reduced by 5%. But everybody is happy now and the relation between '*El campo*' and the government is excellent. Working in the countryside, you can really see the change. The mood changed radically, but the actual situation is not so different. The change is more in the head than in the ground. (Researcher FAUBA and Agricultural Extensionist, Buenos Aires, 24 February 2017)

As expressed in above quote, the business model of soybean production, which is based on leased land, allows for the landowners (here characterized as a person sitting in a rich neighborhood in Buenos Aires) to capture most of the value, not the producers. In this way, the agrarian sector, represented by the big producers' organizations, has good relations with the Macri administration, in spite of the fact that the difference in concrete policies considering the soybean is rather small. However, not only "the countryside" perceived that the Macri administration particularly benefitted the agro-exporting interest, but many urban citizens held the same view:

> The soybean represented at least 80% of all the revenues gathered from export taxes. Macri only reduced it with 5%, so in practice the difference is not very big. However, in the imaginary of the people, Macri removed all taxes from the agribusiness sector. This made the producers happy, but

not the ordinary taxpayers. They associate the soybeans with Monsanto, which is the most hated company in the world, right? I think Macri will pay a high political cost for this. He is now associated with Monsanto and extractivism. I think that there is an emerging social consensus saying that the agrarian sector should pay more taxes. I will give you an illustrative anecdote. Yesterday, I was driving home to Buenos Aires from the beach. I stopped at a gas station to buy some sandwiches for my kids. All of a sudden, a very upset woman in her forties came into the station. 'Ma'am, is everything all right?' I asked her. 'No', she said, 'the guy driving the pickup truck Hilux almost run over me—that soybean-guy', she said. 'I work and I pay 5000 pesos in income tax, and look at the car I can afford, and then look at the new pickup of the guy that Macri took away the export taxes from, and on top of everything he almost ran over me in order to win one minute of extra time'. 'No', she said, 'I am no fan of *the Kirchners*, but these guys deserve to be kicked in the asses'. This illustrates a serious problem of the agrarian sector today; it is associated with extreme inequality, where an important part agrarian generated wealth is spent on giant houses and cars that exist side by side with situations of poverty. There is a very negative vision about the agrarian producers from the people who do not participate in the agrarian sector; a growing tension. This is happening in a context in which the urban society also feels empowered; it knows that it represents the majority of the voters. I mean, there are some 300,000 producers, right, but there are 40 million Argentinians…. (Former Secretary of MAGyP, 2013–2015, Buenos Aires, 2 March 2017)

The former secretary of MAGyP, who is currently the director of the economic, social, and political science research center of INTA, thus argues that the antagonistic positions expressed in the *conflicto del campo* remain in many peoples mind ten years later, and he argues that the population at large takes a critical position toward the agrarian sector. However, the Macri administration seems to have made another analysis of the situation, and since late 2018, the Minister of agroindustry is *Luis Miguel Etchevehere*, a former president of SRA, which means that Argentina's most formidable farming lobby is installed within the government.

When the government launched a new economic program that includes budget cut, increase in revenues, and a USD 57B emergency loan deal with the IMF until 2021, it announced that it would temporally raise the soybean export tax again.[44] [45] In this way, export taxes on soybeans are back on around 30%, which is pretty close to the levy soybeans had when *el campo* protested the most severely against the government of CFK.[46] At the same time, the differential export taxes system for the soybean industrial complex was withdrawn, so that beans were taxed equally as cakes and oil (Bergero et al. 2019). Exports of whole beans rose immediately, while the industrial stages of the soybean chain criticized the reform. Besides decreasing export taxes, Macri has accelerated the pace of approval of new GM events, including a drought-resistant soybean and an herbicide-tolerant soybean from a Chinese company for the first time in history. The Argentine soybean harvest of 2018 was hit by drought, but soybean area has remained high and even increasing the past years. The cattle herd has been continuously increasing since 2011, and reached 54.2 M head in 2018.

Macri's administration has coincided with decreasing commodity prices and economic recession. The government has focused on austerity programs and budget cuts to reduce public expenditures and fiscal deficit. However, Argentina is still a high-income country with a GDP of more than USD 628B in 2017.[47]

There were severe cuts in public spending for agrarian research, innovation, and education as well as to family farming. The staff at the *Secretariat of Family Farming* was dramatically reduced.[48] In terms of environmental regulation, the Macri administration shows important lines of continuity with the former administrations, without any advance in stricter environmental regulation. Quite the opposite, within the realm of austerity policies, the government has cut budgets for several environmental programs and research. Besides financial constraints, his appointment of conservative Rabbi Sergio Bergman as secretary of the *Ministry of the Environment and Sustainable Development* (which in December 2018 was turned into a secretary again) has been questioned. As remarked by one respondent:

> Macri put a rabbi as secretary of the environment, and he obviously does not have a clue about this. For example, he cut the budget for fire prevention. When that decision was questioned, he said that he prays for a fire not to break out. So, the way Argentina combats fires is by praying... This means that this government does not give a shit about environmental issues. (Researcher FAUBA and Agricultural Extensionist, Buenos Aires, 24 February 2017)

While many policies indeed shifted with Macri, there are also important lines of continuity. For example, the national register of family agriculture (RENAF) is still in use. According to a specialist at IICA Argentina, the strongest shifts are at a discursive level, while actual policy lines remain rather intact (Rural Development Specialist IICA, Buenos Aires, 3 March 2017). Family farming remains high on the policy agenda, and many ongoing programs in support for family farming.[49] Here the role of IOs seem important as they finance an important part of the programs, particularly from IFAD, IICA, IDB, and the World Bank. Besides family farming, the idea that the state should facilitate upgrading also remains central in the agrarian policies of the Macri government. The major changes are some institutional modifications changes, less cooperation with local organizations, overall less budget, and focus on entrepreneurialism (Nogueira et al. 2017, 53).

## 5.3   Paraguay—Liberalism or Libertarianism

Compared to Argentina and Uruguay, Paraguay did not experience any defeat of neoliberalism at the turn of the millennium, not even discursively. The 2000s in Paraguay, followed in many respects the same track as during the 1990s. This track was characterized by openness to trade, low taxes, and a small state apparatus (IBRD 2018).[50] While fiscal pressure has been low for all sectors,[51] the agrarian sector has been additionally benefitted by generous tax exemptions from the Value-Added Tax (VAT) on the import of goods, by a complete absence of export taxes, and by a low real estate tax (CEPAL 2018). In addition, the government started to invest heavily in infrastructure, especially highways, bridges, and roads to connect the country in a more efficient way (USDA Foreign Agricultural Service 2017).

Besides domestic policy priorities centered in free enterprising, the "post-Washington consensus" agenda of the international development community has also become influential in Paraguay. IOs form an important part of the agrarian public policy formulation and implementation throughout the region, but particularly in Paraguay, due to severe constraints in the state's administrative, legislative, and funding capacity. At each new government period, the IOs typically meet with the new administration to discuss and set up a so-called framework agreement, with priority lines and programs for the government period (Public Policy Specialist UNDP Paraguay; Innovation Specialist IICA Paraguay, Asunción, 20 and 22 February 2017). The majority of the projects and programs of the *Ministry of Livestock and Agriculture* (MAG) and the *Secretary of Environment* (SEAM),[52] are in this way funded, designed, and coordinated by specialists from the IOs. The deputy minister of agriculture at MAG explained the relation with the IOs in the following way:

> As I said before, this is such an open country that every international organization that want to come and experiment a project or a specific policy can do it without any opportunity cost. To say no to a project means that nothing at all happens, zero, but by saying yes, at least you have something, so you chose something. So, the World Bank comes and says 'now we have to do it this way', and then IDB comes and says 'let us do it that way'. In the end, this is as a huge laboratory where different policy designs are tested. (Paraguay's Deputy Minister of Agriculture, Asunción, 23 February 2017)

As expressed in above quote, the state tends to say yes to all project proposals brought by the IO's. They are also involved in legislative initiatives, state reform, and institutional capacity building. Accordingly, Paraguay's public policies and policy initiatives do not exclusively reflect domestic policy priorities, but wider international development trends. In this way, a look at past and on-going projects of MAG and SEAM (now MADES) show that many focus on climate change mitigation and adaption, natural resource conservation and maintenance of vital ecosystem services, poverty alleviation with a territorial focus, and upgrading of the productive structure. In this way, alongside policies facilitating free enterprising, environmental concerns, indigenous rights, and social inclusion of family farmers have increasingly entered the public policy agenda. In particular, the environmental legislation advanced significantly during the period (Riquelme 2014). Paraguayan policies and regulations are also responding to increasing demands from the international community on reporting, auditing, and monitoring, on development planning with impact assessments, and on participatory decision-making processes. This is materialized in a broad range of development plans for different sectors and for the country as a whole.[53] In this way, many of the specific post-Washington consensus features mentioned in the beginning of this section that swept the world and the region at the dawn of the new millennium are also present in Paraguayan public policies.

The overall policy context in Paraguay combined with rising commodity prices benefitted the agro-exporting sector. The total value of Paraguayan exports rose by six times between 2002 and 2017, at the same time as the relative share of soy-based products remained extremely high, and even rose slightly from 44 to 46%, from USD 0.5B, to 2.99B in absolute export value (see Appendix B). Soybean was thus an extremely dominant export star, offering higher gross margins than any other agrarian activity, at least under normal climate conditions (Public Policy Specialist UNDP Paraguay; Director of DCEA-MAG, Asunción, 20 and 21 February 2017). Beef production and exports have also been rising at a remarkable pace. The value of bovine exports rose by almost fifteen times from 2002 to 2017 (from USD 0.07B to 1.03B), and its relative share of total exports rose from 5.8 to 16% of total exports, over the same period (see Appendix B). Around half of it is produced in the traditional livestock area in Eastern Paraguay (Atlantic forest region) and around half in the Chaco region. Paraguay has also increased its supply of the well-paid premium-chilled cuts (USDA Foreign Agricultural Service 2017).

Paraguay's general economy has also been growing since 2002 and poverty rates have declined. Despite broad fluctuations, Paraguay's GDP expanded by an average of about 5% from 2008 to 2017, benefitting from the country's export of beef and soybeans as well as from successful courting of foreign investment (IBRD 2018). These changes fit well with the three main aims of *The National Plan for Development* (PND 2014–2030): poverty reduction and social development; inclusive economic growth; improved insertion of Paraguay in the international markets (Gobierno Nacional 2014). However, while soybean and beef exports have fueled the economy, they are highly sensitive to climate variability and to a few fluctuating markets. Moreover, they are not generating opportunities to participate in the bonanza for the majority of the 40% of the population living in the countryside (Ezquerro-Cañete and Fogel 2017). Paraguay has among the most extreme unequal land structures in the world and violent conflicts over land are persistent. Large-scale agribusiness farmers have displaced family farmers, and native groups have been particularly badly hit. MAG has had severe budget constraints; support to family agriculture decreased substantially between 2000 and 2010 (Ezquerro-Cañete and Fogel 2017). In a similar way, SEAM has not been able to fulfill many of its activities due to lack of resources (Public Policy Specialist UNDP Paraguay; Senior Technical Advisor UNDP Paraguay; Director of DCEA-MAG, Asunción, February 20 and 21, 2017).

Thus, behind the impressive figures of rising export revenues, soy and pastureland expansion as well as land-use intensification have come at high social-ecological costs. According to a report from the World Bank, the Adjusted Net Savings indicator for Paraguay—which besides per capita wealth incorporates subsoil asset depletion, net forest depletion, and pollution damage—actually shows that the (estimated monetary) loss of value from depletion and damage have been higher than economic growth the past decade (IBRD 2018).[54] The loss of natural wealth indicates that Paraguay's growth model is unsustainable in the medium- to long-term. One of the most spectacular losses of natural resources is the rapid deforestation of the Paraguayan Chaco forests, caused mainly by expanding ranching activities. This sparsely populated area in Western Paraguay was until recently, for the most part, covered by intact forestland and had little agricultural exploitation. During the last one and half decade, however, Chaco has turned into a global hotspot of fast deforestation (Graesser et al. 2015; Hansen et al. 2013; Vallejos et al. 2015, 4). While conversion of forest into pastures is the most significant land-use change in the Chaco region, soybean expansion and livestock intensification are the most important changes in the Atlantic forest region (Eastern region). Of the tropical Atlantic forest, only minor remnants are left and intensive soybean farming is depleting the soils and polluting the water. The Deputy Minister of Agriculture analyzed the underlying causalities in the following way:

> Deforestation comes together with two things; if the soybean is doing well you have to produce more, and if soybeans are not doing good you will not deforest, but if livestock is doing well you need more land for pastures. Sometimes these two trends come together, sometimes separately, but if you analyze the situation and see what really counts—the short-term profit or the long-term sustainability—you'll see that here it is the short-term profit that reigns, nothing else really counts. Soybeans and beef are the two most fundamental pillars of this economy. The soybean business is worth more than USD 3000 million, the meat more than USD 2500 million, approximately some 6000 million taken together. Agribusiness is there, pushing and pushing... There is extreme high dependence on a few commodities. The IDB made a study here. According to them, Paraguay completely depends on eleven commodities. Of course, the ideal is to have much more diversity. Logically, you remember what one always says about putting all the eggs in the same basket.... (Paraguay's Deputy Minister of Agriculture, Asunción, February 21, 2017)

Above quote, synthetizes in many respects agrarian change in Paraguay; ultimately, short-term profits determine its main path. This path creates long-term sustainability costs, high dependence on a few commodities and high exposure to a few external markets, as well as an increasingly powerful agribusiness sector. Actors both inside and outside of the state apparatus are aware of these basic determinates, and they do not seem to perceive that the state has the capacity to change the rules of the game of agrarian change in any radical way. Still, there are an infinite number of decimal places between zero and one, and Paraguay has taken several regulations designed to protect nature and people from the costs of the logic of short-term profit maximization. This section explores the most important of these regulations in detail.

### 5.3.1  Shifts in Social Relations and Regulations

Paraguay has among the most extreme cases of unequal land distribution in the world. According to the last Paraguayan agrarian census from 2008, the country had 264,047 family farmers (less than 50 ha each, according to official definition) of a total number of 289,649 productive units (PU). They thus represent more than 90% of all surveyed farmers, but they only have 6% of total agricultural land. Thus, less than 10% of the farmers have 94% of farmland in their hands (MAG 2009). The biggest two percent of producers manage more than 80% of agricultural land (MAG 2009). As shown in Chapter 2, the historical roots can be found in the huge sell-out of land forced upon the Paraguayan state to repay debt after the devastating *War of the Triple Alliance* (1865–1870). The so-called land reform of Stroessner incorporated millions of hectares of "new" land (forests) to agriculture, but most of it also ended up in the hands of big landowners. However, soy and beef expansion has spurred concentration still. Moreover, most of the soybean land is de facto in the hands of so-called *Brasiguayos*, Paraguayans of Brazilian descent, or Brazilians that have land and live in Paraguay. As mentioned in Chapter 2 and Chapter 4, the first wave of Brazilians entered the Eastern region of Paraguay during the 1970s. After the new millennium, a second wave of more capitalized Brazilian and Argentine farmers and firms arrived and expanded in the departments of Guahory and Caaguazú. It is estimated that around 85 percent of all Paraguayan soybeans are produced by so-called *Brasiguayos* (Lambert 2016). However, property rights are often contested. There has been a lot of corruption and lack of control in the land titling institutions resulting in cases of land with illegitimate titles, false titles, double titles, as well as vast land holdings without any formal titles. Contradictory land legislation has further incited ambiguity and contestedness (see more below about the specific negative consequences for native communities).

Between the agrarian censuses of 1991 and 2008, there was a reduction of 17,000 family farmers. During this period, the amount of PU with more than 10,000 has increased from 350 to 600. They represent less than 0.25% of all PU, but have 41% of all agricultural land (MAG 2009).[55] This shift occurred at the same time as agrarian area increased from 23.82 Mha to 31.09 Mha. Most of the expansion was forest to pastureland conversion in the Chaco region (6 Mha), followed by forestland converted into pastures and crops—mainly beef and soy—in the Atlantic forest region in Eastern Paraguay. Since the last census, based on data from 2008, there has been continuous rapid agrarian change. For example, in 2017, soybean production in Paraguay is estimated in statistical reports from the *Department of Agrarian Statistics and Censuses* (DCEA) of MAG to cover 3.54 Mha, while it covered 2.46 Mha in 2008. In a similar way, pastureland has increased by 0.82 Mha since the census, and the heads of cattle increased from 10.5 million heads to 13.9 million heads in 2016 (IDB 2017). Forest to pastureland conversion accounted for 62% of the new pastureland, almost all in the Chaco (Baumann et al. 2016). However, there is no reliable updated exact data over the actors behind these hectares and heads. While Paraguay is in the midst of rapid and dramatic agrarian transformation, its most recent agrarian census is more than a decade old. It has had a hard time to find external finding and while it is argued important to have an updated census in order to design correct public policies, the opportunity cost for the public budget has been deemed too high (Director of DCEA-MAG; Deputy Minister of Agriculture, Asunción, 21 February 2017). However, IDB announced in 2017 that it would fund a new agrarian census, and the results are expected to be presented in 2020. The IDB-project includes the training of a technical team under DCEA-MAG and FAO pronounced that it would provide additional methodological support (IDB 2017).

While there is a lack of completely reliable and updated data, it is clear that the period after 2008 has been characterized by further land concentration—driven by large-scale beef and soy expansion, and the collapse of the cotton sector (Director of DCEA-MAG; Deputy Minister of Agriculture; Innovation Specialist IICA Paraguay, Asunción, 21 and 22 February 2017). Moreover, the rapid rise of land prices (one hectare of agricultural land in the fertile Eastern region has risen from USD 200 in 2000, to USD 1400 in 2014) fuels concentration and increases barriers to entry.[56] At the same time, the family farming and *campesino* sectors have very low levels of productivity and degraded natural resources (Gobierno Nacional 2014, 58). Moreover, small plots, *minifundios* of 2–3 ha, have been subdivided into smaller and smaller pieces due to demographic reasons (for example, many children in rural areas).

Besides the continuous expulsion of small farmers, there is a dramatic displacement of rural workers in the densely populated Atlantic forest region in Eastern Paraguay (Ezquerro-Cañete 2016). According to the *National Peasant Federation of Paraguay* (FNC), around 900,000 Paraguayans left the countryside and settled in precarious living conditions in the outskirts of Asunción, as mechanized soybean agriculture made them "surplus".[57] The former Minister of Agriculture and Livestock, between 1989 and 2001, expressed this preoccupation in the following way:

The problem with the displacement of the small producers is the tremendous urbanization it is causing. They go to the cities, crime rates soar. Many people say that they do not want to work, but it is not like that, the thing is that there are no jobs. It is complicated. We will never achieve a perfect development model. Of course, the cotton was better in the sense that it demanded much more labor force than the soybean. I do not know. I am in favor of Paraguay using pesticides and genetically modified technologies to produce enough food to feed the world, but we should try to improve our ways of managing this model. (Former Minister of MAG, Asunción, 21 February 2017)

As illustrated in above quote, the social problems of displacement and ultimately urbanization caused by the soybean advancement are acknowledged also by actors that are in favor of the model at large—and in this case have formed part in designing its articulation.

Agrarian change has clearly hit many vulnerable groups in the countryside, but it has been particularly hard on smallholders that lack formal titles to land. About 40% of productive land in Paraguay remains untitled or have incomplete titles. This situation disproportionally affects indigenous communities (IBRD 2018). Both the Paraguayan constitution (from 1992), and the ratified Convention 169 (from 1993) of the *International Labor Organization* (ILO), provide indigenous people with relatively strong communal rights to land in order to be able to preserve and develop their ethnic identity in their own habitat. However, the laws appear to fall short of what is needed to ensure implementation and native groups lack security of their rights to their lands (Human Rights Council 2015, 6).[58] For example, in the sparsely populated Chaco forests live around 40,000 persons of different indigenous communities.[59] They use the forests for food, fiber, shelter, medicine, and spiritual motives (Walcott et al. 2014, 2). Some are hunters and gathers; others are subsistence farmers and/or day laborers at the Mennonites (Chapter 2 in this book). Given their reliance on their lands, they are extremely vulnerable to the land-use change (Sustainable Development Specialist UNDP Paraguay; Public Policy Officer WWF Paraguay, Asunción, 20 and 23 February 2017). However, many of the indigenous populations in the Chaco lack formal property rights to the lands that they occupy (Sustainable Development Specialist UNDP Paraguay, Asunción, 20 February 2017). In spite of the lack of titles and of the fact that most of the Chaco forest is privately owned and found within the big land units, the communal land rights of native communities mentioned above (national constitution and ILO 169), provide legal room for indigenous groups to make claim on land (Py/SEAM/INFONA/ FAPI 2016). There are, however, important legal contradictions, tensions, and confusions between different types of property rights regimes. The private property rights to land (Art. 109 of the constitution), the objectives of agrarian reform (Art. 114), the wildlife protected areas (law 352/95) and the communal land rights often collide with each other, which has given rise to the existence of overlapping ownership deeds that serve as a basis for multiple claims to the same parcels (Santagada 2013, 13–14; Piera Valdés 2016, 33–35). Conflicting claims are often settled in favor of business enterprises (Human Rights Council 2015).

Between 2000 and 2015, approximately 5.5 Mha of forest was lost in the Paraguayan Chaco—almost as much as in the Argentine Chaco during a period of over thirty years—representing a reduction of the forest of more than 20 percent (Py/SEAM/INFONA/FAPI 2016, 9). Big ranchers, while to some degree also by poor peasants using the forests as firewood, drive the lion's share of deforestation (Santagada 2013, 1378). In this way, the livestock expansion violates the rights of indigenous people to their land and threatens their livelihoods (Human Rights Council 2015).

This also means a decline of traditional knowledge and ways of living. Some indigenous groups of the Chaco are traditionally living as nomads, and they cope with periods of local food shortage by moving to other areas, but the privatization of land and the deforestation turn these groups increasingly food insecure (Walcott et al. 2014, 31).[60] The situation of the indigenous population in both the Chaco and in the Atlantic forest region was summed up in the following way:

> Sadly, the indigenous population is ignored in Paraguay. They are suffering rapid deterioration of their land and losing access because of lack of titles. They are more and more unprotected. Agro-exporting firms invade their land, and those who still have land are likewise affected because they lived from what the forest could give, their food came from the forest, but the polluted river contains less and less fish, and the wild fauna of the forest is disappearing, as is the flora, the fruits and the medicine that they used. So, they are basically left without food and medicine. (Public Policy Officer WWF Paraguay, Asunción, 23 February 2017)

While the overall situation is very dark for these peoples, there have been several legal processes regarding indigenous land in the Chaco during the past years. Some cases have been taken to the *Inter-American Court of Human Rights*, which has ruled in favor of the rights of the indigenous groups (Public Policy Officer WWF Paraguay, Asunción, 23 February 2017).

The main driving actors behind the deforestation in the Chaco and displacement of native groups are big Uruguayan, Argentine, and Brazilian ranchers. As land rents skyrocketed in the Pampas region, ranchers from this area have accessed much cheaper land in the remote Paraguayan Chaco (Director of DCEA-MAG, Asunción, 21 February 2017). The newly arrived foreign ranchers have therefore been able to buy bigger amounts of land (Public Policy Officer WWF Paraguay; Paraguay's Deputy Minister of Agriculture, Asunción, 23 February 2017). The Director of the office of Agrarian Censuses and Statistics of MAG, described the situation in the following words:

> The land values continue to rise, because foreign capital is coming and demanding ever more land. The deforestation in the Chaco region right now goes really, but really, fast. All the land bought by Uruguayans and Brazilians goes to deforesting and deforesting. (Director of DCEA-MAG, Asuncion, 21 February 2017)

The foreign actors were besides low land prices, attracted by free enterprising, low taxes, and low salaries. Over 2 Mha of land in the Paraguayan Chaco is estimated to be owned by Uruguayan investors alone (USDA 2017a). However, as PU are big in this sparsely populated area, the number of foreign ranchers is still small. Mennonite communities have been engaged in dairy production in the area for almost one hundred years (Chapter 2 of this book). They have also been expanding the past decades and own several slaughter plants in the area.

While the state lacks capacity to ensure compliance with indigenous' groups legal rights to their lands, big agribusiness companies have managed to get strong protection for their intellectual property rights. In contrast to Argentina, Monsanto succeeded to get the Paraguayan soy complex—including producers' organizations, the state, exporters, and seed companies—to impose a strong private property rights regime for GM seeds.[61] A system in which royalties were charged over harvests upon their sale to crushing industries and trading houses were installed in 2004, when Paraguay finally formally authorized soybeans RR and other GM events (Filomeno 2013, 58; International Service for the Acquisition of Agri-Biotech Applications 2011). Thus, Paraguayan farmers avoid paying the technology premium by saving seeds from harvest (Filomeno 2013, 59–62).

Many social movements—organizations of peasants, *campesinos*, rural workers and native people—have been voicing strong criticisms toward agribusiness agriculture in general and against soybean farmers in particular. The concerns, tensions, and conflicts that had emerged already during the 1990s (Chapter 4 in this book) have escalated along with the accelerating pace of agrarian change. The soybean is perceived as the ultimate symbol for unequal land structure, foreignization, contamination of soils and water, and biodiversity loss (Coordinadora de Derechos Humanos del Paraguay 2007, 19). Many NGOs and social movements claim that the interests of peasant societies and agribusiness are fundamentally irreconcilable, as are sustainability and monoculture (García-López and Arizpe 2010). There is particularly a widespread resentment and anger toward the *Brasiguayos*. The legitimacy of their land titles is often questioned because of the well-known widespread corruption of the Paraguayan agrarian reform programs.[62] Moreover, the *Brasiguayos* are depicted by the peasant movement as the main cause behind displacement of Paraguayan peasants and farmers (Ezquerro-Cañete and Fogel 2017). Some observers find the accusation unjust:

There has been a discourse of demonization of the *Brasiguayos*, which is totally semantic and has no real base to it. This has to do with historical resentment toward both Brazil and Argentina. The narrative that the *Brasiguayo* displaced the Paraguayan in *Alto Paraná* is a fallacy. The truth is that the Brazilian entered in lands where there were no Paraguayans, but in forestland. I worked much in this area in the late 1960s, and I saw that the first Brazilians arriving were very poor and coming from *Rio Grande do Sul*. They did not displace anyone; no one was where they arrived. In the collective imaginary, however, they came to displace the Paraguayans. In the second wave, the most recent one, already in the twenty-first century, then bigger and stronger Brazilians arrived, yes. You know, the king of soybeans, *Tranquilo Favero*, with some 230,000 hectares of land all over the country. But, why do they displace Paraguayans? Because they produce more effectively, that is why. But, the problem is the bad policies of the state, the ineffective support to family farmers, not the presence of Brazilians. (Sustainable Development Specialist UNDP Paraguay, Asunción, 20 February 2017)

As expressed in above quote, there is no agreement on the *Brasiguayo* as a problem. Here the problem is instead framed in terms of the fallacies of the state, which does not provide any solutions for the big majority of the rural population of small or landless Paraguayan peasants. The peasant movement, however, has increasingly voiced a nationalist orientation. This can be a legacy of the Stroessner oppression of any class-based appeal (Ezquerro-Cañete and Fogel 2017). Many peasant organizations, such as FNC, the *Agrarian and Popular Movement* (MAP), and the *Organization for Land Struggle* (OLT) carry out land occupations and fighting against displacements on big *Brasiguayo* farms (García-López and Arizpe 2010). Although titles to the lands could be unclear, the police often violently intervened to protect and support the soy producers and evict the peasants.[63]

Peasant displacement, foreignization, and agribusiness advancement became increasingly discussed under the government of *Nicanor Duarte Frutos* (2003–2008). He passed a bill in 2005 (2532/05), which prohibited foreigners to buy land near the borders (establishing a security zone spanning from the border and 50 km inland) in the name of national sovereignty.[64] Compliance with legislation has nevertheless been put into question and a great deal of land in the border regions that were already owned by foreigners and so remained.[65] Duarte also installed a tax on bigger landowners (IMAGRO) and a progressive tax on personal income

(Ezquerro-Cañete and Fogel 2017).[66] While the role of the state to redistribute wealth was increasingly emphasized, fiscal pressure remained low still; the lowest in the region, with almost half of revenues coming from indirect taxes, mainly VAT (CEPAL 2018).

The polarized situation, with rapid advancement of agribusiness along displacement and urbanization, was increasingly criticized among the population. The skewed land tenure, combined with the well-known corruption involved in the titling processes, was one of the major causes of conflict in the country. Landless peasants, rural workers, and small farmers increasingly organized demonstrations, strikes, and land occupations. Sometimes, urban workers, activist, and environmental organizations joined the struggle (The World Bank 2001). Some peasant resistance struggles ended in executions and forced disappearances (Coordinadora de Derechos Humanos del Paraguay 2007).[67] When cases of arbitrary executions were taken to court, the result has always been impunity, except in three cases where a hired gunman was found to be the only guilty for the deed (Coordinadora de Derechos Humanos del Paraguay 2007, 22).

Six decades of rule by the right-wing Colorado Party was brought to an end when former Catholic bishop Fernando Lugo, influenced by liberation theology, won the elections in 2008, representing the center-left coalition *Patriotic Alliance for Chance* (APC). Lugo, who had worked with peasant and landless organizations for decades, explicitly agonized against the advancement of agribusiness and promised land reform and social redistribution (Ezquerro-Cañete and Fogel 2017). Many environmentalist and peasant-based organizations, and leftist activists supported Lugo's electoral campaign.

The Lugo administration discursively problematized the situation of foreign landowners, talked about reestablishing land sovereignty by limiting foreign ownership, and stressed indigenous' peoples land right. However, in practice there was little change. The APC was an ideologically divided alliance of the pro-market liberal party, which dominated the alliance, and more left-wing parties. The alliance splintered early on (Ezquerro-Cañete and Fogel 2017). The administration also faced opposition in the Chamber of Deputies and the Senate that managed to curb the majority of Lugo's initiatives. In addition, the most important agrarian interest groups—the cattle ranchers in ARP (Chapter 2 in this book); the *Paraguayan Association of Producers of Soybeans, Cereals, and Oilseeds* (APS), representing around 50,000 soybean farmers, the majority *Brasiguayos*, the powerful umbrella organization *Union of Producer's Associations* (UGP)—successfully lobbied to curb all tax and land reforms. Besides lobby, they engaged in demonstrations such as *tractorazos*, blocking roads with tractors and farm equipment (Ezquerro-Cañete and Fogel 2017). In addition, other right-wing forces adopted a diversity of tactics ranging from the use of parliamentary and judiciary activism to military or parliamentary upheavals.

Among the first initiatives of the new administration was to increase funds to the *Land Reform Institute* (INDERT—former IBR) for distribution of land to landless peasants and to make a new Agrarian Census—a minor "revolution" had passed in the countryside during the sixteen years that had passed since the last census (1991). It is very expensive to do a proper Census in a country with hundreds of thousands of small farmers, a lot of untitled and double titled land and a huge forest area (Chaco) with bad roads. The Congress voted to block the proposed funding streams (Ezquerro-Cañete and Fogel 2017). The government finally managed to make the EU pay for the Census, which was published in 2009 (Director of DCEA-MAG, Asuncion, 21 February 2017). The census showed (as discussed above) that concentration rates had soared, and that small farmers were having an increasingly hard time. INDERT, however, continued with a limited budget combined with requirements to compensate owners of expropriated land, which heavily constrained its ability to distribute land. In addition, there were several cases of corruption in INDERT. Lugo also failed to pass a bill in the Congress in order to introduce a 6–10% export tax on unprocessed cereals (soybean, maize, and rapeseed)—at a time when Argentina had a 35% export tax on whole soybeans.

Lugo expanded programs of small credits to family farmers, he appointed peasant leaders to the agrarian reform agency INDERT, but agrarian reform with land distribution did not at all advance under his term, in spite of his electoral promises (Riquelme 2014). At the end of 2009, polls indicated that Lugo had one of the lowest popularity ratings of any leader in the region. The social and peasant movements that had supported his election soon demanded stronger political action. Moreover, under the government of Lugo, laws were passed that by the peasant organizations and environmental groups were seen as agribusiness-friendly, while harming people and nature. One example was the Law regarding *Control of Agrochemical Products* (3742/09), passed in 2009. This law weakened the existing requirements put on agrochemical use by for example reducing the buffer zones separating streams and houses from soy plantations (from 100 to 50 meters), eliminating supervision of aerial spraying and the obligation of prior notice, and reducing sanctions in cases of violation (García-López and Arizpe 2010). It was strongly criticized by civil society groups, but also by the Ministry of Health and SEAM for not adequately protecting health, and because oversight would be only in the hands of SENAVE.

Land conflicts—particularly in soy-area—occupations, and demonstrations increased significantly under Lugo's administration. One of the most violent "responses" to the agrarian situation in Paraguay emerged in 2008 in the form of the guerilla group *Paraguayan People's Army* (EPP). In the name of the peasant population of *campesinos*, EPP has made hundreds of high-profile kidnappings of wealthy Paraguayans, demanding ransom for their return, and killed militaries and polices (Irala and Cardozo 2016; Sánchez 2011).[68] The reaction from the state was to intensify criminalization; passing "state of emergency" legislation with increased control of the peasantry and permanent military presence in vast rural territories (Irala and Cardozo 2016; Martens and Estigarribia 2018). Domestic security was further militarized in response to the large Brazilian criminal organization *First Command of the Capital* (PCC) that started to engage in large-scale rubbery and drug trafficking in the border territory. However, not only EPP and PCC were repressed, but many other peasants and activists suffered increasingly violent state repression. At the same time, many groups were radicalized and there were an increasing amount of cases of violence and assassinations of soybean producing *Brasiguayos*.[69] Violence escalated and Lugo was criticized for lack of control of the situation and driving the country to anarchy.

On 15 June 2012, 11 landless peasants and 6 policies were killed in *Curuguaty* in Eastern Paraguay, in a violent eviction of peasants who had occupied lands. The massacre of Curuguaty led to a human rights and political crisis in the country. It was followed by almost the immediate destitution of president Fernando Lugo, impeached by his opponents, accused of mishandling the deadly clash between farmers and law enforcement. The Vice President, *Federico Franco* (from the Liberal Party), succeeded Lugo to complete the presidential mandate in August 2013. Many organizations and scholars pointed out irregularities in the impeachment against Lugo, and some suggested that the right-wing Congress exploited the violent confrontation between police and landless campesinos to get rid of Lugo and go back to business-as-usual in favor of agribusiness (Ezquerro-Cañete and Fogel 2017). The other Mercosur members claimed that Lugo had been removed from power unfairly and illegitimately. Argentina, Uruguay, and Brazil called home their ambassadors and Paraguay was suspended from Mercosur for a year (2012–2013).[70]

Colorado rule returned with the election of *Horacio Cartes* (2013–2018). His rule was characterized by mainly agribusiness-friendly policies. Taxes remained low (he put a veto on export taxes on soybeans—see below), although he implemented a tax reform, which included a ten percent *tax on agrarian profit* (IRAGRO), which brought a considerable increase in total fiscal income. During Cartes, Paraguay adopted a *National Development Plan* for 2030. This involves many laudable development targets, but it is ultimately showing that economic growth is continuously promoted over all other ends. For example, the special rapporteur on human rights noted that the plan does not refer specifically to indigenous peoples or to their rights (Human Rights Council 2015, 14). In the election 2018, the Colorado candidate won a surprisingly narrow victory over the candidate of the Liberal Party. *Abdo Benitez* was elected new president (2018–2023). He announced that taxes would be kept low, but that the government will do a better job of collecting them to find higher spending on education and healthcare. In a regional perspective, Paraguay remains with the lowest total taxes (IBRD 2018).

The debate over agrarian policies in Paraguay is still vibrant, polarized, and conflictive, however. Whereas the majority of Lugo's reform attempts were blocked (Riquelme 2014), social justice questions have remained on the agenda as one of his legacies. Lugo is still politically active with a seat in the Senate, and his political sector *Frente Guasu* is still formulating legislative proposals for agrarian and social reform. One concrete proposal among Lugo's many blocked initiatives for redistribution of agrarian wealth that has received renewed interest is export taxes on soybeans. Several legislative attempts have been taken to install export taxes, but they were voted down in the Senate. In 2012, the majority of the legislators in the senate actually voted in favor of an export tax, but president Horacio Cartes put his veto against it.[71] In 2017, the *Frente Guasu*, now with six seats in the Senate, made a pact with the government party (Colorados), in which a 15% tax the exports on whole soybeans, wheat and corn was included.[72] After modifying the proposal to tax exclusively whole soybeans and to lower the tax to ten percent, the Senate approved the bill in June 2017.[73] This ten percent export tax on whole soybeans was estimated to bring in USD 300 M/yr in fiscal revenue. Organizations representing producers and traders, such as CAP, APS, UGP, and the *Chamber of Oilseed and Cereal Exporters* (CAPECO) reacted immediately by protesting and blocking several national roads.[74] One of their main arguments was that the export tax will particularly harm small producers and increase concentration.[75] Moreover, these organizations argue the VAT and tax on agrarian profit is more than enough contribution to fiscal revenues from the agrarian sector. Many social movements, quite the opposite, have mobilized in support for the proposal. The Senate decided to postpone the reform and to do previous impact assessments. Since then, however, the process has been slow and the last bid is that the Senate will vote on the final proposal at some moment after May 2019.[76]

Besides the specific case of export taxes, there is an increasing polarization between organizations of peasants, rural workers, and native communities, on one hand, and the producers' and business's sectors, on the other. They all organize big demonstrations to make the politicians listen to their demands. Many organizations, not least the peasant, rural worker, and native communities' umbrella organization *National Inter-sectorial Coordinator* (CNI), demand the government to increase support for family farming and inclusive development, and investigate previous corruption in land distribution, and confiscate land that cannot be proven legitimately accessed. They also question foreignization and ask for clearance of all of their debts at the *National Development Bank* (BNF) and the *Agricultural Credit Agency* (CAH) and for new subsidized lines of credit (IDB 2017). The opposite side, the big producers' organizations and traders, argues that the government lacks control over the repetitive land occupations and violence particularly against *Brasiguayos* and soybean farms. They have organized several manifestations and *tractorazos* demanding that the government takes action to guarantee security and stop land occupations.[77] They also ask the government for tougher methods against violent guerilla groups, such as EPP. However, many observers do not think that violence will be solved by further criminalization and repressive state action, but point at the roots of the problem:

There is the idea that the social tensions and conflicts are transitory and will disappear by themselves, but that is not going to happen. Sadly, here we have EPP staging armed operations, bombings, shootings, and kidnappings since 2008. Against Duarte, against Lugo and against Cartes. They call themselves Green Marxist-Leninists and say that they work in defense of the environment and vulnerable people, but they are only opportunists. They kidnap big farmers in the name of the oppressed, and they receive support from desperate people. The problem is that a lot of people are displaced and desperate, that is the breeding ground. (Public Policy Officer WWF Paraguay, Asunción, 23 February 2017)

While different narratives focus on different blaming targets ranging from Brazilians, to the state, the soybeans, Monsanto, corruption or EPP, it is clear that agrarian change in Paraguay have increased social tensions and conflict over land. In spite of all conflicts and political turbulence, however, the whole period between 2004 and 2017 has been characterized by robust economic growth (4.5% per year on average), based mainly on the continuously expanding agro-exporting sector (see Appendix C). Most of the economic growth has stayed in the hands of the few and income inequalities remain among the highest in the continent. While big soybean and beef farmers are targeted as responsible for social exclusion by many organizations, there is less focus on the even more powerful and concentrated processing and trading firms. Paraguay's top exporters are the transnational grain traders Cargill, ADM and Bunge, followed by the big slaughterhouse *Frigorífico Concepción* (registered as a Paraguayan firm, but capitals are from Brazil).

Paraguay has nevertheless also managed to improve social inclusion and alleviate poverty. Since 2003, total and extreme poverty have fallen by 49 and 65%, respectively (IBRD 2018). At the same time, in 2018, still 18% of the total population is expected to live under the USD 5.5 poverty line—the majority of which live in rural areas. Moreover, the bottom 40% of the population lacks quality and stable access to basic infrastructure and social services (IBRD 2018). As previously mentioned, land has been increasingly concentrated in the hands of a few, while small farms are increasingly fragmented. Policies are in many respects the outcome of distributional struggles, and in Paraguay popular organization is clearly too weak and fragmented to manage to change the balance in any radical way.

### 5.3.2    Shifts in Forms of Production and Regulation

Agrarian change in Paraguay has brought important shifts in forms of production. In the Atlantic forest region, the main shifts have been soybean expansion and land-use intensification in both livestock and crop systems. The arrival of many foreign capitalized agribusiness firms brought technological transfer. In soybeans the transformation started in the 1990s with the RR technology (Chapter 4 in this book), with increased use of agrochemicals, airplane fumigation, and continuous cropping (Ezquerro-Cañete 2016). Soybean production in Paraguay in the twenty-first century has followed the same path of "modern" production forms, while soybean area grew from 1.2 to 3.5 million ha between 2000 and 2017 (USDA 2018).

Technological change in the livestock sector is more recent. It includes incorporation of fences, higher quality genetics, water reservoirs and in-sown pastures, improved nutrition, and health management (Cartes et al. 2015; USDA Foreign Agricultural Service 2017). These shifts have increased the country's average weaning ratio; 10–12 years ago most steers were slaughtered with about 30–40 months of age, they are now slaughtered with roughly 20–24 months of age (USDA 2017a). However, compared to Argentina and Uruguay, the weaning ratio is still around half (USDA Foreign Agricultural Service 2017). Technological change in both livestock and crop systems have increased yields per hectare (land-use intensification) and provoked a rapid rise in land prices, particularly in the Atlantic forest region. These shifts have brought a wide range of problems with pollution, erosion, soil degradation, and weed resistance. Moreover, the agropastoral expansion has turned Paraguay into a world deforestation hotspot. This section will focus on the Paraguayan regulations in relation to deforestation, and briefly mention other environmental regulation.

As mentioned in Chapter 2, Paraguay has historically looked at its vast forests as an untapped resource to exploit and make people settle in. Up until 2004, forestlands were in the 1963 *Agrarian Statute* considered unproductive areas. In accordance with the "social function" of land, landless people could therefore confiscate "unproductive" *latifundio* land, which provided an incentive for forest owners to clear land and put it under "productive use" (Chapter 2 in this book). With the new law No. 2524/2004, forest conservation ceased to be defined as unproductive (Piera Valdés 2016, 29–30).

One of the most radical and important environmental achievements ever in Paraguay is the 2004 *Zero Deforestation* law (2524/04) for conservation, preservation, and management of what is remaining of the Atlantic tropical forest in Eastern Paraguay. After decades of massive deforestation and fragmentation of the Atlantic tropical forest only an estimated 11.7% of the original area remains (Aide et al. 2013; Caldas et al. 2015; Graesser et al. 2015, 5). The *Zero Deforestation* law put a moratorium on all land-use change for agricultural aims on forestland. It was initially set until 2006, but later extended several times and is currently running to 2020.[78] According to the *World Wildlife Foundation* (WWF) in Paraguay, the law was the result of a persistent struggle of many national and IOs as well as local associations, putting pressure on politicians, drafting legislative proposals, and raising general awareness of the dramatic rates of deforestation and their potentially devastating consequences (Public Policy Officer WWF Paraguay, Asunción, 23 February 2017). While more than one Mha has been deforested after the moratorium, the rate of forest loss in the Atlantic forest is still estimated to have slowed down with almost 90% (INFONA 2015; Walcott et al. 2014, 10). However, instead deforestation of the Chaco intensified. In only one year, between 2006 and 2007, deforestation tripled. There is a risk that the strengthened regulatory framework to protect the Atlantic forest fueled the snowballing deforestation of the dry Chaco forests. Extensive cattle-raising activities have been moving away from the eastern region (Atlantic forest) to the western region (Chaco). However, the main actors behind the rapid deforestation of the Paraguayan Chaco is not Paraguayan ranchers going west, but big ranchers from Uruguay, Brazil, and Argentina (Director of DCEA-MAG, in Asuncion, 21st of March 2017). Increased forest protection of the Argentine Chaco and in Brazil (at least before the election of Bolsonaro), and high competition of land with soybeans in Pampas seem to be part of the explanation to the intensified exploitation of the Paraguayan Chaco by foreign ranchers.

SEAM, in partnership with a local NGO and departmental governments, developed an "environmental management plan" for the Paraguayan Chaco. The plan included non-legally binding recommendations against tree removal in certain areas. A "Zero Deforestation law" for the Chaco was proposed and rejected by the chamber of deputies in 2009 (de Waroux et al. 2017). Between 2014 and 2018, more than one million hectares forestland in the Chaco was lost, on average

264,000 ha/yr. According to the *National Service for Animal Quality and Health* (SENACSA), there are, as of December 2018, 8210 livestock PU in the Chaco, owning 6.1 million heads of cattle, representing 44% of total national stock. Ultimately, the main driver is profit. Soybeans are not yet cultivated in Chaco, but several observers note that it is probable that after some years of grazing, the soybean firms will start expanding into the area (Public Policy Specialist UNDP Paraguay, Asunción, 20 February 2017). Paraguay's Deputy Minister of Agriculture had in February 2017, for the first time seen a soybean field in the Chaco region (Paraguay's Deputy Minister of Agriculture, Asunción, 23 February 2017).

While the Paraguayan Chaco still represents one of the world's largest forest reserves and "unexploited" (for large-scale agriculture) fertile land, the pace of deforestation is one of the highest in the world (Vallejos et al. 2015, 7–8; Walcott et al. 2014; Imbach 2016; Hansen et al. 2013).[79] Large cattle agribusiness farms have become the major driver to deforestation, using bulldozers and other heavy machinery to clear land, put cows on and incorporate fences, water reservoirs, and very productive tropical pastures (USDA 2017a). This leads to loss of natural habitats and fragmentation of the landscape with multiple negative social-ecological effects (Chapter 1 of this book).

The most relevant regulative framework for the Chaco forest is still the 1973 *forestry law* (422/73), which declares of public interest and obligatory the protection, conservation, and improvement of forest resources (Piera Valdés 2016, 29, 59). The law stipulates that all landowners (with more than 20 ha of forestland) have to leave at least 25% of the land aside for conservation and ecological rehabilitation. This means that landowners can legally deforest up to 75% of their land.[80] Compliance with the legal framework is, at best, poor (Piera Valdés 2016, 33). As a result, forestlands suffer severe degradation, and illegal conversion of forest is frequent. Illegal and legal forest management efforts have accordingly been very hard to tell apart. Loopholes make possible to transfer the 25% to other owners, in order to further clear 75% of the 25% (Mansourian et al. 2014). There is also a widespread view that there is no penalty associated with not fulfilling the requirements of the law (Py/SEAM/INFONA/FAPI 2016, 36). The remoteness of the Chaco region and the government's lack of resources for monitoring or prosecuting lawbreakers make possible rampant, illegal logging of the slow-growing forest. Once the forest has been degraded, landowners disseminate exotic grass seeds that quickly dominate the area and the authorities declare the land as cattle ranching area. The former Minister of Agriculture and Livestock, 1989–2001, talked about a shift in political culture and the complexity of ecosystems as important reasons to the noncompliance with legislation:

As I said before, Paraguay has in general sufficient and good laws, I think that we have enough legal protection of the forest; the problem is that the laws are not adequately implemented or complied with. I think that an optimistic interpretation would be that around 80% of the farmers actually comply with the forestry law [No. 422/73], but the 20% that of farmers that do not is the problem, particularly in Chaco. You may have heard of how they put a chain in-between two tractors and just take away everything. Of course, the land there is good for pastures and it is hard for people to understand and see the connection that the deforestation changes the rainfall regime, the temperature, the soil. Perhaps, we as human beings are just shortsighted, 'If the yields can be good for at least ten years, then I go for it' [...] In some ways, it was easier to rule before, because the memory of dictatorship was very present, so the Minister was very authoritarian and everybody followed what he said. Regulations were established and then they were followed. Nowadays, this is not so straightforward. In addition, the decentralization and fragmentation of MAG led to less control. I have already said it many times; Paraguay needs that its laws are complied. We have to use much stronger sanctions, not a low penalty fee as we have now. I think illegal logging should lead to prison because we cannot continue destroying our resources, the sanction need to be big, and we need to monitor everything very carefully, we are not doing that today. Here we have impunity. (Former Minister of MAG, Asunción, 21 February 2017)

The ex-minister thus argues that the failing compliance with the legal obligation to leave at least 25% of forestland has multiple causes. One problem he addresses is the difficulties for landowners to see the value of conserving the forest in order to maintain vital long-term ecosystem services at landscape level, rather than just see the short-term yields. Moreover, he sees a general erosion of rule obedience as the memory of the authoritarian dictatorship is vanishing, which is aggravated by the decentralization reforms of the state apparatus and low penalty associated with not fulfilling the requirements of the laws. Several respondents address shifts in political culture after re-democratization and deficient coordination between different public entities responsible for forest policies in the wake of state reform during the 1990s (Chapter 4 in this book), as adding up to poor law compliance and lack of control (Senior consultant at UNDP and senior consultant at IICA, in Asuncion, February 2017).[81] This is aggravated by widespread corruption at all levels (Specialists working at PNUD Paraguay, February 19, 2017; specialists at IICA Paraguay, February 20, 2017, in Asunción). Corruption and lack of monitoring are bigger problems than the level of fines, according to some stakeholders:

> I think the fines are already rather high. That is not the problem. The problem is that there is a culture of bribes. We are talking about powerful economic interest here; they can put the life at risk of the public inspector or the family producer that questions them. The problem is lack of control. Here the vast forest areas are law-less land. There is also a lot of drug production in the forest. They use the trees to hide. You cannot give fines to people in a territory that you don't control with all the risks that it would bring. (Representative IICA in Paraguay, Asunción, February 22, 2017)

The corruption involves inspectors, police, judges, and firms. Moreover, there are occupations by landless peasant farmers (Santagada 2013, 13–14).[82] According to the director of *Agrarian Censuses and Statistics* (DCEA-MAG), however, landless peasants invading forestland was common in the past, but not anymore. Rather it is big business seeing that the chance to get caught is small and that in any case they can pay their way out of it, combined with a small risk to get caught because of lack of inspections (Director of DCEA-MAG, Asuncion, 21 March 2017). The *Secretariat for the Environment* (SEAM, since December 2018 turned into a proper ministry, MADES) is in charge to control the situation, but while monitoring has greatly improved since 2010, it lacks the conditions to do so both in terms of human and financial resources (Director of DCEA-MAG, Asuncion, 21 March 2017).

Besides the forestry law, Paraguay installed a *national system of protected areas* (SINASIP) as a way to protect important ecosystem services. As of 2019, Paraguay has 98 protected areas, covering 14% of total land.[83] The biggest is the National Park *Defensores del Chaco* of 780,000 ha forestland in the Chaco. For this vast region, however, only a single ranger is assigned, and there is a complete lack of logistical and financial support for that ranger (Public Policy Officer WWF Paraguay, Asunción, February 23, 2017). Accordingly, there have been cases of deforestation in the protected areas and the state lacks resources to see that the protected areas are properly respected. In general, a small budget has not allowed SINASIP to secure any larger area of the country's territory. Moreover, several protected areas are not completely state-owned, the borders are not exactly defined and delineated, and the legal authority is ambiguous. This has resulted in deforestation within protected areas and in settlers installing themselves at the borders (Public Policy Officer WWF Paraguay, Asunción, February 23, 2017).

Besides protected areas and the required 25% of forest cover in the forest law, the state has taken initiatives to advance with forest-protection through a series of economic incentives for afforestation, restoration, and conservation.[84] The most fargoing regulation in this respect is the 2006 law for *Valuation and Retribution of Environmental Services* (Law 3001), which was passed after years of lobby from environmental organizations, such as WWF. The framework is based on *Payment Schemes to Ecosystem Services* (PES), in which landowners are given to monetary remuneration through tradable certificates, if they conserve forestland beyond the required minimum of 25%. In this way, an economic value is given to the ecosystem services that forests provide to incentivize landowners to conserve it (Py/SEAM/INFONA/FAPI 2016, 33). This framework emerged within the framework for Paraguay's adoption of the UN *Climate Change Convention* (UNFCCC), including the *Convention on Biological Diversity* (ONU-REDD+/SEAM/INFONA/FAPI 2016, 18). PES is considered a cost-effective way to hinder deforestation since the "costs" of deforestation in the future are estimated higher than the costs of paying landowners to keep the forest. However, less than 25,000 ha (8% of the area of forest annually lost in Paraguay) had received benefit from this regime by 2016 (Py/SEAM/INFONA/FAPI 2016, 35).

One of the constraining factors for implementation of PES is unclear ownership of land and contradictory legal land rights (Piera Valdés 2016, 15–36). A lot of land lacks formal land titles, but even when ownership is clear, it may stand in conflict with communal land rights of indigenous peoples, as mentioned in the previous section about social relations (Human Rights Council 2015, 7). The legal contradictions between different types of property rights regimes constitute an important hinder for all Paraguayan forest conservation initiatives (Santagada 2013, 13–14 and author's interviews with different stakeholders). The biggest constraint for PES implementation, however, seems to be that SEAM lacks resources to be able to pay landowners in parity with alternative land-use, which now is represented by the expected income from cattle ranching in the midst of rising beef prices. One of their public policy officers commented the poor implementation in the following way:

> Sadly, the law that we worked so hard for the legislators to accept, which is the law about valuation of environmental services, is not being implemented because when these services were to be paid for, there were no funds, so no one received anything. (Public Policy Officer WWF Paraguay, Asunción, February 23, 2017).

While the public policy officer lamented the lack of funds, the *National Development Plan of Paraguay 2030* is optimistic about the future of PES as a contributor to increased forest cover, protected biomass area, and restoration of degraded ecosystems (Imbach 2016, 6). There are also high hopes that UN's *Reducing Emissions from Deforestation and Forest Degradation* REDD+, program in Paraguay (working in the country since 2017), which also works with market-based mechanisms such as PES, will bring to an important influx of institutional capacity and capital from abroad (Piera Valdés 2016, 53; Py/SEAM/INFONA/FAPI 2016).[85] Given that almost all of the Chaco forest is in the hands of relatively few big landowners, however, the carbon credits could exacerbate inequality and transfer public resources to the already rich, with no positive impact on poverty alleviation. This works against one of the aims of REDD+, which is to provide social opportunities for the most vulnerable communities, besides reducing emissions from deforestation and forest degradation (ONU-REDD+/ SEAM/INFONA/FAPI 2016). The positive side of the extreme land concentration of the Chaco, however, is that it reduces transaction costs and if the state manages to convince this small group of people it can potentially make efforts to stop deforestation more effective (Py/ SEAM/INFONA/FAPI 2016, 10–13). Paraguay's Deputy Minister of Agriculture, however, expressed less faith in the future of forest conservation:

> So, we designed the law of environmental services [3001/06] to be able to offer landowners profits without deforesting. If the producer wants to produce some thousand ha of soybeans, because soybeans are doing well, the state says: 'Look, if you protect the forest instead, I'll compensate you for the losses'. So, we have that mechanism, in order to protect the soils,

to avoid pastures, to avoid crops. But you have to pay for that, from somewhere you have to find those resources, and here the resources are never sufficient. So, then the state tries to regulate instead, to prohibit logging on certain places such as with the protected areas. But those are isolated attempts, we are still a free market, and even if you try to regulate something over here, the regulation is not working over there... Our tools are weak; there is a lack of strong institutions and a lack of honesty. The biggest problem is probably that SEAM lacks money. It only has funding to cover infrastructure, and they cannot do anything with that. So, when you see that things are not working properly, you turn around and you say that this is impossible, that it is better to let things be the way they are; let's not intervene, come more soybeans and come more beef. (Paraguay's Deputy Minister of Agriculture, Asuncion, 21 February 2017)

As expressed in above quote, the public attempts to regulate exogenous driven agrarian change in more sustainable ways face severe limitations. When the state uses tools of prohibition, it fails to monitor and ensure compliance and when the state uses market-mechanisms of incentives, it lacks resources to make the incentives effective. The overall picture of the Paraguayan attempts to regulate deforestation is that overlaps and gaps in legislation, chronic budget constraints, and widespread corruption and other institutional weaknesses result in lack of control and weak compliance with environmental laws and regulations. At the same time, it is important to remember the relative success of the zero deforestation moratorium in the Atlantic forest region, which indicates that the state still has some regulative capacity. Total prohibition seems in this way to be a more efficient tool. However, the legislative attempts to make a similar moratorium of the Chaco forest has been voted down.

Policies toward the Chaco represent a balancing act between the irreconcilable aims of boosting agrarian-based economic growth and conserving natural resources for long-term sustainability, as well as respecting the rights of peoples living in the forest. Paraguay wants to do both, but the main public arenas in charge of sustainability; SEAM, INFONA, and MAG are chronically underfinanced and weak. The main emphasis of this difficult balancing act shifts across administrations and institutions. For example, in September 2017, president Horacio Cartes passed decree 7702, to the forest law No 422, which allowed for exception of the minimum requirement of 25% of conserved forestland, if the landowner reforest a corresponding amount of land, or buys his way out by purchasing environmental certificates. The president himself used the decree and deforested 2000 ha of land (clearing around two million trees) on his property "San Francisco" in the Chaco in order to expand pastureland.[86] The decree received strong protest from environmental and indigenous NGOs. The new president, Mario Abdo Benitez, derogated the decree in September 2018, upon the argument that the decree contradicts the objectives of sustainable development. This is an illustrative example of the state as an arena of competing interests, including within the dominant *Colorado* party. While there is some degree of variation in ways of conceptualizing legitimate public policy within the dominant Colorado party, stakeholders representing NGOs and IOs say that the environmental knowledge and interest is generally low. Most of the time, it is legislators from oppositional parties that take the lead and push for environmental regulation.

All governments have invested heavily in infrastructure, especially highways, bridges, and roads to connect the Chaco with markets in a more efficient way, for example by reshaping *National Route 9* (Ruta Transchaco) which links el Chaco with the capital, Asunción.[87] This will significantly lower transport costs and time-cost, which in combination with high commodity prices, make exports-oriented agrofood production even more profitable. The *National Development Plan 2030* also establishes that it is a policy priority to improve the roads around Asunción and close to the ports, in order to make access to ports smoother for soy and beef (Gobierno Nacional 2014, 63). Considering the strong economic drivers and the inability of the state to monitor and control the whole territory, it is hard to see that the rapid transformation of forests into cattle production will decrease, if no game-changing shifts emerge.

The limitations of public regulation regarding deforestation are illustrative for environmental regulations at large. Land-use intensification in combination with lack of technical capacity and financial support to set up sustainable production systems has caused a wide range of social-ecological threats, such as degradation of soils, pollution of waterways, and pesticide poisoning. In all these areas, there are insufficient regulation, lack of monitoring systems to supervise compliance with the legislation, lack of coordination between public entities, confuse land titles, corruption, and moreover, very small budgets of SEAM, SENAVE, INFONA, and MAG. Many of these institutional weaknesses in regard to environmental policy—including poor laws, budgets, and management—are recognized flaws in the proper *National Development Plan 2030* (Gobierno Nacional 2014, 60). At the same time, low taxes, investments in infrastructure, and systems that allow strong intellectual property rights to the patent-owning big firms facilitate and support further agribusiness expansion (Human Rights Council 2015, 12).

Many respondents suggest, accordingly, that the most efficient path for enhanced sustainability, in the midst of pressures from beef and soy, is not legislative initiatives, or any type of public policies. Instead, they find that joint partnerships with big agribusiness firms, establishing codes of conduct and "good practices", can be more fruitful. In this way, UNDP Paraguay supports and coordinates the private–public initiative called "green commodities", aiming to protect the biodiversity of the Atlantic rainforest and assure "sustainable land management" throughout the soybean and beef chains (Project coordinator of "green commodities" at UNDP Paraguay, in Asunción, 21/2-2017). This is an arena in which national subsidiaries of the transnational traders (ADM, Cargill and Louise Dreyfus Commodities—LDC), large-scale soy and beef farmers' organizations (ARP, APS, ASP, and FECOPROD), public bodies (SEAM, MAG, INFONA, and the *National Table of Finance*), and IOs (UNDP and the *Global Environmental Fund*—GEF) participate. One of the aims of this arena is to make producer firms adopt more sustainable practices by capacity building, farmers' manuals for sustainable management, and extension services. The underlying idea is that sustainable practices eventually payoff, allowing for long-term production and high yields. Regarding soybeans, for example, the main message is that in the end it is cheaper to do proper rotations than do plant only soybeans:

> We try to show the producers how the soil loses productivity if they don't do good rotations. They already know that many times, but they still find it hard to plant anything else than the crop with the highest margins and many have debts to pay. They pray so that things will work out anyway and they chose to plant whatever provides the highest returns in the short-term. (Senior Technical Advisor UNDP Paraguay, Asunción, February 20, 2017)

Besides the "medicine" of knowledge development, dialogue, capacity building, and diffusion, the idea is that sustainability can also be translated into price premiums by providing access to better-paid market segments. The assumption is that future markets will be more demanding, eventually creating win-win situations in relation to sustainability. Regarding deforestation, for example, it is assumed that as the role of forest area for biodiversity and climate change mitigation becomes more well-known and accepted throughout the world, markets will eventually exclusively demand soy and beef that can show that production has not contributed to deforestation (Technical Advisor UNDP Paraguay, Asunción, February 20, 2017). The same logic is behind the *Round Table of Responsible Soy* (RTRS), in which private firms and traders certify their soy products as "free from deforestation", which is supported by WWF, FAO, and UNDP. Defenders of RTRS claim that it at least may slow down deforestation, but critics argue that certified soybeans may still contribute to deforestation, since when soybean production expands in other areas it pushes other agrarian activities, not least cattle grazing, into forest area (García-López and Arizpe 2010).

In the same way, in order to access as many well-paid meat markets as possible, UNDP supports the *Animal Health Service* (SENACSA) and other public entities work to improve beef safety standards, status, and quality. After a Foot and Mouth Disease outbreak in 2011, Paraguay installed a compulsory traceability system to all cattle operations and herds in the country. This group traceability system tracks all cattle movements within the country, but in order to access demanding markets, for example the 481 beef quota of the EU and the high-quality beef to the Russian Federation, requires individual traceability (USDA Foreign Agricultural Service 2017). Moreover, the markets of the NAFTA countries, Japan, and South Korea are still closed.[88] There is an ongoing discussion in Paraguay if it would be worthwhile to follow Uruguay's sophisticated and expensive traceability system in order to access more highly demanding markets (see the next section about Uruguay). The Paraguayan state, with support from UNDP, is collaborating with the *National Meat Institute* (INAC) from Uruguay to explore this path (Sustainable Development Specialist UNDP Paraguay; Plant and Animal Health Specialist IICA Paraguay, Asunción, 20 and 22 February 2017). There is, however, uncertainty if future markets will, or will not, pay for the costs of detailed information about the whole production process behind a piece of meat, from calving to slaughter:

Everybody talks about the requirements of the markets of the future, but nobody knows. On the one hand, there is the idea that markets are getting more and more demanding, and that well-paying consumers want to be able to follow the whole chain, and be guaranteed no social and environmental costs along the way. On the other hand, world population is growing, land is getting scarcer, and the Asians want to eat more and more beef... There is also a discussion about what is the most environmental-friendly way to produce beef. Here we have the conflict between grass-fed—according to some is more organic, but those cows will emit more greenhouse gas before slaughter as they take more time to grow—and at the other end the feedlot... But, of course, complete traceability would help overcome the enormous deficit in legal compliance. Today there is almost a complete lack of monitoring. The question is who should bear the costs of implementation. The state cannot pay for this alone. (Technical Advisor UNDP Paraguay, Asunción, February 20, 2017)

Meanwhile, the state has not decided if it will take on the costs, a private and voluntary traceability system of individual cattle tracking with tags has been advancing (USDA Foreign Agricultural Service 2017). The biggest meat company and beef exporter, *Frigorífico Concepción*, is one of the companies participating and has made a big number of its advanced control systems and traceability programs. In fact, the company was caught in a smuggling control on May 2, 2018, with 7.3 tons of meat from Brazil, coming in trucks, while it only had registered and received authorization from the *National Service for Animal Quality and Health* (Senacsa) for 3.6 tons of meat. Senacsa suspended the company, which provoked protests from workers and local leaders. On 30 May 2018, the Paraguayan Ministry of Industry and Commerce gave the company a fine of almost USD 3 M for not registering properly the imports, and it was suspended one year for all importing activity. At the time for this smuggling scandal, *Frigorífico Concepción* was in the middle of rapid market expansion, benefitted by the large 2017 corruption scandal of the world's largest meat company JBS SA, in Brazil, which rendered the company a record fine of USD 3.000 million. Therefore, JBS sold its subsidiaries in Paraguay (and in Uruguay and Argentina), which was the second largest beef plant and was bought by another Brazilian beef company; Minerva. As illustrated by this case, corruption is not an exclusive problem of the public arena, and that private governance initiatives may show even less compliance and transparency than the state.

## 5.4    URUGUAY—THE FRENTE AMPLIO
## AND THE COMMODITIES SUPER CYCLE

*Frente Amplio* (FA) won the 2004 elections on a platform that agonized against transnational agribusiness advancement and neoliberal policies, and pledged the return of the interventionist state, rebalancing social relations brought by markets, and guaranteeing long-term sustainability of natural resources (Frente Amplio, 22 December 2003). Since FA, for the first time in its history, entered government in 2005, it has remained there for three consecutive government periods (2005–2010, 2010–2015, and the current administration 2015–2020).[89] This period coincided with the longest uninterrupted period of economic expansion in modern times (Fund. 2019). The annual average GDP growth rate was 4.54% between 2003 and 2016 (IBRD 2015). Despite economic recessions in Argentina and Brazil, Uruguay has managed to sustain growth also under 2017 and 2018 (International Monetary Fund 2019).[90]

The most important social changes brought by FA were a progressive tax reform, with focus on income tax, a public health reform, and the reinstitution of tripartite wage councils and collective labor bargaining, coupled with the launch of several social programs. By tradition, Uruguay stands out in Latin America for being an egalitarian society, with the largest large middle-class, low level of inequality and poverty, and the almost complete absence of extreme poverty. The social reforms of FA further increased social inclusion and reduced inequality; the Gini coefficient decreased from 45.9 in 2006 to 39.7 in 2016 (International Monetary Fund 2019; 28, IBRD 2015).[91] Uruguay occupies the top positions in the region in terms of various measures of well-being.[92] The unemployment rate declined to historically low level (6.6%) in 2014, but rose above 7% in 2017/2018 (International Monetary Fund 2019; IBRD 2015).

While reducing social inequality and improving real wages, these reforms had little implications for the articulation of agrarian change (Antía 2018; Llambi et al. 2016; Martorano 2014). Quite the contrary. In opposition, FA criticized almost all liberalization reforms of *Blancos* and *Colordos*. In government, it has allowed almost all previous policy reforms of the 1980s and 1990s (Chapter 4 of this book) to remain intact, or with only minor modifications. This includes "prudent" macroeconomic policies, proactive utilization of Free Trade Zones, promotion of foreign investments, strong protection of private property rights, no export taxes, and in general a good "business climate" (Baraibar 2014).

In line with neostructuralism from CEPAL, however, FA has argued for export-orientation combined with a proactive "developmentalist" state; planning for long-term change of the productive structure away from reliance on natural resource-based products through "upgrading" for more technologically advanced, diversified, and inclusive growth. In this way, the investment promotion regime was reformed in order to create clearer economic incentives for firms to invest in line with specific development criteria. This included criteria for incorporation of advanced technology, creation of well-paid jobs, contribution to decentralization, and investment in clean energy. Investment projects that meet the most development targets get the largest tax deductions. The combination of an increasing amount of international capital seeking new markets and the benefits provided by the Uruguayan investment regime made Foreign Directs Investments (FDI) arrive in Uruguay as never before, particularly in agriculture (Durán and Salgado 2013; Uruguay XXI 2013, 2016). The average total FDI in Uruguay 2000–2002 was USD 255M, and almost ten times more, on average 2006–2015, USD 2000M (Oyhantçabal and Narbondo 2018). Growth in foreign investment resulted in sharp increases in productivity in agricultural production, processing, and exports. Indeed, a development model based on the intensive use of natural resources has been increasingly pronounced during the FA administrations, and the official slogan for agricultural policies of the Ministry of Livestock, Agriculture and Fishery (MGAP) has been "sustainable intensification". This encapsulates classical productivist aims—to produce more and cheaper—with an increased focus on sustainable natural resource management, particularly regarding soils and water, and mitigation and adaption to climate change (Presidencia de la República Oriental del Uruguay 2017).

Similar to Argentina and Paraguay, Uruguay had a remarkable export growth the first one and a half decade, and the composition of exports shifted toward primary sectors at the expense of manufacturing products (International Monetary Fund 2019, 18). Uruguay has expanded its presence in the world markets even more rapidly than its neighbors, driven by a handful of export commodities—particularly soybeans, bovine meat, and cellulose and other forestry related exports (IBRD 2015). Soy, beef, and forest products together represent more than 50% of total export value in 2017, and more than 73% of the agrarian-based exports, on average 2013–2017 (see Appendix II and CAS 2018). Besides increasing reliance in agrarian commodity exports,

Uruguay is increasingly dependent on China as main buyer. In 2016, 73% of Uruguayan soybeans, 37% of its cellulose, and 35% of its beef was destined to China (Uruguay XXI 2017, 4). These export products in turn reflect important land-use and land-cover changes such as afforestation, crop expansion (led by the soybean "boom") and the land-use intensification through in-sown pastures and other shifts in forms of beef production.

While Uruguay did not participate to any significant extent in the soybean complex until Argentine crop farmers started arriving in 2002/2003 (Chapter 4 of this book), the late insertion was compensated by extraordinary fast growth; within a decade the soybean had become the number one crop (Baraibar 2014). Soybean expansion has mainly been concentrated to the "Littoral" departments of Paysandú, Soriano, Río Negro, San José, and Colonia. The relative increase of soybean area between 2000 and 2017 was over ten times, from 0.09 Mha to more than 1.3 Mha (USDA 2018). In addition, the public policies in support of the forestry sector from the 1990s (Chapter 4 in this book) started to "kick-in, as fast-growing eucalyptus and pine became ready to transform into cellulose exports. While FA reformed the forestry policies and took away some of the support, the complex still benefitted from important tax deductions and new investments have sustained growth levels. Beef—the traditional flagship of Uruguay—has also had a significant increase in terms of production and exports, as well as in quality (Carriquiry 2015; USDA 2017b). These commodities are interlinked in multiple ways, not least because Uruguay has no agrarian frontier and increased competition for land has increased pressures toward intensification of land-use.

Agrarian change has not only brought increasing investments and export revenues, but also increased land concentration, agribusiness advancement, foreignization, land-use intensification, specialization, and significantly increased pressures on natural resources—as in Argentina and Paraguay. Similarities apart, Uruguay's pathway differs in many respects from both its neighbors. For example, it has almost no native forest, no indigenous population, no ambiguous land titles, and no strong movements of peasantry or landless people. Accordingly, deforestation, violation of indigenous rights, and land evictions have not been important policy concerns in Uruguay. Agrarian change has still provoked serious social-ecological concerns regarding pollution, soil degradation, land concentration, foreignization, loss of family farmers, rural depopulation, re-primarization, and dependence on China. As we shall see, FA has implemented some policy regulations in response to the concerns, such as increased land tax, policies for upgrading, differentiated policies for family farmers, and obligatory soil rotations. Some of these measures received critique and resistance from the powerful organizations of the agrarian sector (while the levels of conflict never reached the Argentine levels). Many NGOs and researchers have instead criticized the government for being too permissive with the interests of agribusiness. Thus, while less violent than in Argentina and Paraguay, agrarian change in Uruguay has provoked new tensions and conflicts, with public regulations as their most important arena.

### 5.4.1   Shifts in Social Relations and Regulations

As in Argentina and Paraguay, agrarian change in Uruguay has led to significant shifts in social relations, with growing dominance of big farmers and expulsion of small farmers. Uruguay has not any frontier left; as producers with more than 1000 expanded, all other producer strata retracted, during the period between the agrarian census of 2000 and the last agrarian census of 2011 (DIEA 2014). It is, nevertheless, important to bear in mind that Uruguay has no *campesinos*—no peasants; all family producers are fully integrated into the markets for land, labor, capital, products, and inputs (Piñeiro and Cardeillac 2017). In contrast to Paraguay and Northern Argentina, changes in land property and access have not caused any forced eviction or displacement of previous land-users, or any kind of violent confrontations (Oyhantçabal and Narbondo 2018). Property rights to land are clear and largely undisputed.

The average size of all PU increased from 287 ha in 2000 to 365 ha in 2011, which is among the highest average size per agricultural holding in the world. In 2000, Uruguay had 57 thousand PU, of which 33 thousand were family farmers (less than 500 ha and less than two wage-workers, according to Uruguay's official definition).[93] In 2011, only 45 thousand PU remained. The majority of the 12,000 farmers that abandoned activity where smallholders owning less than 100 ha of land or sharecroppers (MGAP 2014).[94] Of the 45 thousand remaining PU in 2011, 25 thousand, or 63%, were family farming units, of which the majority were livestock producers (73%), owning 15% of the land (DIEA 2014; Paolino 2013). The 2011 agrarian census also showed that the 4167 farms with more than 1000 ha controlled more than 60% of total exploited area, while the 1191 farms over 2500 ha controlled 34% of the same. The sector of industrial crops, such as soybeans, is the most concentrated and has the highest participation of corporate firms; controlling 1.2 Mha out of 1.7 Mha of cropland (MGAP 2014). Only 4% of the family farmers in Uruguay produce grains or oilseeds (DIEA 2014). While there is no new agrarian census, as of yet, agrarian statistics indicate that the amount of family farmers have continued to decline, with 22,187 family farming units left in 2018 (DGDR-MGAP 2018).[95]

One group that was particularly badly hit among crop producers was the traditional sharecroppers. Around half of all sharecroppers in the Litoral disappeared from activity between 2000 and 2009 (Arbeletche and Gutiérrez 2010; Arbeletche and Carballo 2009; Arbeletche et al. 2012). They had typically integrated mixed productive systems in the Littoral and paid for the land with a percentage (around 30–50%) of crop harvest income minus costs. Sharecroppers lost access to land as rents went up and the arriving Argentine crop firms offered landowners to pay for the land in advance without any risk for the owner (Baraibar 2014; Sandoval 2016). In 2015, 30% of soybean production area belonged to landowners, and 70% was done on rented land (Sandoval 2016). Thus, while soybeans offered the highest margins, most ranchers did not start cultivating themselves, but benefited from rising land rents. The rapid expansion of big firms leasing land to plant soybeans was made possible due to the relaxation of rules surrounding land and leasing markets during the 1990s (Chapter 4 in this book). Many "displaced" crop producers, became subcontracted *contratistas* instead—using their machinery to offer drilling, spreading, and harvesting services to the new firms (Baraibar 2014, 181; Manzanal and Schneider 2011, 11; Rossi 2010).

The big Argentine producers thus applied the same technological package and management forms in Uruguay as they had developed in the Argentine Pampas during the 1990s (Chapter 4 in this book). Besides subcontraction, this model involves state-of-the-art technology (satellite monitoring, the most efficient machines), big volumes yielding preferential treatment from suppliers and buyers (better prices and faster response and access to transport and storage requests), and geographical diversification to reduce the risks of spoiled harvests due to local weather conditions (Baraibar 2014). It also relies heavily on "tacit" assets—intangibles, such as knowledge about how to combine different kinds of resources, market information, and know-how (Carriquiry 2015, 30).

The Uruguayan producers soon adopted some of the management forms of the Argentinean producers; to work with professional agronomists when planning and monitoring production, to use specialized subcontracted agrarian services for planting, spraying, and harvesting, and to sell harvest on future markets (Oddone 2015). In general, the Uruguayan producers became more business-oriented and professionalized (Carriquiry 2015, 28; Errea et al. 2011). The farmers that have not been able to keep pace with the capital-intensive forms of production have been out-crowded, while barriers to entry and concentration have risen among crop and cattle producers alike. Some observers argue that the increased concentration is a good thing for Uruguay, since the ones that remain in activity are the ones that can compete, that are professionalized, that can take risks, that have bargaining power and can negotiate prices, as well as take position on future markets (CEO of Marfrig of the Southern Cone, Montevideo, 7 March 2010).

The new agribusiness model, where technological and organizational changes substitute labor for capital, particularly on-farm labor, made many family farmers and rural wageworkers superfluous (Oyhantçabal and Narbondo 2011, 6–7). In total, Uruguay had 68,200 rural wageworkers in 2013, which is a very small number considering the country's 16 Mha of productive land (de Torres Álvarez et al. 2014, 26). On-farm employment generation of the soybean production is lower than for other agrarian sectors, including extensive cattle-raising (Arbeletche et al. 2008). Rural depopulation has been a constant concern in the country since the early twentieth century, when *latifundio* and fencing expulsed people from the countryside (Chapter 2 in this book). The long-term trend of concentration was marginally reverted under the ISI-period with a peak of family farmers during the 1950s. Recent agrarian change has accentuated the trend of depopulation of the countryside since family farmers living on the land is decreasing and most on-farm employment is now made by *contratistas* living in rural towns. Accordingly, rural schools and other social service are closing down (Cirio 2011). While the countryside is becoming increasingly empty on people, the rural towns have become revitalized and become more prosperous, as demand for better-paid jobs, such as agronomists, economists, specialists in human resources, and other specialized professionals increased. The *Technological Oil-Seeds Table* (MTO)—which gathers almost all private actors involved in different stages of the soybean chain as well as the *National Agrarian Research Institute* (INIA), the *Faculty of*

*Agronomy* of the University of the Republic (FAGRO-Udelar), and the *Technological Laboratory of Uruguay* (LATU)[96]—published a study on employment in the soybean complex. According to the study, soybean activities generated 7500 on-farm employments and 8000 employments upstream (input) and downstream (Rosselli 2015).

Together with increased concentration, there has been a rapid process of foreignization of land. In 2000, it was estimated that only 10% (1.8 Mha) of Uruguay's land was foreign owned; by 2006, nearly 25% of all arable land (4.4 Mha) was foreign owned by firms from around the world (Redo et al. 2012, 129). These figures are estimations; it is hard to find reliable exact numbers because the Census only grasp nationality of PU that are registered on physical persons. However, since 2002, the majority of the transacted land had ended up on firms of limited liability (DIEA 2016, 143). The participation of FDI directed to buy land has also been high. It is estimated that 1.53 Mha of land was bought through FDI between 2003 and 2015, representing around 20% of land purchases under this period (Caon 2013; Oyhantçabal and Narbondo 2018). However, in the soybean complex, an important part of foreign firms has mainly leased, not bought, land.

The leasing form was particularly important in the early years of expansion, with the paradigmatic case of the Argentine firm *El Tejar*. The company expanded at an exponential rate on almost exclusively leased between 2003 and 2010, but then changed strategy, listed up at the stock exchange and used the inflow of—mainly Dutch—capital to invest in more fixed assets, such as land. By 2012, El Tejar was the by far biggest agricultural unit in the country. It also participated in societal arenas through "Corporate Social Responsibility" projects, providing funds to rural services and education. Moreover, the CEO, Ismael Turban, was the President of MTO, which since its creation in 2005, rapidly became an important arena for private–public coordination and information sharing regarding soybeans in Uruguay. In 2014, all of a sudden, El Tejar left Uruguay completely and Turban became the CEO of its expanding businesses in Brazil instead (Baraibar 2014; Kassai and Orihuela 2011; Turban 2013). This abrupt exit stood in sharp contrast to Ismael Turban's previous passionate talks about long-term commitment for inclusive sustainable development in Uruguay (CEO of El Tejar, Young, February 23, 2008). However, the combined effects of falling international soybean prices and new harder environmental regulation obligating crop rotations (see the next section), probably made the shareholders redesign the strategy and invest in places with lower labor, land, and transport costs. When El Tejar left Uruguay, the *Union Agriculture Group* (UAG) acquired most of its assets; a company founded in Uruguay with capitals from the US, Canada, and France. UAG became the largest agricultural PU in the country with 181,000 ha of agricultural land, producing diverse agricultural products. UAG is vertically integrated into inputs, logistics, and transport and trading. It has also diversified into energy and oil in several other countries in Latin America. More than 99% of the stock shares of the company are located in the British Virgin Islands.[97] Together with a decrease in international soybean prices, combined with new regulation, there has been a reduction in soybean crop areas and reshuffling of some big transnational actors.

The permanence over time of the new foreign actors in the face of further deteriorating margins is an open question. It is probable that the new crop firms, with lower exit costs due to less fixed assets and shareholders' short-term demands on high profit rates, respond faster to new market signals than the traditional domestic producers. In this way, the financialization of agriculture may have increased volatility. It seems clear that there is a temporal mismatch between time horizons of the stock market and amount of time it would take to repopulate the countrywide when the rural schools have closed down and there is little social service or infrastructure left (Cirio 2011; CNFR 2009; RAP-AL Uruguay 2009).

While short-term leasing contracts were important for the soybean expansion, the foreign companies in the forestry sector received important government support in the 1990s to buy the land they produce. There is a very high concentration of forest industrial companies with the biggest company *Montes del Plata* (a fusion of the Chilean *Arauco* and the Swedish-Finnish *Stora Enso*), owning 250,000 ha all over the country (Bussoni et al. 2019). The next biggest company is Finnish *UPM*, followed by two companies from the US, *Global Forest Partners*, and *Weyerhauser*. The top five forestry companies own 798,000 ha of land (Oyhantçabal and Narbondo 2018). The increased land area owned by foreigners provoked a critical discussion about foreignization; among some NGOs and researchers denoted as "land grabbing". Foreign investments in agriculture and land have nevertheless declined since 2015 (Oyhantçabal and Narbondo 2018). In addition, the recent policy changes and positive outlook in Argentina have accelerated the exit of Argentine investor groups (Sandoval 2016).

FA announced in its electoral platforms and other strategic documents that it aims to strengthen family farming in Uruguay. For the government period 2005–2010, it remarked that "land should be used as a social good, as it is the heritage of all Uruguayans" and this was made equivalent with the necessity to avoid "both underutilization and over-exploitation" (Frente Amplio, 22 December 2003, 52). At the same time, the newly formed *Special Meeting on Family farming* REAF of Mercosur from 2003 stressed the importance of family farming for inclusive rural development, the necessity for the governments to define it properly in accordance to each national context, and to design proper public policies to support it. REAF has continuously been an important arena for policy creation through the political dialogue between the Mercosur governments and with social organizations. Similar to Argentina, the government established definitions and a register for family farmers that made them eligible for specific funds and other targeted policies (Presidencia de la República Oriental del Uruguay 2017, 64).[98] Policies in support for family farming in Uruguay have largely formed part of a wider project for strengthened institutional capacity—involving decentralization, interinstitutional coordination, enhanced participation in decision-making processes, long-term planning, and social inclusion (Presidencia de la República Oriental del Uruguay 2017).

One of the most important institutional initiatives in this respect is the creation of the *General Office of Rural Development* (DGDR) within the *Ministry of Agriculture* (MGAP), to coordinate all public policies and programs in relation to rural development. This was in turn aligned with the 2007 *Decentralization and Coordination of Agrarian* law (No. 18.126), which resulted in three new institutions at different policy levels; National, Regional, and Departmental: the *National Agrarian Council* (CAN), the *Regional Agrarian Councils* and the *Rural Development Tables* (MDR). Moreover, specific sectorial tables were also created. The agrarian councils, at different levels, function as arenas for policy formulation and coordination between the MGAP, the *Agrarian Reform Institute* (INC), the *University of the Republic* (UDELAR), the *National Institute of Agrarian Research* (INIA), the *Agrarian Plan Institute* (IPA), the departmental and municipal governments, local producers' and/or community associations, the private sector, as well as the *National Institutes* of: Meat (INAC); Seeds (INASE); Milk (INALE); Wine (INAVI); Wool (SUL).[99] There are around 40 active MDRs, and they have around 450 civil society organizations participating in monthly meetings together with the other public entities (Villalba Clavijo 2015). Besides the above listed participants that need to participate by law, other public ministries and state enterprises often participate. Most public programs and projects in support of family farmers, rural wageworkers, and small rural communities have been designed and implemented through different local MDRs across the country (Presidencia de la República Oriental del Uruguay 2017, 71). When taking together all MDR projects across the country, the following are the issues most prioritized and worked with: rural housing, health service, electricity, education, clean water, animal health, protected areas and environmental concerns, rural roads, and access to land (Villalba Clavijo 2015).

This model is new in Uruguay, but the country has a long tradition of institutionalized arenas for policy negotiation and coordination between the state and different organized interest groups—not least the traditional producer's organizations in relation to agrarian policies (Chapter 2 in this book). These had, however, been weakened during the neoliberal "interlude" during the late twentieth century. FA reinstituted the state as an arena for bargaining and moderation between different interests. The corporatism of the twenty-first century involves some new elements, however, tainted by a mix of neostructuralist development ideas and the IO's "post-Washington consensus" agenda, with strong emphasis in decentralization and participatory processes, besides collective planning and negotiation, which marks an important break with Uruguay's centralist tradition. In this way, MGAP also created a *Decentralization Unit* to strengthen the Ministry's presence in the various geographical Departments and encourage the participation of local institutions in the design and implementation of agricultural development policies.

Many of the programs oriented toward family farming aim to integrate the producers in a competitive way in different value chains through extension services, credits for technological adoption, and access to new markets.[100] There has also been an increasing amount of funds related to creation of capacities to cope and adapt to climate variability and change, with support from the World Bank. In 2014, Uruguay passed a bill (No. 19.292) that declared family farming to be of national interest and established that in all public procurement of food products and services, the state needs to buy a percentage from family producers (registered as such), as a way to support them.[101] Regarding rural labor, FA has been active in bringing rural labor markets on a par with urban ones by extending most social and labor rights such as collective bargaining, the eight-hour working day, and regulated breaks. It doubled the minimum wage and created a special unit for rural labor within the *Ministry of Labor* (Baraibar 2014, 236).

Uruguay has also launched policies aiming to come to terms with the concentrated land tenure. One of the main tools has since 2005 been to revitalize the *National Institute of Land Reform*, INC, created in 1948 by the *law of Colonization* (No. 11.029), to promote land distribution, increase agrarian production and rural well-being (Diaz 2015). INC has distributed land in mainly two forms: a) by leasing out land to small-holders at low prices, or b) by providing access to credits and selling land to smallholders. From the 1970s and until FA entered the government in 2005, land reform was not a priority and INC had almost no funds or land to distribute (Diaz 2015). FA gave INC new life and more resources so that it could start to buy new land. In 2007, FA passed the law for *the Re-population of the Countryside* (No. 18.187), which changed articles 35 and 70 of the old *law of Colonization*, compelling all sellers of land units larger than 500 ha (of land quality equivalent to Coneat 100) to offer INC first and with preference over any potential buyer. However, despite the preference, INC lacked funds to be able to buy all the land it wanted. The rapidly rising land prices amplified the difficulties. In 2014, the government passed a new law (No. 18.756), to increase the budget of INC, through the creation of a trust fund and access to loans. As of 2018, INC had around 600,000 ha of land.[102] Around 27% of Uruguay's family producers today are *colonos* of INC, of which almost half are dairy farmers.[103] Most *colonos* lease land from INC (below market-value terms) and receive extension services. The *colonos* must live on the land, take care of the soils, and they are not allowed to lease out the land to third parties. Producers with previous experience and women are prioritized (de Torres Álvarez et al. 2014). The idea, according to the former president of INC (and former Minister of Livestock, Agriculture and Fisheries), is to use INC to allow for coexistence between different productive models (agribusiness and family farming) where the state acts as regulator of forces (Former Minister of MGAP, Montevideo, 6 March 2017).

Another FA-initiative to curb land concentration and "foreignization" was the 2007 Law 18.092, which established that anonymous corporations (corporate firms with bearer shares, or no-nominative shares) could not buy or lease land in Uruguay and should change into nominative shares owned by physical persons, who cannot be anonymous. However, this law was designed with generous mechanisms for exceptions (decree 225/07), for example, by presenting a productive project in accordance with the indicators of the investment law. Corporations with bearer shares (the majority of the *pools de siembra*) could thus be excepted from the law by adhering to the tax deduction criteria (show technology transfer, decentralization value-added activities, and employment generation) of the investment regime (Baraibar 2014, 201; 384). This was argued to provide the state with quite good possibilities to incentivize the big agribusiness firms to engage in development projects in line with the broader development interest of the government, and should be considered as one of several legal measures taken to come to terms with the new mega crop-producing companies (Former Minister of MGAP, Montevideo, 6 March 2017).

A progressively higher land tax on large landholdings was passed in 2012, in an explicit response to land concentration. This tax was however taken to Supreme Courte and ruled out in 2013 as "unconstitutional". A new law (Law No. 19088 and Decree No. 293/013) was passed in 2013. Under the new framework, the PU with direct or indirect assets above 12 million indexed Units (representing around USD 1.6 million in September 2013) must pay the tax. However, the value approximation of land is not based on market price, but on the *cadastro* (a national land register), which is estimated to be around one-third of market price. Around 1300 producers/firms owned in 2014 so much land that they must pay this tax, which brings in some USD 60 million a year in tax revenues. The traditional producers' organizations, ARU and FRU, criticized the tax, arguing that the legislation resembles the interventionist policies of the 1970s, which created inefficient and distorted markets. However, the government is also receiving a considerable amount of critique for not acting strong enough to revert the trend of increased concentration and "displacement" of family producers.

Taxes, overall, are an important instrument for social distribution of agrarian wealth. While Uruguay has no export taxes, FA implemented a large overall progressive tax reform, particularly on personal income.[104] At the same time the tax office, DGI, was strengthened and increased control for fiscal compliance in general and within the agrarian sector, it particularly increased the control of employers' contributions to the social security system, BPS, for their rural workers (Director Opypa-MGAP, Montevideo, 8 March 2017). The total fiscal pressure of the agrarian sector has been rather stable around 7–8% (excepting the crises years 2002–2003 when it was lower). Compared to other sectors of the Uruguayan economy the fiscal pressure in agriculture is still low (the average fiscal pressure in the economy is 30%). According to the director and former tax specialist at Opypa-MGAP, the state collects taxes from where it can, and therefore tax the most the things that cannot easily move away to evade paying taxes, which is why income and VAT are important sources of revenue (Director Opypa-MGAP, Montevideo, 8 March 2017). Farms pay a 25% *Corporate Income Tax* (IRAE) on their net taxable income, but the effective rate can be as low as 10–20% with allowable deductions, which cover all necessary expenses to obtain and maintain taxable income. Smaller farms, with presumed annual income below USD 36,466 for 2017, can decide between paying IRAE or *Farming Goods Sale Tax* (IMEBA). IMEBA can sum up to 2.5% depending on the good sold, but without any allowable deductions.[105] This differentiated tax regime for different types of agrarian units was justified in order to stimulate family farming (Tambler 2013). Uruguay lacks asset tax (even for corporations), has low property taxes, and no VAT and no sales tax on most supplies, machinery, and sale of farm products. As mentioned, Uruguay has an investment promotion regime since 1996 that offers important tax benefits for firms that apply for it, as long as investment fulfills the following development criteria, which will be discussed in the next section on upgrading.

It is clear that the governmental programs designed to support small-scale family farmers and to disincentivize concentration did not compensate for the other stressors working in the opposite direction, such as rising soybean and beef prices, rising land values, and the important amounts of economies of scale involved in industrial agriculture, particularly in the soybean RR model. It is, nevertheless, impossible to know how the situation would have looked without these policies.

The *National Commission of Rural Development* (CNFR), representing tenants and small farmers since 1915 (Chapter 2 in this book), is the producers' organization that has voiced the clearest critique against the more concentrated and foreign-owned land tenure structure. The motto of CNFR is "a productive Uruguay WITH people living on the land" and it pledges the state to provide the conditions for family farmers to be able to compete successfully with the big firms and to revert concentration (CNFR 2008, 2010). At the same time, in the wake of decentralization and the institution of development tables, CNFR is more active than ever in various joint arenas with the public sector, defining policy lines. The loudest critical voices against the state in relation to the shift in social relations come from small NGOs advocating agroecology and food sovereignty.

> The role of the state should be to moderate and balance the relations of force, to protect the smaller from the big, the weaker form the strong, that is the way to do it in order to in the end have a more balanced system, which takes into consideration other values than pure market values—social values and ecological values. But the state has largely resigned from this role and instead allowed for the relations emerging from the market to rule alone. (Advisory Director DINAMA-MVOTMA, Montevideo, 6 March 2017)

As mentioned in above quote, critical voices express that the state ought to do much more to protect the "losers" of current agrarian change. Not only are the programs in support for family farmers found insufficient, but also the state is criticized for facilitating for agribusiness expansion, not least through public investments in export-infrastructure (for example the National Plan of Silos and dredging the ports) and in generous tax deductions (for example, the Free Trade Zones and the Investment Promotion Regime).

In the early years of the soybean expansion, the producer's organization the *Rural Federation* (FR) from 1915 also expressed critique to the rapid foreign driven soybean expansion (President of FRU, Montevideo, 3 March 2009). However, after the 2008 financial crisis, combined with increasing polarization between the FA government and "the countryside", FR took on a discourse more similar to the *Rural Association of Uruguay* (ARU) from 1898, against state interventionism and against taxes (Lussich 2009). ARU and FR have organized different forms of resistance against governmental redistributive initiatives. They were for example critical against the law against foreignization of land, against the rural wage councils, the laws strengthening INC, and the progressive land tax. One recurrent argument against these reforms has been that they violate private property rights. The relation between the government and the producer's organization became increasingly tense. However, the strongest "rural" critique emerged outside the traditional arenas. A movement called the *self-convened* (*autoconvocados*) emerged organizing spontaneous manifestations and road blockades against what they express as high taxes on agrarian-generated income, a huge state apparatus, and high production costs.

## 5.4.2    The Strategy of Upgrading Primary Production

During the early years of FA in government, it expressed clearly that the path to development passed through a change of the productive structure away from reliance on a few agro-commodities, and to activities that are more industrial. This echoed the neostructuralist development ideas. However, the more radical critique toward the agribusiness-led agricultural model was increasingly disappearing alongside the years in government and rising commodity prices (Baraibar 2014). Public–private arenas, such as the technological oilseeds table MTO, has also been active in questioning the traditional dichotomy between industry (assumed as more advanced, diversified, and with higher value-added) and agriculture (assumed as "premodern", proto-capitalist, and giving rise to "Dutch disease" problems). However, according to a study on behalf of MTO, the value-added of soybean exports is actually very high, arguing that a lot of technology and know-how is incorporated along the domestic value chain; out of each USD 100 of soybean exports, USD 70 are estimated value added (Rosselli 2015). The idea that development requires industrialization and the a priori understanding of agrarian activities as of low added value has thus started to change. Since the East-Asian rapid growth as a major world manufacture supplier, prices of manufactured goods have declined and become increasingly volatile. At the same time, agrarian commodities include a rising amount of advanced technology (biotechnology, satellite monitoring, and sophisticated software), and sophisticated organizational and management systems. Moreover, commodities have had a long super-cycle of high prices (2003–2014). In this way, the classical divide between agricultural and industrial sectors has become less relevant. Instead, the neo-developmental governments have tried to make use the commodity booms as springboards to promote upgrading; making each productive value chain to incorporate more technology, specialized knowledge, and other forms of value-added. An illustrative description of the main policy line of the FA government in relation to the agribusiness advancement in the country was provided by the president of INASE (at the time), who then became the president of INIA (2010–2012) and currently is the minister of MGAP (2018–2020):

> The multinational corporations are knocking the door every day. We say to the companies; 'we can be your partners if you want to come and produce here, we will not limit your business, but I want you to do the production here, I want you to bring capital, to bring technology and for the

Uruguayans to participate'. The big firms trust that we will respect private property rights, that we have strong investment laws that are very clear, and that we promote foreign investment. We also have mechanisms of extra support if they settle in some other region than Montevideo, if they employ national labor force, if they bring technology. We have a very solid institutional structure in Uruguay and the big firms know that private property rights are respected and so is all laws and deals. This has attracted many investors to Uruguay—and we want that—but not indiscriminately. It is not the same if a company comes and is installed in Montevideo as if it is installed in Tacuarembó; it is not the same if they use domestic work force as if they don't; it is not the same if they bring technology as if they don't. That is our strength as a country, to negotiate that and then our credibility to follow the deals made. Let's see if I can make myself clear, I do not say unconditionally welcome to Monsanto nor to Nidera. I say; 'What do you bring and what do you leave here?' If they only come as firms to produce soybeans and export it, what do they leave us? A piece of eroded land? That is no business, in that shape they would better not come. Now, if they come to produce elite seeds, of first class, to export it to the region and the world, including the developed states, then that is different. That is the basic line. (President of INASE, Montevideo, 6 March 2010)

Above quote synthetizes the official policy line in relation to the new agrarian scenario. It is almost like a schoolbook example of the neostructuralist approach, its combined faith in export-led growth and long-term state planning and interventionism for improved market insertion and development. An important instrument in this respect has been the investment promotion regime. The previously mentioned investment promotion law (16.906) based on generous tax exonerations, including the *Commission for the implementation of the Investment Law* (COMAP), launched by the Colorado government in 1998, has been widely used by the FA administrations. It was nevertheless slightly modified through two decrees (455/007 in November 2007 and 002/12 in January 2012) in order to facilitate more "developmental" investments; that is more labor generating, more technology advanced, and environmentally benign (Former minister of MGAP, in Montevideo, March 2017). In this way, the amount of tax deductions depends on how much the investment project scores on the following criteria list: 1, employment generation (30%); 2, contribution to research and development (R&D) and innovation or use of clean technologies (20%); 3, increase of exports, impact on GDP and local value added (15%); 4, contribution to decentralization (15%); 5, improvement of social indicators (20%).[106] The idea of these indicators is to add value and to incentivize investments that help to develop the productive structure toward a direction that creates knowledge intensive high-quality jobs throughout national territory (MIEM, Cadenas de valor).[107] Tax benefits may include exemption from import duties on fixed assets (as long as these are noncompetitive with the national industry), VAT refund for local purchases of materials and services for building construction, and relief from IRAE.[108] According to a former Minister of Agriculture, the government uses the investment law to make the agribusiness firms contribute more to the social and economic development of the country:

> These conditions aim at generating more work, a better distribution of income, improved levels of technological production and more value added, and the companies coming here to produce need to work in line with these aims. We have some tools, but as with everything, we don't always succeed. The idea is to at least put pressure on the companies so that for example the soy is not only exported as beans, with a very low level of value added and low level of labor generation. (Former Minister of MGAP, Montevideo, 6 March 2017)

As expressed in above quote, the government tries to change the rules of the game, at least slightly, by putting up other than pure market-based conditions, in the benefit of upgrading. However, critical voices, also within FA, claim the possibilities of exonerations are still high for almost any type of investment, including projects that involve harmful environmental practices and generate very little labor opportunities or value-added (Advisory Director DINAMA-MVOTMA, Montevideo, 6 March 2017). Moreover, the lion's share of soybean exports remains as whole beans. No less than 40% of all tax reductions granted by the Investment Promotion Law between 1997 and 2013 were concentrated in the agro-industrial sector (Paolino et al. 2014, 11).

The generous possibilities of tax exonerations through the investment promotion law can be seen to compensate the bigger agrarian firms for their heavier tax burden since the tax reform (Director Opypa-MGAP, Montevideo, 8 March 2017). Besides tax reductions from the investment regime, the use of Free Trade Zones (FTZ) increased substantially under the FA administrations—although FA criticized the FTZ-law when it was launched in 1987. It argues, in line with previous administrations, that the FTZs can become important industrial poles, offering important opportunities for national industry and with strong linkages to the rest of the economy. However, the official statistical figures show that most FTZs are dedicated almost exclusively to warehousing of agro-exports.

In this way, several agribusiness firms have received important tax exonerations. Moreover, through the *state office of budget and planning* (OPP) and COMAP the state has proactively taken contact with the big agribusiness actors and taken several initiatives for public–private clusters for upgrading. Moreover, a new interministerial arena (with participation of OPP), the *National Productive Cabinet*, was created to transform production in ways to promote innovative, sustainable, and inclusive economic development. This cabinet has since 2016 received a clearer mandate (Law No. 19.472) to support a transformation of the country's productive structure through the incorporation of more innovation, technology, and value added (Presidencia de la República Oriental del Uruguay 2017).

In a similar way, the previously mentioned *oil-seeds table* (MTO) has been an important arena for joint private–public collaboration, in order to upgrade and incorporate more value added in the soybean chain (Barrios et al. 2010). This arena is described as vital for the private sector to make the public sector invest in infrastructure and R&D in benefit for the soybean chain. In this way, the relatively high public and private spending on innovation linked to agriculture (public spending on soybean seed innovation is the double of private spending) have for example resulted in that the technology content of soybean export is considered fairly high, in spite of being a primary commodity (Paolino 2015, 43). In the same way, private sector compliance with environmental and other regulation in relation to the soybean business is found to have been benefitted by this public–private space (Baraibar 2014). According to a report from the World Bank, Uruguay has managed to move up the value chain with respect to some of its products and also diversified into new products and markets (IBRD 2015). It also argues that the agriculture sector has experienced an outstanding transformation. Its products are associated with sustainable agriculture practices and clean, green, growth, enabling the country to penetrate high-price niche markets (IBRD 2015).

Uruguayan public policies in support of making agriculture "developmental" is also focused in the creation of "public goods". One of the government's priorities in this respect was to restore international confidence in the high quality and safety of Uruguayan meat. After the outbreak of foot and mouth disease in 2001, many well-paid markets closed down. To build up trust for Uruguayan beef in high quality and value markets, a wide range of public programs and regulations were launched, with a strong traceability system as the flagship policy.[109] Between 2007 and 2011, Uruguay, through its *National Meat Institute* (INAC), incorporated all cattle into a system called the *Electronic Information System of the Meat Industry* (SNIG). Uruguay became the first country in the world with a complete and mandatory traceable meat process (the *System of Identification and Animal Registration Law*, No. 17,997 and Decree 266/2008). The state delivers the identification devices to producers free of charge. The system involves all types of livestock producers, livestock auction locations, and slaughter plants in the country (more than 75,000 units), and through it each beef cut can be traced back to the animal from which it was originated. SNIG collects detailed information about each animal through an ear tag that all calves receive at weaning. This collects information about all the places where the animal has been since birth (including data on productivity of land and types of pastures); all the other animals it has encountered during its lifetime, and vast socioeconomic information about owners (Osorio 2009). The statistical information is in real time and a web application gives different involved actors (breeders, brokers, industry, veterinarian services, and police) information related to their activities, providing transparency to the system.[110] In 2016, 33 million heads of cattle had received ear tags since the system was launched. The ex-minister of agriculture and livestock (2009–2010) and deputy of the senate talked about the system in the following terms:

> Uruguay is an expensive country; everything is expensive here. We have the highest salaries in the region, and we want to have the highest salaries. So we cannot compete with price in the world market; we have to compete with quality. But we can't go out on the world market and just say that Uruguayan meat is best, which it is, because the advanced markets don't trust us, we have to prove in the most transparent and rigorous way possible that we are doing the right things all along the productive chain. It costs money to prove your innocence, so here the Uruguayan state has

carried the main part of that costs, in order to allow for our producers to access the best markets and the best prices. (Former Minister of MGAP, Montevideo, 6 March 2017)

As illustrated in above quote, while the system cost the state a lot of money, it is expected to generate higher income and is part of a broader development idea of inserting itself in the world market in such a way that it allows for well-paid jobs. In concrete, Uruguay uses the traceability to differentiate its product from other global competitors and receive price premiums (Presidencia de la República Oriental del Uruguay 2017, 82–89). The system has allowed for access to the *EU 481* quota for feedlot cattle meat (Chapter 3 in this book),[111] which requires individual traceability, and the *Hilton Quota* of high value, fresh and chilled, boneless cuts of beef that comes from grass-fed and free-range systems (USDA 2017b). Moreover, in 2018, the most demanding market, Japan, reopened its market to Uruguayan beef.[112] China is the number one market for Uruguayan beef. After the implementation of SNIG, Uruguay has a USD 100/t price premium over world beef prices, which is argued to has been facilitated by the introduction of the traceability technology that enables access to high-end markets (IBRD 2015). In short, the traceability is believed to have helped to reinforce the country's reputation as the world's premier supplier and producer of quality beef with good sanitary status. Besides the implementation of SNIG, the Uruguayan government has worked to open up third markets through promotion travels, diplomatic means, and trade negotiations.

The "upgrading" policies of FA seem to have yielded access to markets that pay better prices for higher quality goods (MGAP 2017). Within the wide label of "quality", the state hopes that both safety, taste, and environmental aspects count, and will eventually payoff through access to the best-paid markets (Presidencia de la República Oriental del Uruguay 2017, 66). In this way, the government argues that in the end, there would not necessarily be any tension between agrarian growth and sustainability. At the same time, there is awareness of the fact that most consumers in the world are not willing to pay a higher price for guarantees of strong environmental protection along the production chain. In addition, even the consumers who are willing to pay a higher price, do not trace the whole chain, and there is almost no consumption pressure on "inputs"—such as soybeans for feed.

Many of the upgrading policies have benefitted agribusiness expansion. The regime for investment promotion, the extended number of free trade zones, the increased targeted blends for production of biodiesel; the public investments in infrastructure, the authorization of new GM events, and the construction of a strong intellectual property right framework, are a few examples of public policies—and often of public funds—benefitting disproportionately the big firms. Moreover, the government has maintained a relatively low fiscal pressure on agriculture (vis-à-vis other sectors), and not least maintained the absence of export taxes (Baraibar 2014; MGAP 2017).

### 5.4.3   Shifts in Forms of Production and Regulation

As one of the world's richest grasslands, Uruguay has undergone profound land-use change in the last decades. Some changes share many features with the Argentine Pampas *sojization* model, but as mentioned, Uruguay has also had a process of afforestation and pastureland intensification. While there are many important interacting effects between the different productive orientations and their respective technological packages (so-called cocktail effects), the most heated societal debate has been about the soybean model and its massive use of glyphosate, which largely reflects the same conflict lines as expressed in Argentina and Paraguay.[113]

The pace of soybean expansion was spectacular between 2003 and 2015, with a short dip after the financial crisis in 2008. The increased competition for land drove up land prices. Between 2000 and 2015, land prices multiplied by seven (Oyhantçabal and Narbondo 2018).[114] In a corresponding way, prices of land leasing have increased from an average of USD 476/ha in 2007, to USD 721/ha in 2014 (DIEA 2016, 144, 150). As mentioned, rising land prices tend to increase concentration, but another equally important effect is that land becomes perceived a scarce resource that needs to be used more effectively than in the past (Carriquiry 2015, 28). The higher land values increased the incentives to invest more capital and technology in the land, and changed the relations between livestock and crops, as expressed by the former CEO of El Tejar in Uruguay:

In this region, land was worth USD 200 and leasing between 15 and 17. So, the farmer preferred to but more land or lease more land rather than improving it. [...] In this way, it was easy to grow outwardly. That is what happened historically. The cultivations were always perceived as a necessary evil as we identified ourselves as ranchers. Historically, what happened was that the rancher rented out some cultivable land to crop producers at the end of the period of pastures. It implied some income, but more importantly, it prepared the land for livestock again, and that was the main objective. Now the value relations have changed. You can no longer use crops as a mere supplement to livestock when you have to pay five times more for the land. (CEO El Tejar, Young, February 23, 2008)

Thus, as expressed in above quote, the traditional, extensive, low technology, experience-based, and risk-adverse production model (Chapter 2 in this book) started to disappear, as well as the traditional subordination of crops to livestock. The dominant system of the fertile *Littoral* went from mixed crop-pasture systems to continuous cropping. This shift was driven by changes in price relation, but also by the fact that the Argentine crop firms that led the soybean expansion were used of doing continuous crops in the more fertile Pampa Húmeda (where mixed systems already had started to disappear more than a decade earlier). In the early years of the expansion, some producers did not even rotate soybeans with a winter crop, but planted soybeans over soybeans (Arbeletche and Carballo 2009). This provoked a rapid response of erosion, resulting in lower yields, which made most producers rotate soybeans with other crops, albeit seldom with pastures. Uruguayan researchers soon came to see, however, that even continuous crops systems under no-tillage created erosion (Ernst and Siri-Prieto 2009; García-Préchac et al. 2004; Bidegain et al. 2010; Prieto and Ernst 2010; Salvo et al. 2010).[115]

Besides erosion, the soybean expansion in Uruguay also led to an exponential rise in the use of glyphosate and other agrochemicals. As natural enemies and other biological control agents became wiped out by insecticides, more resistant and secondary plagues show up on the fields and more toxic agrochemicals have to be applied to the crops (Bruno 2007). The agrochemicals are found to reach nontarget species and provoke hazardous effects on flora, fauna, and aquatic systems (Oyhantçabal and Narbondo 2011; Ríos et al. 2010). The use of fertilizers and pesticides also contribute to erosion as they kill the organisms that bind the soil together. Loss of biological diversity and weed resistance from the wide use of glyphosate are other concerns (Ríos et al. 2008, Ríos 2011). More recently, the use of 2, 4-D to eliminate Glyphosate-resistant weeds have also shown significant increases (GEF 2015).

In a similar way as in the Argentine Pampas, a strong debate emerged about the health effects on animal and human life nearby the sprayed fields, as well as due to high chemical residues in food. The *Producers' Organization of Small and Medium Size Farmers* (CNFR) engaged in the critical discussion on the massive use of agrochemicals, pinpointing the soybean production as the biggest problem. The *Uruguayan Association of Beekeepers* (SAU) demanded stricter regulation on glyphosate and several insecticides, both because of direct harm on the bees and because of high levels of agrochemical residues in the honey, making it impossible to sell.[116] An important volume of exports of honey to Germany has been rejected and sent back to Uruguay on various occasions because the levels of glyphosate residues were several times higher the accepted levels in the EU. Besides harm from the agrochemicals, poorer habitat indirectly harms the bees, as soybeans have displaced grasslands of high floral resources. Moreover, it is increasingly difficult for beekeepers to access land since no one is on the land to coordinate about how and when they could come with their hives. Thus, similarly to the Argentine Pampas, *Sojization* in the Littoral provoked a strong societal debate about pollution of water, biodiversity loss, glyphosate residues in food, and other health hazards caused by the skyrocketing use of agrochemicals.

The firms and farmers themselves involved in different stages of the soybean chain, represented in the *Uruguayan oilseeds table* (MTO), have also engaged in the debate. They acknowledge some of the expressed concerns over polluted waterways, cases of intoxication and risks of erosion, but they argue that these problems basically emerge from "bad practices" by some farmers, which in turn depend on information deficit, rather than inherent problems to "the model" itself. At the same time, many soybean farmers of all strata also defend "the model". For them, the criticism against agrochemical spraying comes from urban sectors that do not know about the rural "reality". In this way, the discussion about the soybean productive model has tapped into other discussions and divisions, where the strong rural–urban divide, and love–hate relationship, in a similar way in Uruguay as in Argentina.

As mentioned, Uruguay has not only had a dramatic soybean expansion during the past one and a half decade, but also a substantial increase in beef production, in spite of less pastureland.[117] Livestock area shrunk from 10.2 to 8.4 Mha, due to loss of land to crops and afforestation, between 2000 and 2013, while cattle stocking rates increased on average by 8.6% in this period (Bussoni et al. 2019). An important part of the

Uruguayan beef production moved away from its traditional extensive ranching model, based almost completely on natural or seminatural pastures (Barrán and Nahum 1986), to greater incorporation of improved pastures, use of sophisticated large-scale machinery, irrigation, fertilizers, and supplement feed.[118] Moreover, some big meat companies made strategic alliances with the "network" crop firms, in which contract farmers linked to the meat companies get access to feed below market price (CEO El Tejar, Young, February 23, 2008). In a similar way, there was a rapid modernization and intensification of dairy production, counteracting the effects of area reduction (Carriquiry 2015, 27).

However, as mentioned, land-use intensification also brought new concerns. The augmented intensity, with more animal on less land, has increased pressures on both soil and water resources. A substantial proportion of livestock manure is wasted or discharged in the environment. Animal feces from effluents and from livestock leaving manure in the water cause water quality impairment and contamination of drinking water, particularly through high levels of nitrogen and phosphorous (Researcher in limnology, CURE-Udelar, in Punta del Este, March 2017). Cattle moving freely in river and lake areas also destroy a stream bank and the streamside vegetation, as they walk down the bank to enter the river. It increases the sediment load into the river along with all of the manure and urine. Sometimes dead animals in the water contribute to the problem. During the past years, there are increased societal debates and concerns over poisoned waterways from agrochemicals and waste (GEF 2015). At the same time, the pulp mill plants from the forestry sector also emit important amounts of phosphorous. In addition, afforestation has altered water cycles, decreased the volume of the water flows, which has led to a higher concentration levels of phosphorous. Moreover, from agriculture, both agrochemicals end up in the water, spread by the wind and processes of erosion and sedimentation (MVOTMA 2017, 132). Uruguay has severe problems with Cyanobacteria growth, threatening to contaminate the drinking water. In 2015, the drinking water from Río *Santa Lucía*, whereof more than half of the Uruguayan population get its drinking water, started to have outbreaks of bad taste and smell. This created a scandal among the population and water issues became high on the agenda. The problem was attributed to a process of accumulation of nutrients—particularly nitrogen and phosphorus—steaming from both agricultural and industrial activity (MGAP 2017). While soybean expansion, afforestation, and livestock intensification, each contributes with specific new social-eco-logical concerns, the combined effects—the cocktail effect—put snow-balling pressures on natural resources (Avila-Vazquez et al. 2018; Demetrio 2012; Díaz-Zorita et al. 2002; Modernel et al. 2016; Nardo 2011; Novelli et al. 2017; Oliveira and Hecht 2016; Primost et al. 2017; Ronco et al. 2016). Moreover, since agrarian activities are so important in Uruguay, at the same time as the country lacks a strong industry sec-tor, agriculture represents 80% of the National greenhouse gas emis-sions—compared to the world average of 13% (MGAP 2017).

Public policy responses to concerns over erosion, biodiversity loss, con-tamination, and generally in relation to natural resources in the wake of agrarian change have been uneven, with some areas more characterized by laissez-faire and others of innovative regulations. In general there have been important advances in legislation regarding biodiversity, climate change mitigation and adaptation, agrochemical regulation, food safety, and sustainable management of soils and water resources (MGAP 2017). As in Argentina and Paraguay, Uruguay has implemented a national sys-tem of protected areas (SNAP). As of 2019, the country has 29 protected areas covering 3% of national territory.[119] While environmental concerns have increased in policy importance, "sustainable intensification" is the overall lead motto of Uruguayan public policies in regard to agrarian change. The government has said that it has taken the challenge to make Uruguay—a small country with a population of 3 million persons—in 2050 produce food to 50 million people. In this way, public policies have focused on productivity increase, while combined it with regulation for sustainable management of the main productive factors (water and soils).

There have also been important shifts in the institutional structure and policy-formulating processes regarding environmental protection, with the creation of many new arenas and new forms of decision-mak-ing. As in the case of social and distributive policies, these are character-ized by a strong belief in decentralization, broad participatory processes, and interinstitutional coordination and negotiation. Politicians, pub-lic officials, researchers, producers, civil society, and citizens are con-sulted in broad calls with workshops for capacity building, negotiation, and consensus-creation. While some arenas are formed ad hoc to spe-cific projects, typically engaging thousands of actors participating in a consultative process, many transversal new arenas have also been estab-lished and institutionalized. For example, a new *National Environmental*

*Cabinet* has been established with participation of the Presidency of the Republic and leaders from the all relevant ministries to create integrated and equitable environmental policies for "a sustainable national development that is territorially equilibrated" (Presidencia de la República Oriental del Uruguay 2017).

Uruguay has a rather strict legal regulation for protection of the soils in *Law of Use and Conservation of water and soils* (No. 15.239),[120] which provides the state with clear responsibility to prevent and control the erosion and degradation of the soils. This law was passed already in 1981, and at the time, it marked a break with previous notions of landowners as entitled to do whatever they wanted with their land in the name of private property rights. The new legislation, however, gave the state authority to limit the power of the property holders in the name of soil conservation and erosion prevention, which still makes Uruguay to stand out in relation to its neighbors. The bill, perhaps counterintuitively, was passed under the military regime (1973–1985), but probably it would have been harder for a democratic government to pass a law that by many were seen to violate basic private property rights (Director of SARAS, Punta del Este, 10 March 2017). The content of the law, however, was not built on the expertise of the military regime. Quite the contrary, it was built on the extremely ambitious work of the agrarian commission under CIDE in the 1960–1970s, generating deep site-specific knowledge about its soils (Petraglia et al. 1982). As mentioned in Chapter 2, this gave a new empirically based understanding of the alarming rates of erosion, threatening the natural resource dependent economy as a whole. While the 1981 law provided the state with strong tools to prevent erosion and land degradation, it was not implemented properly. Moreover, an important amount of the research of CIDE was gradually forgotten, as conservation of natural resources was not prioritized. Many of the researchers that had been involved in the work left the country under the dictatorship, and there was likewise an important brain drain in the universities. After re-democratization, the neoliberal governments did not take an interest in safeguarding the previous work, as the state in general was taking a step back from agrarian development planning (Former researcher at CONEAT, Punta del Este, 11 March 2017. Thus, decades passed by without much consideration of the soils. MGAP had almost no resources, or staff with knowledge and capacity to control compliance.

However, as a new wave of soil degradation and risk of erosion emerged when soybeans broke the mixed farming systems and replaced it with continuous cropping, while under no-tillage, the long-lasting preoccupation of the soils became revoked and erosion returned to the policy agenda (Paolino 2013). In 2008, the government passed a decree (No. 450/008—*Sustainable and responsible use of the soils*) that was an attempt to put the 1981 *Law on Use and Conservation of water and soil* into force. It gave the state authority to demand producers to make plans for land-use and land management for sustainable production systems with proper use, conservation and recovery of soils while minimizing soil erosion (Hill and Clérici 2013a; Bidegain et al. 2010). An inspection team was installed and fines were increased for producers who did not take care of the soils (Law No. 18.564). At the same time, the state together with the *Faculty of Agronomy* (FAGRO), started to rebuild research and capacity to increase the understanding of the situation of Uruguayan soils and the need to mitigate erosion. Moreover, during several years, the state, researchers, and producers engaged in a participatory process for the developing and testing of a new policy tool for law enforcement.

This work culminated in the government's flagship agricultural policy regarding the protection and conservation of soils is since 2013, the mandatory *Soil Use and Management Plans* for croplands (Presidencia de la República Oriental del Uruguay 2017, 82). All producers must present a soil management plan, including a plan for rotations for five years ahead, to the *general division of Natural Renewable Resources* of MGAP (RENARE), where they are geo-referenced and a traced through a Satellite-based monitoring system. The plans are filled online by an accredited agronomist, using a software called Erosion 6.0. The software has an equation built in it, based on a mathematical model; the *Universal Soil Loss Equation* (USLE), which estimates the soil erosion that occur under defined situations (MGAP 2017, 67–68).[121] USLE builds on international research, but is adapted and calibrated for Uruguay's soils based on previous research of CIDE, as well on new research from FAGRO, INIA, and MGAP (Member and researcher of the CONEAT team and current consultant to the soils division of MGAP, Punta del Este, March 2017). The software matches the submitted data with soil quality maps for that particular piece of land, based on drones and satellite imagery as well as previous research (Hill and Clérici 2013b). The equation tells if soil loss rates are acceptable, that is below the stipulated tolerable level, or if land-use and management need to change in order to preserve, restore or increase soil organic matter, and minimize erosion (Hill et al. 2016). Thus, the system is designed to make the producers diversify their productive systems to prevent losses of soil over a preestablished tolerated threshold.

In 2016, 14,944 plans were submitted corresponding to an area of more than 1.5 Mha of cropland, representing 96% of total cropland (MGAP 2017). Compliance is controlled and monitored through satellite images, and noncompliance is sanctioned with a fine or the suspension of the license to practice agriculture (de Torres Álvarez et al. 2014). In this way, the country moved toward fully implementing for the first time, the 1981 *Law of Use and Conservation of water and soils*. It is estimated to have contributed to a slight reduction of soybean area, mainly outside the Lítoral (since these traditional livestock soils are in general less apt for crop cultivations), which now have converted back to pastures (Sandoval 2016). While the soil loss rate is the exclusive parameter considered, the plans are argued to provide data that will be used for the development of ports, road transport logistics, the use of water basins and other areas and for planning the country's natural resources. The explicit aim of the regulation is "sustainable intensification" in line with the broader aims of MGAP (Bidegain et al. 2010).

The obligatory plans for rotation constitute a flagship of Uruguayan public regulation. Government authorities often stress it as an illustration of the capacity of the Uruguayan state to create institutional solutions to the growing pressures on natural resources that emerged in the wake of the crop expansion (Paolino et al. 2014). According to the Minister of MGAP, Uruguay is the only country in the world with a science-based regulation that guarantees to avoid soil loss. One often-stressed reason to that Uruguay managed to implement the policy successfully is the participatory processes preceding the law. For example, the Oilseeds table (MTO) with members from the all big crop firms, and representatives from FAGRO and INIA, had already been discussed the problem of erosion and soil degradation resulting from continuous cropping, and a consensus-made understanding for the need of more diversified rotation schemes had started to emerge. MTO had also published manual for "good agricultural practices" pledging for more diversified rotations. In this way, the majority of the private actors accepted and in much agreed with the obligatory plans, and only a few voices criticized the reform for putting on unnecessary costs and work on producers, and violating private property rights. Probably, the long tradition of research on soils and the long tradition of putting faith to science also played a role.

Public policies have prioritized crop productions systems. Water resources have been less considered. However, the regulations to curb erosion are not only important for the soil, but also for water resources, as erosion ends up in the water and is the principle source of contamination from agrarian activities (MGAP 2017). Moreover, the *Soil Use and Management Plans* have also been made mandatory for dairy farmers in the watershed of the Santa Lucía River, including a plan for both rotation (to curb erosion and sedimentation) the use of nutrients, in order to make sure that phosphorous levels in the water are below a predefined tolerated level and to protect the water quality (MVOTMA 2017). Producers are also mandated to make an obligatory declaration and plan for management of effluents. Family producers can receive economic support, covering up to 80% of the costs, for installing technological improvements to manage effluents (MGAP 2017). There are also installed buffer zones areas in the *Santa Lucía* basin, to restrict cattle from entering the water, and/or manure to flow down into the water with the rain. In 2019, the *Ministry of Housing, Land Management and Environment* (MVOTMA) announced that it wishes to extend the buffer zones to all waterways. Institutionally, policies to protect the water resources have been strengthened Uruguay through the 2013 *Law concerning hydro biological resources* (No. 19.175), which declared conservation, investigation, and sustainable development of water resources of general interest. Moreover, the 2009 *National Water Policy Act* (Law No.18 610) created a roadmap for integrated water policies and establishes that the country need to develop a *National Water Plan* based on three long-term commitments: water for sustainable development; access to water and sanitation as a human right; and the management of the risk of floods and droughts (MVOTMA 2017). The work with the plan started in 2010, following a participatory process involving some 2000 people and led by the newly established *National Directorate of Waters* (DINAGUA). The work culminated with Executive Branch passing the plan (Decree No. 205/017) in 2017 (Presidencia de la República Oriental del Uruguay 2017, 63).[122] The plan gathers all commitments related to water undertaken by different public institutions and unite them under three major objectives, and establishes guidelines for the development of regional and local plans. It also includes an analysis of the current situation of all water resources in Uruguay, and points out how they should be solved (MVOTMA 2017). The increasing pressure from the intensification of agriculture and livestock activities, as well as forestry is highlighted, and there are several policy lines to prevent agrochemicals, effluents, and hydrological erosion to contaminate the water. At the same time, MGAP has created policies to promote irrigation (Paolino 2013), which may increase stress on water resources.

Regarding agrochemical spraying, Uruguay has, compared to Argentina and Paraguay, rather strict rules with demands on a safety distance between urban centers or rural schools of at least 500 meters from aerial agrochemical application and 300 meters from terrestrial applications, as well as a "buffer area" of 50 meters distance from any waterway (MGAP 2017).[123] However, compliance with regulations of safety distance and buffer zones is put on doubt by many observers. The main mechanism to for compliance (as in Argentina and Paraguay) is that any citizen can file a complaint about violation of the agrochemical rules to the *Division of Agricultural Services* (DGSA) of MGAP, which in turn can fine producers that do not follow the rules.[124] The same mechanism is used for all types on environmental crimes; in line with the idea that it is inefficient and costly for the state to do inspections, it instead relies on citizens that fill in complaints. However, nested and dependent relations in the countryside often hamper people from lodging a formal complaint despite violation of rules (Advisory Director DINAMA-MVOTMA, Montevideo, 6 March 2017; Researcher FAGRO-Udelar; Stockholm, 23 June 2018). In addition, large productive areas in the Uruguayan countryside has become practically empty of people, and thus there are no people watching, or able to witness about broken environmental rules. In this way, there have also been many cases of animal cadavers lying for weeks in a river without anyone noticing, and causing severe problems in the water. For example, the politicians in the department of Canelones, Uruguay, prohibited all pesticide use in the *Laguna del Cisne*.[125] At a national level, Uruguay is currently implementing a system of satellite monitoring of agrochemical use, using technology that links agrochemical machines, using geo-referenced data, in real time to DGSA, for full control of compliance (MGAP 2017). In this way, the Uruguayan state seeks technological solutions that can oversee compliance without need for costly inspections or people living in the countryside that can fill in complaints.

In December 2018, Uruguay also passed a Law for a *National Plan for Agroecology*, for the promotion and development of agrarian production and products based on agroecological principles (Presidencia de la República Oriental del Uruguay 2017, 63). The law has been discussed for many years and is an initiative from several ecologist organizations mobilizing against agribusiness and industrial agriculture, as well as interdisciplinary networks of environmental researchers.[126] A new commission has been formed under MGAP in charge of coordinating more concrete policy lines during 2019, through a participatory and interinstitutional process (including stakeholders from the Ministries, INC, INIA, municipalities, the University of the Republic, civil society and producers' organizations). The basic philosophy of agroecology is based on ecological knowledge and stands in sharp contrast with the dominant productivist paradigm of agribusiness led agriculture. The objectives are to strengthen food sovereignty, based on local, small-scale and biodiverse systems with family farmers and urban food producers at the front, and based on the view of high quality, nutritious, and poison-free food as a human right. While it is too early to say what the National Plan for Agroecology will end up meaning for Uruguayan agriculture as a whole, the approval of the law could perhaps illustrate a trend shift for Uruguayan agrarian public policies, away from the almost exclusive consideration of market values.

However, other decisions rather point at the opposite direction. In December 2017, Uruguay authorized four new genetically modified events for maize and soybeans. The approval of the events was made in spite of both the *National Directorate of the Environment* (DINAMA) of and the *Ministry of Public Health* (MSP) voted against the authorization of the events due to understudied risks. MGAP, the *Ministry of Industry* and the *Ministry of Foreign Relations* voted for approval.[127] This is just one, of many cases, where there are differences in priorities and analysis between different Ministries regarding agrarian regulation. In particular, there are recurrent cases of MVOTMA favoring stricter environmental regulations, while MGAP generally argues for policies that are closer to the interests of the organized agrarian sector itself (Advisory Director DINAMA-MVOTMA, Montevideo, 6 March 2017).

Notwithstanding the diversity of regulations and the differences between different state offices, it is safe to say that the main Uruguayan policy line clearly supports a development model based on agro-industrial exports (already representing 75% of total exports). This is clear in the bulk of funded public programs in relation to agriculture deals with improved management of different public goods (water, soils) aiming to increase international competitiveness, product differentiation, and technological upgrading, responsible soils management, meat traceability, research and capacity building (Paolino 2015, 39–41). The motto of "sustainable intensification" and the targets to increase agrofood production to feed 50 million people in the world (MGAP 2017), encapsulates the main thread of Uruguayan public policies in relation to agrarian change, and it illustrates how the government wants to communicate that it is possible to reconcile the aim of productivity growth with the aim of sustainability.

## 5.5   CONCLUDING ANALYSIS OF REGULATIVE SHIFTS AND AGRARIAN CHANGE OF THE TWENTY-FIRST CENTURY

This chapter has explored the different ways Argentina, Paraguay, and Uruguay acted through public policies and regulations co-shaping the accelerating agrarian change of the twenty-first century. As shown, the public regulations in these countries during this period show some common traits. They all included a significant policy reorientation away from the neoliberal reforms of the 1980s and 1990s (Chapter 4 in this book), and were influenced by the neostructuralist winds sweeping the region and the "Post-Washington Consensus" agenda of the international community. This was centered in bringing the state back in as a central unit for strategic development planning, to make export-oriented commodity chains more "developmental" through redistribution, upgrading, and improved insertion in the world market. It also involves higher priorities given to poverty alleviation, participatory processes, transparency, and decentralization of the state apparatus. Environmental policies have also appeared higher on the agenda compared to previous periods, with all countries building up ministries, offices and agencies for the environment. However, it is clear that productivity gains and economic growth still take precedence over long-term care of natural resources. In Uruguay and Argentina, the *"Pink Tide"* was strong and implied a higher priority given to social redistribution and inclusion. The *"Pink Tide"* governments did nevertheless not radically challenge the new forms of production and social relations brought by the agro-commodity "booms". While the political orientation of the governments naturally plays an important role, it is clear that an essential part of national regulative frameworks did not change with a shift in political force in government. Regulations are slowly evolving historical constructs involving a great deal of path dependency and inertia. There are thus strong lines of regulative continuity, even in times of rapidly changing policy discourse. In all three countries, agribusiness advancement and agro-commodity exports increased significantly throughout the period, while family farmers and natural resources such as biodiversity and water and soils quality were facing important losses.

In Argentina, during the 12-year period (2003–2015) of rather heterodox rule of *Kirchnerismo*, state interference in the agrofood market and redistributive policies increased substantially. The government, however, continued on the path of export oriented, agribusiness dominated, agrarian change. At the same time, Argentina's export item per excellence—the soybean—evolved into a highly polarized battlefield with the government and the "countryside" at opposite sides. At the core of the conflict was the struggle over the surplus generated by the soybean business, materialized in the discussion about export taxes, but there was also an infected debate about export restrictions on beef. *Kirchnerismo* ended rather abruptly with the 2015 election of the economically liberal (orthodox) and socially conservative government of Mauricio Macri (2015–2019).

Uruguay in many ways followed a similar path as of Argentina. It moved from a neoliberal period in the 1990s, to a more neostructuralist, or neo-developmental rule with FA in government since 2005 and a progressive tax reform in as the most far-reaching redistributive reform. Shifts in Uruguay were nevertheless more gradual. While tensions largely followed the same fault lines as in Argentina, positions were less polarized and conflicts less violent. FA was reelected in 2009 and 2015, and is governing for the third time in arrow (2015–2020). Both Argentina and Uruguay have upcoming general elections by the end of this year, 2019.

In Paraguay, the *"Pink Tide"* Lugo administration (2008–2012) was the exception to the rule of having the economically liberal and socially conservative Colorado party in government.[128] However, Lugo faced broad opposition in the Colorado-dominated Paraguayan Congress, with frequent calls for his impeachment, and he did not manage to impose any substantial changes or reforms regarding forms of production or social relations in agriculture (Riquelme 2014). After Lugo's impeachment the former Vice President Federico Franco (2012–2013) completed the presidential mandate, followed by the election of Horacio Cartes (2013–2018) and then the election of the sitting administration of Abdo Benitez (2018–2023).[129] Throughout this period, independently of rulers, the conflicts over land, mostly articulated between *Brasiguayo* agribusiness farmers and poor small farmers, peasants and indigenous populations were severe, and often violent. Thus, each country, albeit to varying degrees, had important tensions and sometimes conflicts between the governments and different actors of the agrarian sector. The different ways Argentina, Paraguay, and Uruguay have coped with agrarian change and the following social-ecological concerns are partly the result of differences in policy goals and visions, but they are also due to differentiated state capacity and autonomy; that is to effectively, independently and coherently formulate and implement chosen paths. The abilities of the state to do so are constrained, both by domestic factors (including legacies from the past) and by features inherent to contemporary agrofood globalization. In the next, final and concluding chapter, these aspects are further explored.

## Notes

1. Corn, dairy products, wheat, sunflowers, and forestry products are also important export products, albeit to diverging degrees in the three countries.
2. While pressures from increased international demands for soy and beef are similar throughout the region, it is clear that agrarian change has still been articulated slightly differently over the territory depending on local, historically formed, social-ecological conditions.
3. The forestry sector has also received important amount of direct public subsidies and support in the region.
4. The value of agrarian-based exports represents over 60% of all exports in Argentina, Paraguay, and Uruguay (60, 63, and 73% respectively), and these in turn rely heavily on soybeans and beef in all three countries (51, 76, and 48% respectively). Moreover, these exports mainly end up in one destination: China (CAS 2018).
5. Other important "*Pink Tide*" governments: Ricardo Lagos (2000–2006) and Michelle Bachelet (2006–2018) in Chile; Evo Morales in Bolivia (2006–present); Hugo Chávez (1999–2013) and Nicolás Maduro (2013–present) in Venezuela; Rafael Correa (2007–2017) and Lenín Moreno (2017–present) in Ecuador; Alan García (2006–2011) and Ollanta Humala (2011–2016) in Peru.
6. RTD is particularly driven by IADB, IFAD, the Food and Agriculture Organization of the United Nations (FAO), and IICA throughout the region.

7. Most South American countries had experienced a deterioration in TOT between 1998 and 2002. But the situation improved from 2003 through 2008. In 2008, the TOT in the region, mainly linked to agricultural commodities, showed an impressive 70% rise compared to their 1990s average.

8. In 2002, family farming represented 218,868 PU, representing 66% of total PU, managing 23,519,642 ha of land, representing 13.5% of productive land (Alianza Internacional de las organizaciones de productores familiares de soya 2008, 8).

9. Soy products represent a remarkable 47% of the total value of agrarian-based exports, which in turn represent 60% of all Argentine exports (CAS 2018). If considering the whole soybean value chain (which includes activities outside of agriculture), the share of the soybean in the Argentine GDP has been estimated to 5.5% in 2014 (FAO 2017).

10. The World Bank, GDP growth (annual percent), https://data.worldbank.org/indicator/NY.GDP.MKTP.KD.ZG?locations=AR.

11. An exceptional drought in 2018 reduced exports, exacerbating the large current account deficit. The government asked the IMF for help and received big emergency bailout loan of USD 57B.

12. The subnational governments (Provinces and Municipalities) in Federal Argentina on average collect 49% of total government revenues.

13. Fiscal revenues from the export taxes accounted for 14–18% of total tax collection in the period 2005–2007 (FAO 2017, 28).

14. The government also provided subsidies to food processors (wheat and maize mills) for low consumption price on bread and pasta. The government targeted a price for domestic sales of wheat flour, and paid millers for an eventual difference with market price.

15. Argentine domestic consumption of soybean meal is thus still relatively low, compared with, for example, Brazil who produced 31.66 mmt soybean meal, in 2016/2017 of which 13.76 mmt was exported (USDA 2018), where the availability of cheap soybean meal played a pivotal role in the development of a competitive and dynamic broiler and pork industry.

16. For example, funds were provided livestock farmers that implemented plans for sanitary improvement as well as for purchase of machinery, input and new technologies for increased productivity, through the *National Service of Agri-Food Health and Quality* (SENASA).

17. Decree 192/2011, Infoleg, Ministerio de Justicia y derechos humanos, Presidencia de la Nación, Bs. As., 24 February 2011, http://servicios.infoleg.gob.ar/infolegInternet/anexos/175000-179999/179598/norma.htm. Accessed 4 March 2019.

18. Argentine beef imports were banned in more than 60 countries, including the US and Canada. After an aggressive vaccination program, Argentina regained the status as "foot and mouth free with vaccination".

19. In November 2007, CFK raised export taxes from 27.5 to 35% on whole soybeans, from 24 to 32% on soybean oil, while biodiesel remained at "only" 12%. Export taxes on sunflower were sat at 32%; wheat at 28%; maize at 25%. Resolution 368/2007. The total tax revenue from grains exports in 2008 was approximately USD 7.4B (Richardson 2009).

20. Resolution No. 125. Meanwhile, the export tax on soy-based biodiesel was set at 15.2%, less than half the tax rate on soybeans and soy coproducts (Naylor and Higgins 2017).

21. For example, *the Argentine Association of the Soybean Chain* (ACSOJA) and the *Chamber of the vegetable oil industry, la Cámara de la Industria Aceitera.*

22. The term *Gorila* emerged in the 1950s, referring to an anti-peronist person. Its use has been extended to a right-wring reactionary person.

23. Old "tropes" about the countryside were revoked; it was portrayed as backward and barbaric, in contrast to the urban enlightenment and the dream of unlimited progress.

24. Native groups used the forests for hunting and gathering, subsistence cattle ranching, timber harvest, and small-scale charcoal production.

25. FONAF organized several mega-meetings, with over 900 organizations participating, diagnosing the situation for family farmers and suggesting policy interventions.

26. CFK remained unpopular among the traditional producers' organizations and many other actors linked to agricultural and agro-industrial activities, but after a significant decline in 2008, the support for her was steadily improving up until the 2011 general elections. CFK, representing the *Front for Victory* (FpV), managed to be re-elected by the historically strong support of more than 54% of the votes.

27. The most important programs were the *Agrarian Social Program* (PSA), *the project for development of small farmers* (PROINDER), *Mohair, Rural Change* (Cambio Rural) the development projects PRODERNOA, PRODERNEA y PRODERPA, special projects of INTA, to name a few.

28. More than USD 50 million was allocated to 250,000 families by SSDRAF under a period of four years.

29. Access the law via FONAF's website: http://www.fonaf.org.ar/documentos/Ley_27118_Reparacion_historica_AF.pdf. Accessed 22 December 2018.

30. As described in Chapter 4, the Argentine Chaco forest suffered intensified deforestation in the wake of soybeans expanding in the area as well as livestock ranching ("displaced" from the Pampas).

31. World Database on Protected Areas, Protected Planet: Argentina, https://protectedplanet.net/country/AR. Accessed 8 February 2019.

32. Access Law 26331 (in Spanish) at http://servicios.infoleg.gob.ar/info-legInternet/anexos/135000-139999/136125/norma.htm. Accessed 10 October 2018.
33. El 9 de Julio, "Las abejas argentinas están en peligro (y nadie hace nada)" http://www.diarioel9dejulio.com.ar/noticia/90982; Conclusión "Agroquímicos: los apicultores le contestan a Etchevehere por el genocidio de abejas en Córdoba", https://www.conclusion.com.ar/sin-categoria/agroquimicos-los-apicultores-le-contestan-a-etcheve-here-por-el-genocidio-de-abejas-en-cordoba/04/2018/. Accessed 10 September 2019.
34. In Argentina, national regulations stipulate that communities should be informed no later than 48 hours before fumigation, so that they can close the windows and take other precaution, but there are no national-level rules for security distance between actual fumigation site and waterways or communities. Thus, the security distance may vary significantly between different products and different localities.
35. INTA, "Extracción de nutrientes en la agricultura Argentina". See report: http://inta.gob.ar/documentos/extraccion-de-nutrientes-en-la-agricultura-argentina/at_multi_download/file/Extraccion_de_nutrientes.pdf. Accessed 22 June 2018.
36. It should however be mentioned that the government resisted Monsanto's attempt to make it harder for farmers to save seeds. CFK also increased significantly the budget for environmental research, and some restrictions on agrochemical spraying and safety zones were imposed.
37. Two important income transfer programs were the Unemployed Household Heads program, and from 2010 the Universal Allowance per Child. Both transferred funds to the unemployed or to families with income below the statutory minimum wage (FAO 2017, 28).
38. Over 500 meetings and over 7000 persons from academia, the state, Provinces, public agencies, producers' organizations, NGOs, chambers of commerce, and private actors participated in the drafting of the plan, which was facilitated and supported by CEPAL, IICA, FAO, and UNDP.
39. See *La Nacion*: https://www.lanacion.com.ar/2137527-retenciones-a-la-soja-el-campo-le-pedira-una-audiencia-urgente-a-macri. Accessed 12 January 2019.
40. In September 2018, Macri reformed the ministries, and agroindustry was (again) turned into a secretariat under the Ministry of Production.
41. Eduardo Duhalde was the President of Argentina in the turmoil period 2002–2003.
42. The respondent has a long trajectory within the agrarian sector, both as high-level public official, researcher, and private consultant for a number of IOs and private agribusiness firms.

43. The respondent is referring to the big landowners that lease out their land to the soybean producers, see Chapter 4 in this book, while they live in *Recoleta*—a rich neighborhood in Buenos Aires.
44. Word Bank, Argentina—overview: http://www.worldbank.org/en/country/argentina/overview. Accessed 12 January 2019.
45. Clarín Rural, 2018, "Medidas del Gobierno: La soja volvería a tributar cerca de 30% de retenciones", 03/09-2018 https://www.clarin.com/rural/soja-gran-ganadora-medidas-gobierno_0_H1CL-a5D7.html. Accessed 2 November 2018.
46. At the same time, corn and wheat, with no taxes since December 2015, will now pay USD 4, equivalent to 11% of export tax. Processed grains (soybean, corn and wheat) will pay USD 3, almost 9%. The export taxes are estimated to generate USD 6,838 M in 2019 in fiscal revenues Clarín Rural, 2018, "Medidas del Gobierno: La soja volvería a tributar cerca de 30% de retenciones", 3 September 2018, https://www.clarin.com/rural/soja-gran-ganadora-medidas-gobierno_0_H1CL-a5D7.html. Accessed 2 November 2018.
47. World Development Indicators database. The classification changed since July 1, 2018, from "Upper-middle income" country to "High-income" country, https://blogs.worldbank.org/opendata/new-country-classifications-income-level-2018-2019. Accessed 5 May 2019.
48. It was also renamed to the *Secretariat of Family farming, Coordination and Territorial Development* (SAFCyDT).
49. See "Dirección de Ejecución de Programas y Proyectos destinados a Pequeños Productores", http://www.agroindustria.gob.ar/sitio/areas/d_desarrollo_territorial_rural/d_ejecucion_programas_proyectos/ and http://www.agroindustria.gob.ar/sitio/areas/agroemprende/. Accessed 10 February 2018.
50. The share of revenues and expenditures as percentage of GDP is well below the *Organization of Economic Cooperation and Development* (OECD) average and that of regional peers.
51. The tax burden, measured as tax revenue relative to GDP, was in 2007 around 8%. The revenues rely heavily on indirect taxes such as Value Added Tax (VAT), which generates 50% of the total.
52. SEAM was created in in 2000 in order to advance, coordinate, monitor, and implement all environmental policy. In late 2018, it changed into the *Ministry of Environment and Sustainable Development* (MADES).
53. See websites and annual reports from the IOs and MAG, SEAM, INFONA. The UNFCCC, and particularly the REDD+ initiative and the WWF have been very dominant in developing Paraguay's new

regulative framework regarding forests. IDB is supporting so that the country can do a proper agrarian Census in 2019–2020 (the last Census—CAN—is from 2008), with methodological support from FAO and streamlined with the SDGs of Agenda 2030. IFAD and World Bank are very influential in policies regarding family farming. CEPAL is particularly present in social policy formulation.

54. Adjusted net savings measures the true rate of saving in an economy after taking into account depletion of natural resources and damages caused by pollution. This provides a relatively simple indicator of how sustainable their country's investment policies are https://development-data-hub-s3-public.s3.amazonaws.com/ddhfiles/143151/ans-methodology-january-30-2018_2_0_0.pdf. Accessed 29 February 2019.

55. The total amount of soybean farmers has been rather stable between 1991 and 2008, around 27.000, but they managed four times more land in 2008 compared to 1991 (from 0.55 Mha to 2.46 Mha). The herd of cattle expanded with 38% between 1991 and 2008; the summed up to 10,496,641 heads in 2008 (MAG 2009).

56. See El País, "La codicia por la tierra en Paraguay", Published 2 March 2017, https://elpais.com/elpais/2017/02/07/planeta_futuro/1486488199_675583.html. Accessed 10 March 2018.

57. See El País, "La codicia por la tierra en Paraguay", Published 2 March 2017, https://elpais.com/elpais/2017/02/07/planeta_futuro/1486488199_675583.html. Accessed 10 March 2018. Between 1991 and 2008, 797,366 rural laborers left activity, a contraction of 75%.

58. There are, according to the last Indigenous Census (2012) 117,150 indigenous persons (1.81% of population), of which 52% live in the Eastern region (Atlantic forest) and 48% in the Chaco region (IDB 2017). Mainly subsistence farmers with 2–3 ha of land in Eastern Paraguay. According to official sources, the expropriation process that should be followed in the case of indigenous land awards is often unworkable because of high land prices, and this situation often redounds to the benefit of private landowners. As a result, the issuance of land titles is frequently limited to separated parcels that are then broken up into lots which are often too small to be of practical use (Human Rights Council 2015, 6).

59. The Paraguayan Chaco includes the department of Boquerón, Alto Paraguay, and Presidente Hayes. The Chaco peoples are Ayoreo, Chamacoco, Enxet, Nivakle, Manjuy, Maka'a, Toba Qom, Nandeva, and the Guarayo. The Chaco region represents a third of national territory, but less than 3% of the country's population lives there

(Walcott et al. 2014). While population is sparse, it increased between 1962 and 2002, from 74,129 to 135,186 people (Caldas et al. 2015).

60. Some communities of the native groups living in the Paraguayan Chaco have chosen voluntary isolation from the outside world and live in regions containing the ample supply of water and the richest biodiversity in the region, although it has remained poorly known (Yanosky 2013, 115–118).

61. Besides Monsanto, the system was supported by the big producers' organizations (the *Association of Producers of Paraguay*—APS, the *Coordinator of Rural Producers' Organizations of Paraguay*—CAP, and the *Federation of Cooperatives of Agrarian Production*—FECOPROD), the seed companies, the *Association of seed producers of Paraguay* (APROSEMP) and the *Chamber of Grain and Oilseed Exporters and Traders* (CAPECO).

62. Brazilian producers acquired the land by illegal purchase with the support of officers related to the state institutions in charge if land titles; IBR and later INDERT.

63. See various reports from NGOs and social movements; La Coordinadora Latinoamericana de Organizaciones del Campo (CLOC)—Vía Campesina, Capítulo Paraguay, https://clocvcparaguay.wordpress. com/2014/01/22/el-reino-del-latifundio/; Base IS, http://www. baseis.org.py/wp-content/uploads/2017/02/guahory-vale.com-pressed.pdf; Biodiversidad, Paraguay, http://www.biodiversidadla. org/Paises/Paraguay/(offset)/27 and the Environmental Justice Atlas, https://www.ejatlas.org/print/the-guahory-crisis, and Farm Land Grab, https://www.farmlandgrab.org/post/view/20455. Accessed 21 October 2018.

64. See the webpage of the land reform institute, Indert, http://www. indert.gov.py/index.php/institucion.

65. See El País, "La codicia por la tierra en Paraguay", Published 2 March 2017, https://elpais.com/elpais/2017/02/07/planeta_futuro/ 1486488199_675583.html. Accessed 10 March 2018.

66. IMAGRO was in 2013 replaced by IRAGRO which instead of area tax agrarian rent. Fiscal revenue increased with the shift from IMAGRO to IRAGRO. http://www.leyes.com.py/disposiciones/categoria/4/52/ impuesto-a-las-rentas-de-las-actividades-agropecuarias-iragro-.html. Accessed 19 January 2019.

67. Between 1989 and 2005, at least 75 arbitrary executions of members of organizations of peasants and rural workers.

68. In 2009, the Paraguayan state declared EPP as a terrorist group through decree No. 363/2009.

69. See several news articles in national press: Ultima hora 2018, https://www.ultimahora.com/disputa-parcela-tierra-deja-un-muerto-y-dos-heridos-n2704854.html; Ñandutí 2016, http://www.nanduti.com.py/2016/12/27/nuevo-desalojo-en-guajhory-un-muerto-y-otro-en-grave-estado1/; Vanguardia 2016, http://www.vanguardia.com.py/2016/10/31/matan-a-productor-cuando-realizaba-trabajos-de-fumigacion/; ABC 2004, http://www.abc.com.py/edicion-impresa/policiales/campesinos-asesinan-a-brasiguayo-y-a-su-guardia-a-escopetazos-en-itakyry-767248.html; ABC 2011, http://www.abc.com.py/edicion-impresa/policiales/degellan-a-colono-brasiguayo-206200.html; and Noticde 2016, http://www.noticde.com/2016/10/itakyry-asesinan-al-hijo-de-un.html. All accessed 22 October 2018.

70. The suspension of Paraguay allowed for the others to admit Venezuela to the bloc, without Paraguay's new center-right government could block it.

71. Agromeat, 2012, "Paraguay le pone retenciones a la soja", http://www.agromeat.com/95447/paraguay-le-pone-retenciones-a-la-soja. Accessed 22 May 2018.

72. Asaga, 2017, "Paraguay: impulsan un 15% de retenciones para la exportación de granos", http://www.asaga.org.ar/index.php/es/cap/16-noticias/internacionales/1339-paraguay-impulsan-un-15-de-retenciones-para-la-exportaci%C3%B3n-de-granos.html. Accessed 22 May 2018.

73. La Nacion, 2017, "El senado paraguayo aprobó un diez porciento de retenciones a la soja y cortan ruta en respuesta", https://www.lanacion.com.ar/2036044-el-senado-paraguayo-aprobo-un-10-de-retenciones-a-la-soja-y-cortan-rutas-en-protesta; Cronista, 2017, "Media sanción en Paraguay a retención del 10% a la soja", https://www.cronista.com/internacionales/Media-sancion-en-Paraguay-a-retencion-del-10-a-la-soja-20170623-0058.html. Accessed 23 May 2018.

74. La red del Campo—Chacra, "Tractorazo a la Argentina por las retenciones en Paraguay", http://www.revistachacra.com.ar/nota/13937/. Accessed 12 January 2019.

75. Agrolatam, "Retenciones de 10% a la soja sera nefasto", http://www.agrolatam.com/nota/34442-retenciones-de-10-a-la-soja-sera-nefasto-dicen-en-paraguay/. Accessed 23 January 2019.

76. Base IS, "Reforma fiscal planteada por el gobierno no incluye impuesto a la soja", http://www.baseis.org.py/reforma-fiscal-planteada-por-el-gobierno-no-incluye-impuesto-a-la-soja/. Accessed 2 May 2019; ABC, 15th of November 2018, "Postergan impuesto a exportación de granos," http://www.abc.com.py/nacionales/dilatan-aumento-del-impuesto-a-la-soja-1759730.html. Accessed 22 May 2019.

77. La Nación, 21st of December 2018, "Productores amenazan con trac-torazo a nivel país", https://www.lanacion.com.py/pais/2018/12/21/productores-amenazan-con-tractorazo-a-nivel-pais/. Accessed 10 January 2019.

78. The first law (2524/04) was a total moratorium for two years, but the period has been extended many times, first through law 3139/06, later by 3663/08, by 5045/18, and in 2018 law 5288/18 further extended the period until 2020. See the National Forestry Institute, Infona, http://www.infona.gov.py/application/files/1714/2902/4900/Ley_N_2524_-_Deforestacion_cero.pdf. Accessed 20 December 2018.

79. According to the national forest Institute, INFONA, of the Paraguayan government, the dry Chaco forest in Paraguay covered 11.561.519 hectares of land in Paraguay, representing 60% of all Paraguayan forest land, and 28,4% of total Paraguayan land area. See Infona, http://www.infona.gov.py/application/files/2114/3093/5539/BNB2011_6ESTRATOS.jpg. Accessed 11 October 2018.

80. Complementary forestry legislation from the 2000s were: increased fines for deeds against the environment; obligatory previous Environmental Impact Analysis for large-scale projects; the prohibition of exports and traffic of Rolls, timber pieces and wooden beams; the installation and expansion of Protected Wild life areas, and other types of protected forest area (Piera Valdés 2016, 25–33, 57–62).

81. An illustrative quote was provided by the specialist on innovation and biotechnology at IICA Paraguay: *Since re-democratization there is such a high frequency in change of ministers and public units and authorities that it became impossible to follow-up norms and laws. There is no continuity. Moreover, there has been a clear deterioration in technical capacity among public officials. This is linked to an educational crisis, where many so-called universities have emerged, offering cheap academic titles, but without teaching the necessary knowledge and skills,* (Innovation Specialist IICA Paraguay, Asunción, 22 February 2017).

82. See webpage of the Paraguayan Senate: http://www.diputados.gov.py/ww5/index.php/noticias/introducen-modificaciones-al-proyecto-que-prohibe-la-deforestacion-en-la-region-oriental?ccm_paging_p=2. Accessed 20 November 2018.

83. World Database on Protected Areas: "Protected Planet: Paraguay", https://protectedplanet.net/country/PY. Accessed 8 February 2019.

84. Law No. 536 from 1995 created tax exonerations and cost reductions for afforestation on soils that had been classified as suitable for forests SEAM, Law 536: http://www.seam.gov.py/sites/default/files/ley_536.pdf. Accessed 11 September 2018). Law No. 1639 from 2000

complemented the initiative by prohibiting change of land-use after having received benefits for afforestation (Piera Valdés 2016, 29–30; 63).

85. The high price on carbon seems to potentially offer landowners almost equal rents per hectare as provided by the livestock activities in the Paraguayan Chaco (Py/SEAM/INFONA/FAPI 2016, 15–17, 23–24, 28–29).

86. Resúmen de Noticias, "Mediante decreto Cartes taló 2 millones de árboles", Published 17 November 2017, https://www.rdn.com. py/2017/11/15/mediante-decreto-cartes-talo-2-millones-de-arboles/; WWF Paraguay, "El Poder Ejecutivo derogó el DECRETO 7702", 20th of August, 2018, http://www.wwf.org.py/?uNewsID=333411. Accessed 21 December 2018.

87. See information on the website for the Ministry of public constructions and communications, http://www.mopc.gov.py/siguen-las-tareas-de-mantenimiento-de-la-ruta-9-n2773. Accessed 12 October 2018.

88. The Chinese market, as in the case of soybeans, continues to be closed due to diplomatic reasons. The livestock sector is pushing hard to make the government try to improve relations with China.

89. See the electoral platforms for the government period 2005–2010, 2010–2015, and 2015–2020, downloadable at the website of FA, https://frenteamplio.uy. Accessed 2 December 2018.

90. According to IMF, Uruguay has managed to be relative successful despite the deteriorating external environment due to its strong institutions, prudent policies, and large buffers (International Monetary Fund 2019). Uruguay's economy in 2019 is still negatively affected by lower investment, higher oil prices, lower exports to neighboring countries, and lower agricultural exports due to a 2018 drought, which particularly hit the soybean harvest (International Monetary Fund 2019).

91. Uruguay has managed to attain a high level of equality of opportunity in terms of access to basic services such as education, running water, electricity, and sanitation. Uruguay's middle class represents 60% of its population. Moderate poverty went from 32.5% in 2006 to 9.4% in 2016, while extreme poverty has practically disappeared: it went down from 2.5 to 0.2% in the same period. In terms of equity, income levels among the poorest 40% of the Uruguayan population increased much faster that the average growth rate of income levels for the entire population. Around 87% of the over-65 population is covered by the pension system (IBRD 2015). Only 1.4% of the population lives on USD 3.20 or less per day at 2011 PPP (International Monetary Fund 2019, 28).

92. It scores among the highest in for example the Human Development Index, the Human Opportunity Index and the Economic Freedom Index. Institutional stability and low levels of corruption are reflected

in the high level of confidence that citizens have on the government. According to the World Bank's Human Opportunity Index.

93. Definition from MGAP through 2008, Resolución 527/008 of 2917/08.

94. Around half of all sharecroppers in the Uruguayan Litoral that had been registered as active crop producers in the agrarian census of 2000, had abandoned the activity in 2009 (Arbeletche and Gutiérrez 2010; Arbeletche et al. 2012).

95. A family producer in Uruguay is defined by MGAP as one who operates not more than 500 ha, employs up to two waged workers, resides on the farm or at a distance that is no further than 50 km, and has farm activity as its main source of income.

96. The private actor members of MTO were in 2009 only 17 but represented 80% of the total amount of exported and processed soybeans. The Department of Livestock and agriculture (MGAP) is nor a member of the board but participates in many events.

97. El País, "El gigante del campo uruguayo se tambalea," Buenos Aires, 25 April 2017. www.elpais.com/economia/2017/04/21/actualidad/1492772304_849739.html. Accessed 6 March 2019.

98. Family farmers must be registered as such to access funds from DGDR, DIGEGRA, MEVIR, and benefits from the Bank of the Republic and BPS. The official definition of a family farmer in Uruguay (manage less than 500 ha and to live on the land, or at least within the frame of 50 km from the land) was taken through a MGAP resolution (No. 1013/2016). The total amount of public support to the agrarian sector has yearly (2009–2013) been around USD 206 M, of which USD 80 M has been to producers, representing 1.3% of annual total value of agrarian production (Presidencia de la República Oriental del Uruguay 2017, 80).

99. There has also been established public–private "tables" for specific agricultural orientations, aiming to construct a bridge between the public sector and the producers through joint capacity building and policy formulation (Presidencia de la República Oriental del Uruguay 2017, 71)

100. For example, the projects: "family ranchers and climate change", "Rural Inclusion", "More value and more technology", "Uruguay Rural" (with long-term support from IFAD), "the livestock project", and "Responsible Production", "the rural microcredit program".

101. Declárase de interés general la producción familiar agropecuaria y la pesca artesanal: https://legislativo.parlamento.gub.uy/temporales/leytemp4430371.htm. Accessed 12 January 2019.

102. It is established that when land plots bigger than 500 ha is put on sale on the land market, INC has to be offered to buy it (for market price).

See the official website: https://www.colonizacion.com.uy/adjudica-cion-de-tierras-2005-2017. Accessed 12 January 2019.

103. See the official INC website: https://www.colonizacion.com.uy/adjudi-cacion-segun-superficie. Accessed 11 October 2018.

104. Personal Income Tax (IRPF) from work is taxed at progressive rates (ranging from 10 to 36%). Moreover, holding income and capital gains are taxed at a flat 12% rate, with almost no deductions. Taxable income for IRAE purposes is basically net income shown by the company's books after making various tax adjustments, including an adjustment for inflation.

105. See the website of the tax office, DGI, https://www.dgi.gub.uy/wdgi/page?2,principal,_Ampliacion,O,es,0,PAG;CONC;1187;1;D;opcion-irae-imeba;19;PAG. Accessed 12 January 2019; Report from Deloitte, 2018, Uruguay: Tax and Investment Profile, The way to new oppor-tunities, https://www2.deloitte.com/content/dam/Deloitte/uy/Documents/tax/Tax%20and%20Investment%20ProfileFY17.pdf. Accessed 21 December 2018.

106. Depending on how much investment projects score on these indicators and the size of their project, investors can be exonerated from corporate income tax (25%) of total investment (up to 100%) for up until 25 years (Durán and Salgado 2013).

107. The project is evaluated by a special Commission and receives a score based on fulfillment of development criteria, and in accordance with the size of the project and amount invested.

108. Exemption from IRAE (which is directly from the tax and not from the taxable income) can be of up to 100% of the amount invested from 3 to 25 years. Each year, the amount deducted from IRAE cannot exceed 60% of the payable tax. In addition to this general regime, there are specific tax incentives applicable to certain activities such as renewable energy.

109. Uruguay's beef cattle herd is the largest in the world in terms of animals per inhabitant (about 40 per capita). It is composed of 12 million head, mostly of British breeds, such as Hereford and Aberdeen Angus.

110. An authorized operator notifies any movement of cattle or change in holding. Each livestock transaction must be preauthorized and recorded in real time. This allows SNIG to receive information about all opera-tions, verify that the actors are registered and do not have any sanitary or legal prohibitions, and establish and register the origin and desti-nation of each movement. Each participating actor has different privi-leges that give authorized access to distinct data and functionality. For example, breeders may access maps of their farmlands and data on activ-ities or animal species through their respective registration numbers.

Veterinarian services may access maps showing rings of neighboring farms, data used in case of animal diseases. See the SNIG/MGAP website, https://www.snig.gub.uy/. Accessed 22 December 2018.

111. The EU 481 quota demands cattle to be less than 26 months of age and be finished the last 100 days with grain. The EU ranks as the number two destination for Uruguay's beef exports, but the stalled negotiations between EU and Mercosur puts the future of the beef quotas in high uncertainty.

112. USDA, https://www.fas.usda.gov/data/uruguay-livestock-and-products-annual-2. Accessed 20 December 2018.

113. Glyphosate is also used for many other crops, such as corn and cotton; it is used in gardening, in citrus plantations and in forestry. Given the dominance of soybeans in the Pampas, however, the wide use of glyphosate is mostly explained by these cultivations.

114. Evolution of average price per ha in USD: 2000—448/ha; 2002—385/ha; 2004—664/ha; 2006—1132/ha; 2008—1844/ha; 2010—2633/ha; 2012—3473/ha; 2014—3934/ha (DIEA 2016, 138).

115. Once extracted from earth the soybean leaves almost no residue, which increase vulnerability for rain induced erosion.

116. The most harmful insecticides Endosulfan and Ffonil were recurrently used in soybean production between 2003 and 2008, when MGAP restricted use. Studies still find high residues of Endosulfan in the water.

117. Annual bovine meat production augmented from 0.453 mmt in 2000, to 0.593 mmt in 2017 (ECLAC 2017, 108).

118. The traditional constraint of the seasonality of pastures is overcome with increased grain complement. One of the benefits of new culture, knowledge, available grains, and infrastructure (silos everywhere) is that livestock farmers now cope with droughts by buying forage, while previous droughts sometimes resulted in cattle death.

119. World Database on Protected Areas, Protected Planet: Uruguay, https://protectedplanet.net/country/UY. Accessed 8 February 2019.

120. See https://legislativo.parlamento.gub.uy/temporales/leytemp2183552.htm. Accessed 23 January 2019.

121. MGAP has an agreement with the *Faculty of Agronomy* (FAGRO) of the Udelar, and the *Uruguayan Association of Agronomists* (AIA), for the accreditation system, which guarantees that the agronomists are knowledgeable about sustainable development (Hill et al. 2016). See the guidelines for the plan, Renare-MGAP 2013 "Instructivo para elaborar un plan de uso y manejo responsable del suelo," www.cebra.com.uy/renare/media/INSTRUCTIVO-PARA-ELABORAR-UN-PLAN-DE-USO-Y-MANEJO-RESPONSABLE-DEL-SUELO-11-09-2013.pdf. Accessed 2 October 2018.

122. The plan contributes directly to the National Environmental Plan (in consultation process), to the National Climate Change Policy, and to the National Biodiversity Strategy.
123. DGSA, "Denuncaias por mal uso de productos fitosanitarios," http://www.mgap.gub.uy/unidad-organizativa/direccion-general-de-servicios-agricolas/tramites-y-servicios/servicios/denuncias-mal-uso-fitosanitarios. Accessed 1 October 2018.
124. DGSA is in charge of main register (online), but since 2013, DINAMA is also responsible for the enforcement of sound pesticide and chemical waste management (MGAP 2011) DGSA-MGAP, Denuncias, Manejo Seguro de Productos Fitosanitarios, http://www2.mgap.gub.uy/portal/page.aspx?2,dgsa,dgsa-denuncias,O,es,0. Accessed 1 October 2018. Decree 152/2013 enacts Environmentally Appropriate Management of Waste resulting from agrochemicals. This includes containers of chemicals or biological products used in crop production, and it mandates the manufacturers and importers to submit management plans and defines the requirements for such plans with regards to the management of stocks of obsolete pesticides and empty containers.
125. El País, "Aire tóxico: acusaciones cruzadas sobre plaguicidas", https://www.elpais.com.uy/que-pasa/aire-toxico-acusaciones-cruzadas-plaguicidas.html. Accessed 10 September 2018.
126. See the website of Redes—Friends of the Earth Uruguay, https://www.redes.org.uy/2018/12/12/ante-la-aprobacion-del-proyecto-de-plan-nacional-de-agroecologia/. Accessed 12 January 2019.
127. La Diaria, "Gobierno aprobó la liberación comercial de cuatro nuevos eventos transgénicos de maíz y soja con el voto en contra de los ministerios de Salud y Ambiente," 20 December 2017, https://ladiaria.com.uy/articulo/2017/12/gobierno-aprobo-la-liberacion-comercial-de-cuatro-nuevos-eventos-transgenicos-de-maiz-y-soja-con-el-voto-en-contra-de-los-ministerios-de-salud-y-ambiente/. Accessed 12 January 2019.
128. The Colorado party involves, nevertheless, different factions, representing slightly divergent ideological orientations and that have from time to time been engaged in severe power struggles against each other. According to the only text about Colorado ideology that the party has published on its official website, the "Colorado doctrine" is close to the social doctrine of the church, such as loving your neighbor. At the same time, the text remarks that the Colorado party makes a clear distinction between politics and economics; "The Colorado party does not intervene in the economic life, except in the most essential to correct social inequalities, reflecting our faith in universal values without oppressed classes and well-being for all. This is why until today the services of electricity, water, waste and communications are in the

hands of the state to serve free individuals" Cf. "Conclusiones de la disertación del Dr. Leandro Prieto Yegros acerca de la Ideología de la Asociación Nacional Republicana Partido Colorado," http://www.anr.org.py/paginas.php?cod=53. The sitting president, Abdo Benitez, also quoted this text in a post on his personal Facebook page the 11th of September 2016, https://www.facebook.com/MaritoAbdo2018/posts/1300512773314916. Accessed 1 December 2018.

129. Abdo's father, Mario Abdo Benítez, was private secretary for 25 years of Alfredo Stroessner.

## References

Aide, T. M., D. Clark, H. R. Grau, M. Levy, D. Carr, D. Redo, and M. Andrade. 2013. "Deforestation and Forest Expansion in Latin America: 2001–2010." *Biotropica* 45 (262): e271.

Alianza Internacional de las organizaciones de productores familiares de soya. 2008. *Declaración de Asunción*. IV Encuentro Internacional de Pequeños Productores de Soya y la Sociedad Civil, 6, 7, y 8 de Febrero del 2008, Asunción: Tekokatu, Ser, Probioma and Solidaridad.

Amdan, M. L., R. Aragón, E. G. Jobbágy, J. N. Volante, and J. M. Paruelo. 2013. "Onset of Deep Drainage and Salt Mobilization Following Forest Clearing and Cultivation in the Chaco Plains (Argentina)." *Water Resources Research* 49 (10): 6601–6612.

Anichini, F., D. Borda, F. Masi, J. Ramirez, B. Servin, and G. Setrini. 2013. *Estudio de Potencialidad de Desarrollo de las Cadenas de Valor*. Edited by Centro de Analisis y Difusion de la Economia Paraguaya. Asunción: Agencia de Cooperación Internacional del Japón (JICA).

Antía, Florencia. 2018. "The Political Dynamic of Redistribution in Unequal Democracies: The Center-Left Governments of Chile and Uruguay in Comparative Perspective." *Latin American Perspectives* 46 (1): 152–166. https://doi.org/10.1177/0094582X18806827.

Arancibia, Florencia. 2016. "Regulatory Science and Social Movements: The Trial Against the Use of Pesticides in Argentina." *Theory in Action* 9 (4): 1–21.

Arbeletche, Pedro, and Carolina Carballo. 2009. "La expansión agrícola en Uruguay: Algunas de sus principales consecuencias." *Revista de Desarrollo Rural y Cooperativismo Agrario* 12: 7–20.

Arbeletche, Pedro, and Gonzalo Gutiérrez. 2010. "Crecimiento de la agricultura en Uruguay: exclusión social o integración económica en redes." *PAMPA, Revista Interuniversitaria de Estudios Territoriales* 6: 113–138.

Arbeletche, Pedro, J. M. Ferrari, and G. Souto. 2008. *Impactos socioeconómico de la soja en Uruguay*. Edited by Jorge Escudero. Montevideo: MTO.

Arbeletche, Pedro, Macarena Coppola, and Cintia Paladino. 2012. "Análisis del agro-negocio como forma de gestión empresarial en América del Sur: el caso uruguayo." *Agrociencia Uruguay* 16: 110–119.

Avila-Vazquez, Medardo, Flavia S Difilippo, Bryan Mac Lean, Eduardo Maturano, and Agustina Etchegoyen. 2018. "Environmental Exposure to Glyphosate and Reproductive Health Impacts in Agricultural Population of Argentina." *Journal of Environmental Protection* 9 (03): 241.

Baraibar, Matilda. 2014. "Green Deserts or New Opportunities?: Competing and Complementary Views on the Soybean Expansion in Uruguay, 2002–2013." Doctoral thesis Monograph, Economic History, Stockholm University, Stockholm Studies in Economic History 64.

Bárcena Ibarra, Alicia, and Antonio Prado. 2015. "Introducción." In *Neoestructuralismo y corrientes heterodoxas en América Latina y el Caribe a inicios del siglo XXI*, edited by Alicia Bárcena Ibarra and Antonio Prado. Santiago de Chile: CEPAL.

Barrán, José Pedro, and Benjamín Nahum. 1986. *Batlle, los estancieros y el imperio británico*. Vol. 1. El Uruguay del novecientos. Montevideo: Ediciones de la Banda Oriental.

Barrios, Juan José, Nestor Gandelman, and Gustavo Michelin. 2010. "Analysis of Several Productive Development Policies in Uruguay." IDB Working Paper Series. Montevideo: Universidad ORT Uruguay.

Baumann, Matthias, Ignacio Gasparri, María Piquer-Rodríguez, Gregorio Gavier Pizarro, Patrick Griffiths, Patrick Hostert, and Tobias Kuemmerle. 2016. "Carbon Emissions from Agricultural Expansion and Intensification in the Chaco." *Global Change Biology* 23 (5): 1902–1916.

Bergero, Patricia, Julio Calzada, Federico Di Yenno, and Emilce Terré. 2019. "The Removal of Differential Export Taxes Partially Discourages Soybean Value Chain in Argentina." In *BCR Weekly Review*, Buenos Aires: Bolsa de Comercio de Rosario—BCR.

Bidegain, Mario Pérez, Federico García Préchac, Mariana Hill, and C. Clérci. 2010. "La erosión de suelos en sistemas agrícolas." In *Intensificación agrícola: oportunidades y amenazas para un país productivo y natural*, 67–88. Montevideo: Udelar.

Blois, María Paula. 2016. "Ciencia y glifosato: interpelando órdenes. Una investigación en la prensa en el contexto argentino." *Cuadernos de antropología social* (43): 73–93. http://www.redalyc.org/pdf/1809/180948645007.pdf.

Bruno, Alfredo. 2007. *Plaguicidas usados en el cultivo de soja: Evolución de su uso y estimación de su impacto ambiental*. Río Negro: Coyuntura Agropecuaria.

Burns, Sarah L., and Lukas Giessen. 2016. "Dismantling Comprehensive Forest Bureaucracies: Direct Access, the World Bank, Agricultural Interests, and Neoliberal Administrative Reform of Forest Policy in Argentina." *Society & Natural Resources* 29 (4): 493–508. https://doi.org/10.1080/08941920.2015.1089608.

Bussoni, Adriana, Jorge Alvarez, Frederick Cubbage, Gustavo Ferreira, and Valentin Picasso. 2019. "Diverse Strategies for Integration of Forestry and Livestock Production." *Agroforestry Systems* 93 (1): 333–344. https://doi.org/10.1007/s10457-017-0092-7.

Caldas, Marcellus M., Douglas Goodin, Steven Sherwood, Juan M. Campos Krauer, and Samantha M. Wisely. 2015. Land-Cover Change in the Paraguayan Chaco: 2000–2011. *Journal of Land Use Science* 10 (1): 1–18.

Caligaris, Gastón. 2017. "Las grandes empresas agropecuarias en Argentina: los casos de Cresud y El Tejar." *Cuadernos de Economía* 36 (71): 469–488.

Caon, Lucrezia. 2013. "Land Management Style and Soil Erosion in the Western Area of Uruguay: Local Farmers vs. Foreign Investors." Masters thesis, International Land and Water Management Land Degradation and Development Group, Wagenagen University (880730-154-070).

Carriquiry, Florencia. 2015. "El papel del agro en el desarrollo económico nacional." In *El desarrollo agropecuario y agroindustrial de Uruguay: Reflexiones en el 50 aniversario de la Oficina de Programación y Política Agropecuaria (OPYPA-MGAP)*, edited by Unidad de Comunicación Organizacional y Difusión, 26–35. Montevideo: OPYPA-MGAP.

Cartes, José L, J. J. Thompson, and A. Yanosky. 2015. "El Chaco paraguayo como uno de los últimos refugios para los mamíferos amenazados del Cono Sur." *Paraquaria Natural* 3 (2): 37–47.

CAS. 2018. *Anuario de Comercio Exterior de base agraria de los países CAS 2013–2017*. Edited by M. Ackermann and L. Gorga. Montevideo: Consejo Agropecuario del Sur.

Castellano, Andrés, and Mercedes Elida Goizueta. 2011. *Agregado de valor en la cadena de la soja: alternativa de upgrading para productores primarios*. EEA Balcarce Área de Economía y Sociología Rural. Balcarce: INTA.

CEPAL, ECLAC. 2018. "Paraguay." In *Economic Survey of Latin America and the Caribbean*, edited by Economic Commission for Latin America and the Caribbean (ECLAC). Santiago: CEPAL.

Cirio, Ignacio. 2011. *Efectos Colaterales: Testimonios de afectados y afectadas por el agrongeocio en Uruguay 2011*. Edited by Amigos de la Tierra - Redes. Montevideo: Astra.

CLAEH, FIDA Mercosur. 2015. *Institucionalidad responsable por la agricultura familiar y las políticas públicas diferenciadas en el Mercosur. Marco de actuación de la REAF*. Montevideo, Uruguay: Programa FIDA Mercosur CLAEH.

CNFR. 2008. "Hacia dónde vamos con este modelo productivo?" In *Rap-AL Uruguay, Transgénicos*. Montevideo: Comisión Nacional de Fomento Rural.

CNFR. 2009. *Propuesta de políticas públicas diferenciadas para el desarrollo de la agricultura familiar*. Comunicado de Prensa. Montevideo: Comision Nacional de Fomento Rural.

CNFR. 2010. "Diálogo Nacional sobre la Funcion Social de la Tierra: Por Políticas de Acceso a Tierra y Desarrollo Rural sostenible." *El Noticiero* 19: 4–8.

Colla, Julia Lucía. 2017. "La territorialidad campesina indígena y la disputa por el territorio en el Chaco (Argentina)." *Geograficando* 13 (2): 2–16.

Coordinadora de Derechos Humanos del Paraguay. 2007. *Informe Chokokue: Informe al Relator Especial sobre las ejecuciones extrajudiciales, sumarias o arbitrarias del Consejo de Derechos Humanos de Naciones Unidas sobre las violaciones al derecho a la vida en contra de miembors y dirigentes de las organizaciones campesinas en el contexto de reforma agaria en Paraguay (1989–2005).* Asunción: Codehupy.

Dabezies, Martín. 2009. "Área Cadenas Agroindustriales: Informe final de la consultoría sobre Cadenas Agroindustriales en el marco del Plan Estratégico Nacional en Ciencia, Tecnología e Innovación." In *Plan Estratégico Nacional de Ciencia, Tecnología e Innovación (PENCTI)*, edited by Equipo Operativo Gabinete Ministerial de la Innovación. Montevideo: PENCTI.

Davila, Mabel, Daniel Iglesias, Carlos Alonso, and Laura Rodríguez Silvera. 2017. "Regulation for Wheat and Soybean Exports in Argentina, 2004–2014." *International Journal for Advanced Research (IJAR)*, 5 (6): 883–899.

Demetrio, Pablo Martín. 2012. "Estudio de efectos biológicos de plaguicidas utilizados en cultivos de soja RR y evaluación de impactos adversos en ambientes acuáticos de agroecosistemas de la región pampeana." Doctoral dissertation, Facultad de Ciencias Exactas. Universidad Nacional de la Plata, Argentina.

de Torres Álvarez, M. F., P. Arbeletche, E. Sabourin, J. Cardelliac Gula, and G. Massardier. 2014. "La agricultura familiar entre proyectos nacionales de desarrollo." *Eutopía* (6): 25–40.

de Waroux, Yann le Polain, Rachael D. Garrett, Jordan Graesser, Christoph Nolte, Christopher White, and Eric F. Lambin. 2017. "The Restructuring of South American Soy and Beef Production and Trade Under Changing Environmental Regulations." *World Development* 121: 188–202. https://doi.org/10.1016/j.worlddev.2017.05.034.

DGDR-MGAP. 2018. *Tramites y servicios de la producción familiar, Unidad organizativa, Dirección General de Desarrollo Rural.* Montevideo: MGAP.

Diaz, Pablo. 2015. "Legislación sobre acceso a la tierra en Uruguay." Declaracion Final del VI Congreso CLOC-Via Campesina, Buenos Aires, 18 April 2015.

Díaz-Zorita, Martín, Gustavo A. Duarte, and John H. Grove. 2002. "A Review of No-Till Systems and Soil Management for Sustainable Crop Production in the Subhumid and Semiarid Pampas of Argentina." *Soil and Tillage Research* 65 (1): 1–18. https://doi.org/10.1016/S0167-1987(01)00274-4.

DIEA. 2014. *Anuario Estadístico Agropecuario 2013.* Edited by Ministerio de Ganadería, Agricultura y Pesca - Dirección de Investigaciones Económicas Agropecuarias. Montevideo: MGAP.

DIEA. 2016. *Anuario Estadístico Agropecuario 2015*. Edited by Ministerio de Ganadería, Agricultura y Pesca - Dirección de Investigaciones Económicas Agropecuarias. Montevideo: MGAP.

Dominguez, Diego, and Pablo Sabatino. 2008. "La conflicitividad en los espacios rurales de Argentina." *Lavboratorio, Cambio Estructural y Desigualdad Social (CEyDS)/Facultad de Ciencias Soicales, UBA* 1m0 (22): 38–52.

Durán, Verónica, and Lucía Salgado. 2013. "Avances en el régimen de promoción de inversionen (COMAP)." In *Anuario Opypa 2013*, edited by MGAP-OPYPA. Montevideo: Agropecuaria Hemisferio Sur.

ECLAC, FAO, IICA. 2017. *The Outlook for Agriculture and Rural Development in the Americas: A Perspective on Latin America and the Caribbean 2017–2018*. San José, Costa Rica: Economic Commission for Latin America and the Caribbean (ECLAC), Food and Agriculture Organization of the United Nations (FAO), and Inter-American Institute for Cooperation on Agriculture (IICA).

Ernst, Oswaldo, and Guillermo Siri-Prieto. 2009. "Impact of Perennial Pasture and Tillage Systems on Carbon Input and Soil Quality Indicators." *Soil and Tillage Research* 105 (2): 260–268.

Errea, Eduardo, Juan Peyrou, Joaquín Secco, and Gonzalo Souto. 2011. *Transformaciones en el agro uruguayo - Nuevas instituciones y modelos de orgnaización empresarial*. Edited by Dámaso Antonio Larrañaga Universidad Católica del Uruguay, *Universidad Católica*. Montevideo: Facutad de Ciencias Empresariales, Programa de Agronegocios.

Ezquerro-Cañete, Arturo. 2016. "Poisoned, Dispossessed and Excluded: A Critique of the Neoliberal Soy Regime in Paraguay." *Journal of Agrarian Change* 16 (4): 702–710.

Ezquerro-Cañete, Arturo, and Ramón Fogel. 2017. "A Coup Foretold: Fernando Lugo and the Lost Promise of Agrarian Reform in Paraguay." *Journal of Agrarian Change* 17 (2): 279–295.

FAO. 2017. "Soybean Prices, Economic Growth and Poverty in Argentina and Brazil." In *Background paper to the UNCTAD-FAO Commodities and Development Report 2017 Commodity Markets, Economic Growth and Development*. Rome: Food and Agriculture Organization of the United Nations.

Ffrench-Davis, Ricardo. 2005. *Reformas para América Latina después del fundamentalismo neoliberal*. Santiago: CEPAL.

Filomeno, Felipe Amin. 2013. "How Argentine Farmers Overpowered Monsanto: The Mobilization of Knowledge-Users and Intellectual Property Regimes." *Journal of Politics in Latin America* 5 (3): 35–71.

FONAF. 2008. *Foro nacional de agricultura familiar: Propuestas para un plan estratégico de desarrollo rural*. Buenos Aires: FONAF-SAGyP.

Frente Amplio. 2003, December 22. *Grandes lineamientos programáticos para el gobierno 2005–2009 - Porque entre todos otro Urugauy es posible.* Edited by Frente Amplio. Montevideo: IV Congreso extraordinario del Frente Amplio.

García-López, A. Gustavo, and Nancy Arizpe. 2010. "Participatory Processes in the Soy Conflicts in Paraguay and Argentina." *Ecological Economics* 70 (2): 196–206. https://doi.org/10.1016/j.ecolecon.2010.06.013.

García-Préchac, F., O. Ernst, G. Siri-Prieto, and J. A. Terra. 2004. "Integrating No-Till into Crop-Pasture Rotations in Uruguay." *Soil and Tillage Research* 77 (1): 1–13.

GEF, FAO. 2015. *Project: Strengthening Capacities for the Sound Management of Pesticides Including POPs (MSP) in Uruguay.* Food and Agriculture Organization, Global Environment Facility.

Giarracca, Norma. 2008. "La Argentina y la democratización de la tierra." *Lavboratorio, Cambio Estructural y Desigualdad Social (CEyDS)/Facultad de Ciencias Soicales, UBA.* 10 (22): 18–21.

Gisclard, Marie, Gilles Allaire, and Roberto Cittadini. 2015. "Proceso de institucionalización de la agricultura familiar y nuevo referencial para el desarrollo rural en la Argentina." *Mundo agrario* 16 (31): 1–18.

Gobierno Nacional. 2014. *Plan Nacional de Desarrollo—Paraguay 2030.* Asunción: Secretaría Técnica de Planificaión del Desarrollo Económico y Social.

Gomez, Marcelo. 2008. "La soja de la discordia. Los sentidos y estrategias en la movilización de la pequeña buruesía." *Lavboratorio, Cambio Estructural y Desigualdad Social (CEyDS)/Facultad de Ciencias Soicales, UBA* 10 (22): 22–35.

Gonzaga Belluzo, Luis. 2015. "La reciente internacionalización del régimen del capital." In *Neoestructuralismo y corrientes heterodoxas en América Latina y el Caribe a inicios del siglo XXI*, edited by Alicia Bárcena and Antonio Prado, 112–125. Santiago de Chile: CEPAL.

Graesser, Jordan, T. Mitchell Aide, H. Ricardo Grau, and Navin Ramankutty. 2015. "Cropland/Pastureland Dynamics and the Slowdown of Deforestation in Latin America." *Environmental Research Letters* 10 (3). https://doi.org/10.1088/1748-9326/10/3/034017.

Gras, Carla, and Valera Hernández. 2013. *El agro como negocio: producción, sociedad y territorios en la globalización.* Buenos Aires: Editorial Biblos.

Hansen, Matthew C., Peter V. Potapov, Rebecca Moore, Matt Hancher, S. A. Turubanova, Alexandra Tyukavina, David Thau, S. V. Stehman, S. J. Goetz, and T.R. Loveland. 2013. "High-Resolution Global Maps of 21st-Century Forest Cover Change." *Science* 342 (6160): 850–853.

Hernández, René. 2015. "Transformación del Estado y paradigmas de desarrollo en América Latina." In *Neoestructuralismo y corrientes heterodoxas en América*

*Latina y el Caribe a inicios del siglo XXI*, edited by Alicia Bárcena Ibarra and Antonio Prado. Santiago: CEPAL.

Hill, Mariana, and Carlos Clérici. 2013a. "Avances en la política de conservación de suelos." In *Anuario OPYPA-MGAP*, edited by OPYAPA-MGAP. Montevideo: MGAP.

Hill, Mariana, and Carlos Clérici. 2013b. "Avances en la política de conservación de suelos." In *Anuario de OPYPA*, edited by Opypa-MGAP. Montevideo: MAGAP.

Hill, Mariana, Kate Kennedy Freeman, and Raquel Orejas. 2016. "Plotting Productivity: Soil Use Management Plans in Uruguay." 2016 World Bank Conference on Land and Poverty, Washington, DC. 15 March 2016.

Human Rights Council. 2015. "Report of the Special Rapporteur on the Rights of Indigenous Peoples, Victoria Tauli-Corpuz, Addendum. The Situation of Indigenous Peoples in Paraguay." *Human Rights Council, Thirtieth session, Agenda Item 3*, United Nations—General Assembly.

IBRD, IFC, MIGA. 2014. *Country Partnership Framework for the Argentine Republic for the Period 2015–2018*. Report Number: 131971.

IBRD, IFC, MIGA. 2015. *Country Partnership Framework for the Oriental Republic of Uruguay for the Period 2016–2020*. Report Number: 97063.

IBRD, IFC, MIGA. 2018. *Country Partnership Framework for the Republic of Paraguay for the Period 2019–2023*. Report Number: 131046.

IDB, DCEA-MAG. 2017. *Proyecto de implementación del sistema de censo y encuestas agropecuarias—Análisis Socio Cultural—Informe Final*. Edited by John Renshaw. San Lorenzo-Asunción: Inter-American Development Bank.

Imbach, P., J. Robalino, J. Zamora, C. Brenes, C. Sandoval, E. Pacay, M. Cifuentes-Jara, and G. Labbate. 2016. *Escenarios de deforestación futura en Paraguay*. Edited by FAO/PNUD/PNUMA. Asunción: PNC ONU-REDD+Py/SEAM/INFONA/FAPI.

INFONA. 2015. *Inventario Forestal Nacional*. Edited by INFONA Programa ONU-REDD, SEAM, FAPI, FAO, PNUD, and PNUMA. Asuncion: Dirección de sistema nacional de información forestal.

International Monetary Fund. 2019. "Uruguay: 2018 Article IV Consultation-Press Release; Staff Report; and Statement by the Executive Director for Republic of Uruguay." In *Country Report*. Washington: IMF.

International Service for the Acquisition of Agri-Biotech Applications, ISAAA. 2011. "GM Approval Database, Soybean, GTS 40-3-2." ISAAA. Accessed 12 May 2011. http://www.isaaa.org/gmapprovaldatabase/events/default.asp?EventID=94.

Irala, Abel Enrique, and Hugo Javier Pereira Cardozo. 2016. "Violencia armada y avance de la soja en el norte del Paraguay." *Conflicto Social* 9 (16): 180–208.

Jobbágy, E. G., H. R. Grau, J. M. Paruelo, and E. F. Viglizzo. 2015. "Farming the Chaco: Tales from Both Sides of the Fence." *Journal of Arid Environments* 123: 1–2. https://doi.org/10.1016/j.jaridenv.2015.07.011.

Kassai, Lucia, and Rodrigo Orihuela. 2011. "Hedge-Fund Backed El Tejar Tests IPO Appetite with $300 Million Bond Sale." *Bloomberg.* www.bloomberg. com/news/print/2011-05-09/hedge-fund-backed-el-tejar-tests-ipo-appetite-with-300-million-bond-sale.html.

Kroes, Joop, Jos van Dam, Iwan Supit, Diego de Abelleyra, Santiago Verón, Allard de Wit, Hendrik Boogaard, Marcos Angelini, Francisco Damiano, Piet Groenendijk, Jan Wesseling, and Ab Veldhuizen. 2019. "Agrohydrological Analysis of Groundwater Recharge and Land Use Changes in the Pampas of Argentina." *Agricultural Water Management* 213: 843–857. https://doi. org/10.1016/j.agwat.2018.12.008.

Lambert, Peter. 2016. "The Myth of the Good Neighbour: Paraguay's Uneasy Relationship with Brazil." *Bulletin of Latin American Research* 35 (1): 34–48. https://doi.org/10.1111/blar.12410.

Lapegna, Pablo. 2016. "Genetically Modified Soybeans, Agrochemical Exposure, and Everyday Forms of Peasant Collaboration in Argentina." *The Journal of Peasant Studies* 43 (2): 517–536. https://doi.org/10.1080/03066150.201 5.1041519.

Leguizamon, Amalia. 2014. "Roundup Ready Nation: The Political Ecology of Genetically Modified Soy in Argentina." PhD, Sociology, City University of New York.

Lende, Sebastián Gómez. 2017. "Usos del territorio, acumulación por desposesión y derecho a la salud en la Argentina contemporánea: el caso de la soja transgénica." *GEOgraphia* 19 (39): 3–15.

Lindenboim, Javier. 2008. "Distribuir y redistribuir: he ahí la cuestión." *Lavboratorio, Cambio Estructural y Desigualdad Social (CEyDS)/Facultad de Ciencias Soicales, UBA* 10 (22): 13–17.

Llambi, Cecilia, Silvia Laens, and Marcelo Perera. 2016. "Assessing the Impacts of a Major Tax Reform: A CGE-Microsimulation Analysis for Uruguay." *International Journal of Microsimulation* 9 (1): 134–166.

Lussich, Manuel. 2009. "Discurso del Presidente de la Asociación Rural del Uruguay." *Revista ARU.* Octubre 2009. Montevideo: Duplex Comunicación.

MAG. 2009. "Censo agropecuario nacional de 2008." In *CAN*, edited by DCEA-MAG. San Lorenzo: Ministerio de Agricultura y Ganadería.

Magliano, Patricio N., Roberto J. Fernández, Eva L. Florio, Francisco Murray, and Esteban G. Jobbágy. 2017. "Soil Physical Changes After Conversion of Woodlands to Pastures in Dry Chaco Rangelands (Argentina)." *Rangeland Ecology & Management* 70 (2): 225–229.

Mansourian, Stephanie, Lucy Aquino, Thomas Erdmann, and Francisco Pereira. 2014. "A Comparison of Governance Challenges in Forest Restoration in

Paraguay's Privately-Owned Forests and Madagascar's Co-Managed State Forests." *Forests* 5 (4): 763–783.

Manzanal, Mabel, and Sergio Schneider. 2011. "Agricultura Familiar y Políticas de Desarrollo Rural en Argentina y Brasil (análisis comparativo, 1990–2010)." *Revista Interdisciplinaria de Estudios Agrarios* 34 (CIEA, FCE, UBA): 35–71.

Manzetti, Luigi. 1992. "The Evolution of Agricultural Interest Groups in Argentina." *Journal of Latin American Studies* 24 (3): 585–616.

Martens, Juan, and Rodrigo Estigarribia. 2018. "Eficacia de las políticas de mano dura en Paraguay." *Revista Jurídica Investigaciòn en Ciencias Jurìdicas y Sociales* (7): 43–135. https://ojs.ministeriopublico.gov.py/index.php/rjmp/issue/view/8.

Martorano, Bruno. 2014. "The Impact of Uruguay's 2007 Tax Reform on Equity and Efficiency." *Development Policy Review* 32 (6): 701–714.

MAyDS. 2010. *Argentina Líder Agroalimentario. Plan Estratégico Agroalimentario y Agroindustrial Participativo y Federal 2010–2020, Metas 2020 para el Sector Agroalimentario y Agroindustrial argentino.* Edited by Ganadería y Pesca Ministerio de Agricultura. Buenos Aires: Presidencia de la Nación.

MAyDS. 2017. "Informe de estado de implementación 2010–2015: Ordenamiento Territorial de Bosques Nativos y planes alcanzados por el Fondo Nacional para el Enriquecimiento y la Conservación de los Bosques Nativos." Ley No. 26. 331 de Presupuestos Mínimos de Protección Ambiental de los Bosques Nativos, Ministro de Ambiente y Desarrollo Sustentable de la Nación.

MGAP. 2017. *Uruguay Agrointeligente: Los desafíos para un desarrollo sostenible.* Montevideo: Ministerio de Ganadería, Agricultura y Pesca—MGAP.

MGAP, DGSA. 2011. IMPORTACIÓN DE PRODUCTOS FITOSANITARIOS—Resumen estadístico Ejercicio 2010. MGAP Servicios Agrícolas.

MGAP, DIEA. 2014. *Censo General Agropecuario 2011—Resultados definitivos.* Edited by MGAP. Montevideo: Estadísticas Agropecuarias (DIEA).

Modernel, Pablo, Walter A. H. Rossing, Marc Corbeels, Santiago Dogliotti, Valentin Picasso, and Pablo Tittonell. 2016. "Land Use Change and Ecosystem Service Provision in Pampas and Campos Grasslands of Southern South America." *Environmental Research Letters* 11 (11): 113002.

Muñoz, Roberto. 2016. "Organizaciones campesinas en la provincia de Chaco, Argentina. Una aproximación a su composición social a partir de sus acciones de protesta: el caso de la Unión Campesina de Chaco (UCC), 2002–2011." *el@ tina. Revista electrónica de estudios latinoamericanos* 14 (55): 22–43.

MVOTMA. 2017. *Plan Nacional de Aguas.* Edited by D. Greif and E. Fierro. Montevideo: Presidencia de la República.

Nardo, Daniela. 2011. "Estudio del impacto de plaguicidas utilizados en el cultivo de soja y en otras actividades agrícolas sobre las especies acuáticas de consumo humano en el Área Protegida Laguna de Rocha." Maestría en Nutrición, Universidad católica del Uruguay.

Naylor, Rosamond L., and Matthew M. Higgins. 2017. "The Political Economy of Biodiesel in an Era of Low Oil Prices." *Renewable and Sustainable Energy Reviews* 77: 695–705. https://doi.org/10.1016/j.rser.2017.04.026.

Nogueira, María Elena, Marcos Andrés Urcola, and Mario Lattuada. 2017. "La gestión estatal del desarrollo rural y la agricultura familiar en Argentina: estilos de gestión y análisis de coyuntura 2004–2014 y 2015–2017." *Revista Latinoamericana de Estudios Rurales* 2 (4): 23–59.

Nogués, Julio J. 2011. "Agricultural Export Barriers and Domestic Prices: Argentina During the Last Decade." *Report Prepared for FAO.*

Novelli, Leonardo E., Octavio P. Caviglia, and Gervasio Piñeiro. 2017. "Increased Cropping Intensity Improves Crop Residue Inputs to the Soil and Aggregate-Associated Soil Organic Carbon Stocks." *Soil and Tillage Research* 165: 128–136.

Ocampo, José Antonio. 2004. "Latin America's Growth and Equity Frustrations During Structural Reforms." *Journal of Economic Perspectives* 18 (2): 67–88.

Ocampo, José Antonio. 2015. "América Latina frente a la turbulencia económica mundial." In *Neoestructuralismo y corrientes heterodoxas en América Latina y el Caribe a inicios del siglo XXI*, edited by Alicia Bárcena and Antonio Prado. Santiago de Chile: CEPAL.

Oddone, Gabriel. 2015. "Sustento del fuerte dinamismo del sector agropecuario." In *El desarrollo agropecuario y agroindustrial de Uruguay: Reflexiones en el 50 aniversario de la Oficina de Programación y Política Agropecuaria (OPYPA-MGAP)*, edited by Unidad de Comunicación Organizacional y Difusión, 32–36. Montevideo: MGAP.

Oliveira, Gustavo, and Susanna Hecht. 2016. "Sacred Groves, Sacrifice Zones and Soy Production: Globalization, Intensification and Neo-Nature in South America." *The Journal of Peasant Studies* 43 (2): 251–285. https://doi.org/1 0.1080/03066150.2016.1146705.

ONU-REDD+/SEAM/INFONA/FAPI, PNC. 2016. Mapeo de los benificios múltiples de REDD+en Paraguay: analisís adicionales para orientar la toma de decisiones sobre políticas y medidas REDD+. Edited by Programa de las Naciones Unidas para el Medio Ambiente. Asunción: FAO/PNUD/PNUMA.

Osorio, Gillard Gabriel. 2009. "Individual Cattle Traceability." In *Un nodo de cooperación sobre la experiencia de Uruguay en trazabilidad bovina*, edited by Inter-American Institute for Cooperation on Agriculture (IICA). Montevideo: MGAP-INAC-IICA.

Oyhantçabal, Gabriel, and Ignacio Narbondo. 2011. *Radiografía del agronegocio sojero—Descripción de los principales actores y los impactos socio-económicos en Uruguay. Datos actualizados a 2010.* 2nd ed. (first 2008) ed. Montevideo: Redes-AT.

Oyhantçabal, Gabriel, and Ignacio Narbondo. 2018. "Land Grabbing in Uruguay: New Forms of Land Concentration." *Canadian Journal of Development Studies/Revue canadienne d'études du développement*: 1–19. https://doi.org/10.1080/02255189.2018.1524749.

Pagnussatt, Eva. 2018. "El derecho a una alimentación sana. Inseguridad alimentaria y salud humana." *Dilemata* (26): 169–177.

Paolino, Carlos. 2013. "Uruguay: estrategia de desarrollo agropecuario." In *Políticas para la agricultura en América Latina y el Caribe: competitividad, sostenibilidad e inclusión social*, edited by CEPAL, 47–54. Santiago de Chile: CEPAL.

Paolino, Carlos. 2015. "La política púbica y el apoyo al sector agropecuario." In *El desarrollo agropecuario y agroindustrial de Uruguay: Reflexiones en el 50 aniversario de la Oficina de Programación y Política Agropecuaria (OPYPA-MGAP)*, edited by Unidad de Comunicación Organizacional y Difusión and Diego Campoy, 37–45. Montevideo: MGAP.

Paolino, Carlos, Lucía Pittaluga, and Mario Mondelli. 2014. "Cambios en la dinámica agropecuaria y agroindustrial del Uruguay y las políticas públicas." LC/L.3821. *Estudios y Perspectivas*, 15. Santiago de Chile: CEPAL.

Papotto, Daniel. 2013. "Argentina: el Plan Estratégico Agroalimentario y Agroindustrial Participativo y Federal 2010–2020." In *Políticas para la agricultura en América Latina y el Caribe: competitividad, sostenibilidad e inclusión social*, edited by CEPAL, 29–34. Santiago de Chile: CEPAL.

Petraglia, Cecilia, Ruben Puentes, Ricardo Cayssials, José Barrios, and José P. Lucas. 1982. *Avances en conservación de suelos en el Uruguay*. Montevideo: Departamento de Uso, Manejo y Conservación de Suelos, Ministerio de Agricultura y Pesca, and Instituto Interamericano de Cooperación para la Agricultura.

Piera Valdés, A. 2016. "Consultoría Nacional: Análisis del Marco Legal e Institucional Vigente para la Implementación de REDD+ en Paraguay." In *Program ONU-REDD+ Paraguay*. Asunción: FAO, PNUD, PUMA, INFONA, SEAM, and FAPI.

Piñeiro, Diego E., and Joaquín Cardeillac. 2017. "The Frente Amplio and Agrarian Policy in Uruguay." *Journal of Agrarian Change* 17 (2): 365–380.

Presidencia de la República Oriental del Uruguay, Oficina de Planeamiento y Presupuesto. 2017. *Objetivos de Desarrollo Sostenible: Informe Nacional Voluntario—Uruguay*. Montevideo: Presidencia, OPP, AGEV, INE, and AUCI.

Prieto, G. S., and O. Ernst. 2010. "Manejo del suelo y rotación con pasturas: Efecto sobre la calidad del suelo, el rendimiento de los cultivos y el uso de insumos." *Informaciones Agronómicas del Cono Sur* (45): 22–26.

Primost, Jezabel E., Damián J. G. Marino, Virginia C. Aparicio, José Luis Costa, and Pedro Carriquiriborde. 2017. "Glyphosate and AMPA, 'Pseudo-Persistent' Pollutants Under Real-World Agricultural Management Practices in the Mesopotamic Pampas Agroecosystem, Argentina." *Environmental Pollution* 229: 771–779.

Py/SEAM/INFONA/FAPI, PNC ONU-REDD+. 2016. *Paraguay: cambio de uso e suelo y costos de oportunidad. Sinergias entre REDD+y la Ley de Valoración y Retribución de Servicios Ambientales.* Ciudad de Panama: FAO/PNUD/PNUMA.

RAP-AL Uruguay. 2009. "Escuelas rurales contaminadas y en peligro de desaparición." *RAP-AL.* Accessed 30 August 2012. http://www.rapaluruguay.org/Comunicados/Escuelas_rurales.html.

Redo, Daniel J., T. Mitchell Aide, Matthew L. Clark, and María José Andrade-Núñez. 2012. "Impacts of Internal and External Policies on Land Change in Uruguay, 2001–2009." *Environmental Conservation* 39 (2): 122–131. https://doi.org/10.1017/s0376892911000658.

Richardson, Neal P. 2009. "Export-Oriented Populism: Commodities and Coalitions in Argentina." *Studies in Comparative International Development* 44 (3): 228.

Ríos, Amalia. 2011. "El riesgo de resistencia a glifosato en Uruguay." In *Jornada Cultivos de Invierno 2011, INIA LA Estanzuela,* edited by INIA. Flores, Uruguay: INIA LA Estanzuela.

Ríos, A., M. A. García, A. Belgeri, M. P. Caulin, V. Mailhos, and G. San Roman. 2008. "Comunidades florísticas asociadas a los sistemas de siembra directa en Uruguay—Factores que pueden afectar la efectividad del herbicida glifosato." *Serie de Actividades de Diffusión INIA* (554): 96.

Ríos, Mariana, Natalia Zaldúa, and Sabrina Cupeiro. 2010. *Evaluación participativa de plaguicidas en el sitio RAMSAR, parque Nacional Esteros de Farrapos e ISlas del Río Uruguay.* Montevideo: Vida Silvestre.

Riquelme, Quintin. 2014. "El derecho a la tierra desde la concepción de los movimientos campesinos." In *La tierra en el Paraguay: de la desigualdad al ejercicio de derechos,* edited by Patricio Dobrée, 47–62. Asunción: Programa Democratización y Construcción de la Paz—Paraguay.

Ronco, A. E., D. J. Marino, M. Abelando, P. Almada, and C. D. Apartin. 2016. "Water Quality of the Main Tributaries of the Paraná Basin: Glyphosate and AMPA in Surface Water and Bottom Sediments." *Environmental Monitoring and Assessment* 188 (8): 1–13.

Rosselli, Pablo. 2015. *Observatorio de Oleaginosos Uruguay: Indicadores sectoriales y escenarios futuros.* Edited by Deloitte, MTO and Oleaginosos Uruguay. Montevideo: Mesa Tecnológica dr Oleaginosos.

Rossi, Virginia. 2010. "La producción familiar en la cuestión agraria uruguaya." *REVISTA NERA* 13 (16): 63–80.

Salvo, L., J. Hernández, and O. Ernst. 2010. "Distribution of Soil Organic Carbon in Different Size Fractions, Under Pasture and Crop Rotations with Conventional Tillage and No-Till Systems." *Soil and Tillage Research* 109 (2): 116–122.

Sánchez, W. Alejandro. 2011. "Sangre Joven: Understanding the New Wave of Armed Groups in Latin America." *Security and Defense Studies Review* 12 (1): 135–155.

Sandoval, Lazaro 2016. "Oilseeds and Products Annual: Uruguay." In *GAIN Report*, edited by Global Agricultural Information Network. USDA.

Santagada, Ezequiel. 2013. *Reporte sobre la problemática de la tenencia de la tierra en el Paraguay de cara a la implementación del Programa REDD+.* Edited by ONU-REDD PNC Paraguay. Asunción: Programa de Naciones Unidas para el Medio Ambiente (PNUMA).

Saylor, Ryan. 2014. "Commodity Booms, Coalitional Politics and Government Intervention in Credit Markets." *Review of International Political Economy* 21 (3): 640–669. https://doi.org/10.1080/09692290.2013.806271.

Shurtleff, William, and Akiko Aoyagi. 2009. *History of Soybeans and Souyfoods in South America (1882–2009): Extensively Annotated Bibliography and Sourcebook.* Lafayette: Soyinfo Center.

Svampa, Maristella Noemi. 2013. "Consenso de los Commodities y lenguajes de valoración en América Latina." *Nueva Sociedad* 244 (4): 30–46.

Tambler, Adrián. 2013. "Recaudación y presión fiscal en el agro." In *Anuario 2013*, edited by Opypa. Montevideo: Opypa—MGAP.

The World Bank. 2001. *Paraguay Country Assistance Evaluation.* Edited by Operations Evaluation Department. Washington, DC: World Bank.

Tkachuk, Maximiliano, and Marina Dossi. 2014. *Dinámica de la producción ganadera Argentina: Análisis de variables intervinientes y de escenarios futuros.* Buenos Aires: FAUBA, Apuntes Agroeconomicos.

Toledo Lopez, Virginia. 2017. "La política agraria del kirchnerismo. Entre el espejismo de la coexistencia y el predominio del agronegocio." *Mundo Agrario* 18 (37): 25.

Torrella, Sebastián A., María Piquer-Rodríguez, Christian Levers, Rubén Ginzburg, Gregorio Gavier-Pizarro, and Tobias Kuemmerle. 2018. "Multiscale Spatial Planning to Maintain Forest Connectivity in the Argentine Chaco in the Face of Deforestation." *Ecology and Society* 23 (4). https://doi.org/10.5751/es-10546-230437.

Turban, Ismael. 2013. *El Tejar: Ismael Turban.* SoundCloud: Blasina y Asociados: Agronegocios y comunicación.

Uruguay XXI. 2013. *Uruguay: A Place to Invest, Work and Live.* Montevideo: Uruguay XXI.

Uruguay XXI. 2016. *Encuesta Inversores Extranjeros.* Montevideo: Uruguay XXI.

Uruguay XXI. 2017. Foreign Trade Report 2016. Montevideo: Uruguay XXI.

USDA. 2017a. "Paraguay: Livestock and Products Annual." In *GAIN Report*, edited by USDA *Foreign Agricultural Services.* Global Agricultural Information Network—GAIN. Washington, DC: USDA.

USDA. 2017b. "Uruguay: Livestock and Products Annual." In *GAIN Report*, edited by USDA Foreign Agricultural Service: Global Agricultural Information Network. Washington, DC: USDA.

USDA. 2018. "Oilseeds: World Market and Trade." In *World Agricultural Reports*. United States Department of Agriculture, Foreign Agricultural Services.

USDA Foreign Agricultural Service. 2017. "Paraguay: Livestock and Products Annual, 2016." In *GAIN Report*, edited by GAIN Publications. USDA.

Walcott, J., J. Thorley, G. Casco, L. Coronel, V. Kapos, L. Miles, R. Blaney, and S. Woroniecki. 2014. *Mapeo de los beneficios múltiples de REDD+en Paraguay: el uso de la información espacial para apoyar la planificación del uso de la tierra*. Cambridge: Programa de las Naciones Unidas para el Medio Ambiente.

Vallejos, María, José N. Volante, María J. Mosciaro, Laura M. Vale, M. Laura Bustamante, and José M. Paruelo. 2015. "Transformation Dynamics of the Natural Cover in the Dry Chaco Ecoregion: A Plot Level Geo-Database from 1976 to 2012." *Journal of Arid Environments* 123: 3–11. https://doi.org/10.1016/j.jaridenv.2014.11.009.

Vergara-Camus, Leandro, and Cristobal Kay. 2017a. "New Agrarian Democracies: The Pink Tide's Lost Opportunity." *Socialist Register* 54 (54): 211–230.

Vergara-Camus, Leandro, and Cristobal Kay. 2017b. "Agribusiness, Peasants, Left-Wing Governments, and the State in Latin America: An Overview and Theoretical Reflections." *Journal of Agrarian Change* (17): 239–257.

Villalba Clavijo, Clara. 2015. *Estudio de las mesas de desarrollo rural en Uruguay como innovación institucional para la participación y la inclusión*. Edited by IICA. Montevideo: IICA, DGDR.

Volante, José Norberto, and Lucas Seghezzo. 2018. "Can't See the Forest for the Trees: Can Declining Deforestation Trends in the Argentinian Chaco Region be Ascribed to Efficient Law Enforcement?" *Ecological Economics* 146: 408–413. doi:https://doi.org/10.1016/j.ecolecon.2017.12.007.

Yanosky, Alberto. 2013. "Paraguay's, Challenge of Conserving Natural Habitats and Biodiversity with Global Markets Demanding for Products." In *Conservation Biology: Voices from the Tropics*, edited by Luke Gibson Navjot S. Sodhi and Peter H. Raven, 113–119. Asunción: Wiley and Sons.

# CHAPTER 6

# Conclusion: State Autonomy and Capacity in a Comparative Light

As underlined many times in this book, agrarian change in Latin America during the past two decades has been truly dramatic. Argentina, Paraguay, and Uruguay are at the heart of the transformation—beef and soy are the main protagonists of change—but the main driving forces of these commodity chains are for the most part to be found far away from Latin America. Agrarian change is ultimately driven by increased international demand on food, particularly meat, and thus on pastureland and cropland for feed. A plethora of interacting shifts in the political economy of the international agrofood system has in turn co-constructed and fueled the international demand. Some of the most significant shifts and trends since the 1980s are trade liberalization and deregulation (albeit uneven), stronger intellectual private property rights, economic growth, and national policy shifts in Asia (particularly China), population growth and new technologies. These shifts have in turn allowed for growing dominance of increasingly concentrated and vertically integrated giant transnational agribusiness actors (Chapter 3).

While exogenous drivers have been decisive, endogenous features also played an important role, co-shaping agrarian change in Latin America. Clearly, availability of relatively cheap and productive land was an important precondition, but macroeconomic policies, social and environmental legislation, tax systems, land tenure, and so forth also influence the ways pressures from agrofood globalization have played out. For example, the restructuring reforms during the most neoliberal period of

© The Author(s) 2020

M. Baraibar Norberg, *The Political Economy of Agrarian Change in Latin America*, Governance, Development, and Social Inclusion in Latin America, https://doi.org/10.1007/978-3-030-24586-3_6

the late twentieth century (Chapter 4) facilitated significantly the later agribusiness-led agrarian change in the region, notwithstanding country-specific differences. How states have acted through public policies and regulations in relation to the transformative pressures of agrofood globalization, mediating between the global driven market forces and the domestic economic, social and political forces, have been thoroughly explored in this book.

Besides considering the macro and micro linkages, this book has situated contemporary agrarian change in a much longer history of distant drivers, policy regulations, and agrarian transformations. By exploring the political economy of agrarian change in the region since the first wave of massive trade globalization—from the 1870s and until the crack of the Bretton-Woods system a century later (Chapter 2)—up to the neoliberal reorientation of the last decades of the twentieth century (Chapter 4), historical legacies in present-day articulations have been identified. "You have to know the past to understand the present", as Carl Sagan once famously put it,[1] and this inquiry has made it very clear that it is impossible to understand contemporary patterns of change in Latin America without a thorough comprehension of historically formed institutions, of which many have proven to be extraordinary persistent over time. At the same time, the historical backdrop also shows that in the midst of strong lines of continuities, path dependency and inertia, rather abrupt changes have from time to time occurred. Thus, historical processes have created paths, consolidated power-relations, institutionalized privileges, and so on, that can be self-reinforcing and difficult to reverse, but they alone are not determining the present-day struggles or the future.

This final and concluding chapter builds on all previous chapters and provides a comparative analysis of the differentiated and historically formed capacities of Argentina, Paraguay, and Uruguay to regulate distantly driven contemporary agrarian change in ways that boost the benefits and mitigate real and perceived risks to human well-being and the environment. More concretely, the three states have during the twenty-first century explicitly expressed that they aim to use public policies in order to make agrarian change: (1) more "developmental", through value-added incorporation and upgrading of the productive structure; (2) more socially inclusive, through redistribution, such as targeted support programs to family producers and other vulnerable groups; and (3) more sustainable, through long-term care of natural resources. Similarities and differences in the main ways of the states to try to achieve these ends are discussed. However, the states are not the only game in town. Social groups and organizations (including producers' organizations, environmental organizations, social movements, international organizations, and agribusiness actors) strive to put their interests and concerns on the agenda. These power struggles and their implications for the regulatory frameworks are also discussed in a comparative light. Moreover, the regulations, implicit norms, and power relations of the wider agrofood system also influence, and put some clear limits to, the potential success of the states' regulative attempts to transform contemporary agrarian change in ways that are aligned with national development aims. Ultimately, this taps into the discussion about state capacity and autonomy of semi-peripheral states in contemporary agrofood globalization.

The organization of the chapter is the following: The first section outlines the main agrarian development agenda of the twenty-first century in the region, and the possibilities and constraints the states come across when trying to fulfill their objectives (Sect. 6.1). Concretely, the subsections explore thematically how the states have acted to slightly reform and "improve" the articulation of agrarian change in order to make it more: "developmental"/upgraded (Sect. 6.1.1), socially inclusive (Sect. 6.1.2), and environmentally sustainable (Sect. 6.1.3), and discuss the country-specific similarities and differences, as well as the limitations of adopted actions. The second, and final, section discusses what the possibilities and constraints of regulations in relation to agrarian change and development in Argentina, Paraguay, and Uruguay can teach about differentiated state autonomy and capacity in the face of distantly driven transformative pressures (Sect. 6.2). The section highlights both shared and country-specific limitations in state capacity and state autonomy and discusses their wider theoretical implications for the understanding of constraints facing semi-peripheral states in contemporary agrofood globalization.

## 6.1   Common Trends in Dealing with Agrarian Change

There is significant disagreement about what exactly the state ought to do in response to the pressures from agrofood globalization among different groups, ranging from merely facilitating agribusiness expansion to imposing restrictive regulations in order to protect natural resources from exploitation and/or to protect vulnerable groups from either exploitation or exclusion. This disagreement reflects a wider contestedness about the severity of social-ecological risks, the causal mechanisms involved, and what is a fair distribution of the benefits and "costs" involved. Ultimately, it is about what is perceived as sustainable and just. Notwithstanding the disagreement and notwithstanding the differences between Argentina, Paraguay, and Uruguay, in broad brushes, the overall policy orientation of the states moved toward liberalization, stabilization, privatization, and investment promotion in the late twentieth century, reflecting the overall trend of "Washington-Consensus" (Chapter 4 in this book). A few years into the new millennium, however, when international prices on agrofood commodities started to rise rapidly, there was a rather pronounced policy break with neoliberal development ideals in Latin America. Rising inequality, disappointing growth rates, and the full-blown 2001 economic crisis were important reasons to the increasing unpopularity of neoliberal "solutions". Instead, the agrarian development agenda for the twenty-first century of all three countries involves a more proactive role for the states in creating poverty reduction, social development, inclusive economic growth, sustainable use of natural resources, and improved insertion in the international markets. This policy agenda relies on neostructuralist development ideals and on a "post-Washington Consensus" agenda.

*Neostructuralism*
Neostructuralism swept the region. The *Economic Commission for Latin America and the Caribbean* (CEPAL) once again—as under the era of Latin American structuralism—played a leading role in formulating and diffusing its intellectual underpinnings. Many of the tenants of "Classical" Latin American structuralism of the 1950s and 1960s remain in neostructuralism, for example the belief that the roots to Latin American relative "underdevelopment" are to be found in the power relations between center and periphery, as well as in the domestic productive heterogeneity and inequality. The solutions are not to be found in "comparative advantage" and unleashed market forces, but require a strong, active, and "developmental" state that plans and invests in redistributive reforms, in industrial policies as well as in public education and health. Compared to its predecessor, however, neostructuralism takes world market integration and financialization for granted, and abandoned the more inward-looking orientation (Chapter 2 in this book). Exports are thus still considered important for economic development, but the state was "brought back in" to procure

for improved world-market insertion, meaning to sell a diversified set of goods, with high content of technology and knowledge, to high price markets. This is assumed to make possible the generation of well-paid high-quality jobs, benign "spill-overs" to the rest of the economy and reduce social-environmental harm. While not all countries embraced neostructuralism equally full heartedly, agrarian policies across the region involved important tenants of this stream of development thought. For example, all countries drafted long-term planning programs for agrarian development. These were designed to boost the benefits of contemporary dynamic agro-commodity exports by using them as springboards for value-added incorporation, export market diversification, and long-term upgrading of the productive structure.

The social dimensions of the neostructuralist development agenda were particularly present in the countries where the so-called *Pink Tide* was strong (Chapter 5 in this book). This was the case of Argentina under *Kirchnerismo* (2003–2015) and of Uruguay under the tree consecutive *Frente Amplio* administrations (2005–2020). As many times before throughout history, the path of Paraguay differs from that of Argentina and Uruguay. The Paraguayan policies of the new millennium were mainly characterized by *Colorado*-led "business as usual"; openness to trade, low taxes, and a small state apparatus. However, neostructuralist development ideas and a general wave of critique against neoliberalism also influenced Paraguayan policies throughout the period. Issues concerning social equality became high on the policy agenda, culminating in a *"Pink Tide"*-government with the election of Fernando Lugo (2008–2012). These ideas probably also spread via regional arenas such as Mercosur, CEPAL, and other spaces for regional exchange. For example, the *Mercosur Arena for Family Farmers*, REAF, took a leading role in designing public policies and was particularly important in pushing the member states to clearly define the group of family farmers in each country, in order to establish national registers and design specific support programs. REAF expressed that the states have a responsibility to create conditions for family farmers that allow them to continue their way of living, which in turn was argued vital for rural development, social cohesion, national food security and sovereignty, sound localized ecosystem stewardship, rootedness, and local knowledge production. The member countries adopted this vision to a large extent and followed many of the policy proposals (CLAEH 2015).

The Neostructuralist approach also emphasizes democratization, participation, and decentralization of power. A renewed focus on "territoriality", "participatory processes", and "democratic accountability" have spread across the region (Public officials at MAG, MGAP and MAGPyA, and specialists from IICA and UNDP, Paraguay, Argentina and Uruguay, in Asunción, Buenos Aires and Montevideo, February–March 2017). These ideals are not exclusive of neostructuralism. On the contrary, they are shared by most of the contemporary influential development-approaches, and are at the core of many programs of the International Organizations (IO's). Decentralization was already a central part of the "Washington-consensus" agenda, but back then it was combined with a general restructuring of the state and with austerity policies (Chapter 4 in this book). In this way, during the 1990s, Menem in Argentina increased the power of the Provinces and Municipalities pulling over the responsibility for budget cuts on local authorities. In Paraguay, the end of the long Stroessner regime in 1989, created an impetus for radical restructuring of the state, in which decentralization was central. In the post-Stroessner political turmoil and power vacuum, the influential IOs redesigned the organization of the state apparatus in accordance with ideals of decentralization and fragmentation for improved efficiency and responsiveness, creating small independent offices that manage their own budgets. Under Argentine *Kirchnerismo*, decentralization was combined with empowerment of the municipalities and grassroots' participation, while the already powerful Provinces received less attention. In Uruguay, decentralization did not become an important part of public policy until FA entered the government in 2005. In this small country of very strong centralist tradition, decentralization has thus come to be associated with "leftist policies".

*Post-Washington Consensus*
The international community of cooperation and financial organizations also moved away from neoliberal orthodoxy and over to a "post-Washington consensus" after the turn of the millennium. This meant less focus on free movement of capital, and more focus on active development planning, support to domestic industries, and construction of social safety net programs. This new consensus acknowledges the persistence of "market failures" in agricultural activities, which leads to a significant externalization of social-ecological costs; such as pollution, biodiversity loss, land degradation, and marginalization of vulnerable groups. At the

same time, the IOs often promote market based (but state created) tools for full cost/benefit internalization, to incentivize "correct" behavior (for example by pricing ecosystem services, creating tradeable carbon credits, and by making pollution cost). The concrete policy lines promoted by the IOs are for the most part aligned with the broader international agenda for sustainable development, not least the influential declarations of the UN-system, such as the 2030 Agenda and the Sustainable Development Goals. As mentioned in Chapter 5, IOs are important in designing, funding, and implementing agrarian policies throughout the region. While they are supposed to support the policy agenda set by the governments, in line with the 2005 *Paris Declaration*,[2] and while many projects are performed in collaboration with public entities, it is clear that their own aims and ways of doing things shine through. It is for example noteworthy that many of the programs and projects in Argentina, Paraguay, and Uruguay share more or less the same aims, priorities, and design.[3] In this way, many of the public programs and projects funded by the IOs the past decades have followed one and the same post-Washington agenda, thus creating significant policy coherence across the region and across different government administrations. The concrete content is often concerned with issues such as poverty alleviation, establishment of protected areas, mitigation/adaptation of climate change, food security, support of family farmers, rural development, and decentralization. There is also significant focus on the "legitimate" procedures for public decision-making, centered in broad stakeholder participation for capacity building and consensus making, as well as the establishment of transdisciplinary and public–private arenas. These arenas are often seen as complementary to legal regulations, and oftentimes stressed as more accountable, efficient, and leading to less conflicts than legislation (Public officials of MAGPyA, MVOTMA, MGAP & MAG in Argentina, Paraguay, and Uruguay, and employees of IICA Paraguay and Argentina, UNDP Paraguay, in Buenos Aires, Asunción and Montevideo, February–March 2017). In this way, the IOs working in the region have often coordinated (and funded) broad decision-making processes for participatory policy creation and implementation.

While IOs are important throughout the region, they have divergent amount of influence over public policies in different countries. The arenas for joint policy articulation are often initiated and coordinated by the IOs in Paraguay, while there is a long and institutionalized tradition of state-led negotiation and bargaining in Argentina and Uruguay (not least from the corporativist *Peronísmo* and *neo-Batllismo* governments). Overall, the IOs are extremely important for agrarian policies in Paraguay. The weak institutions and low taxes of this relatively poor country have rendered low capabilities and chronic scarcity of public funds. Indeed, Paraguay can be seen as a laboratory of different new policy ideas and designs, since it says "yes" to all proposals; "since the opportunity cost is zero", as expressed by the Deputy Minister of Agriculture (Asunción, 23 February 2017). While the development agendas of the IO's are influential, their new and "trendy" development ideas and concepts may often turn out to be mere old wine in new bottles. Thus, under new labels such as "creating resilience", "reducing vulnerability to climate change", or "strengthen ecosystem services", many of the same development activities that have been implemented for decades often remain, such as to provide farmers with credits, extension services, and inputs (Specialists at IICA Argentina. IICA Paraguay, UNDP Paraguay; public official at MVOTMA, in Asunción, Buenos Aires and Montevideo, February–March 2017).

*A Revived Belief in the State, But also Inertia, Power Struggles, and Constrained Capacities*

All the new development trends in the region after the turn of the millennium—Neostructuralist thought, *"Pink Tide"* governments and post-Washington Consensus—thus have one thing in common; a revived belief in the role of the state for long-term, sustainable, and inclusive economic development. However, regulative frameworks are slowly evolving historical constructs involving a great deal of path dependency and inertia. Thus, shifts in policy trends and of political forces in government have for the most part not implied game changing shifts for agrarian change. Sometimes shifts in discourse seem to have been wider than in actual policies. For example, the newly installed center-left governments in the early 2000s agonized against the rapidly advancing agribusiness firms, financialization of agriculture, and displacement of family farmers. At the same time, most of the policies of their neoliberal predecessors were not rolled back and no land reforms were implemented. Instead foreign direct investments continued to receive generous tax exemptions, currencies were held down to support the export sectors, public investments were made in export infrastructure, and agribusiness advanced as never before.

At first glance, this seems to indicate that the anti-neoliberal agenda was a mere discursive construct, and that the relation between the agribusiness complex and the states is nothing but a long-lasting love affair. However, a deeper investigation into the plethora of regulations and policies in each country (Chapter 5 in this book) provides a more complex picture. For example, *Kirchnerismo* in Argentina had a long-lasting and severe conflict with the agro-exporting interests, centered in the high export taxes. While less violent than in Argentina, some of the agrarian policies of the FA-administrations in Uruguay did also provoke articulated protests from influential producers' organizations. Even in the general liberal and agribusiness-friendly Paraguay, the producers' have been organizing *"tractorazos"* against policies and regulations that oppose their interests. At the same time, NGOs, small farmers' organizations, and indigenous groups in all three countries, but particularly in Paraguay and Northern Argentina, have criticized the governments for too much *laissez-faire* and facilitation of agribusiness advancement.

State action has effects on power-relations between different social actors in multiple ways; while some may benefit some actors, others (intentionally or unintentionally) may harm the same. National regulatory frameworks do not represent one stable and coherent interest, but rather the negotiated results emanating from power struggles in an arena full of dispute. It is thus not surprising that some policy lines seem to go in one direction and others in other directions. At the same time, many other stressors than public policies and regulations shape the reality of agrarian change. Thus, while the states aim to transform their agrarian sectors so that they become more upgraded and inserted in improved ways in the international markets; more socially inclusive and developmental; more environmentally sustainable, and so on, they may still fail to do so. As mentioned in Chapter 1, the states are not the only game in town. The coming sections explore in depth the differentiated capacities of the states to regulate agrarian change in ways that boost the benefits and mitigate real and perceived risks to human well-being and the environment, in line with the abovementioned development objectives. This includes an analysis of both country-specific and shared constraints facing Argentina, Paraguay, and Uruguay to fulfill their agendas. Ultimately, this taps into a discussion about state capacity and autonomy of semi-peripheral states in contemporary agrofood globalization.

### 6.1.1   The Possibilities and Constraints to Achieve "Upgrading" of Agrarian Change

An important part of the agrarian policies in Argentina, Paraguay, and Uruguay have sought to incorporate more technology and knowledge into specific value chains to increase access to well-paid markets, to generate more high quality jobs, and to strengthen backward and forward linkages; ultimately to upgrade the production structure. Accordingly, the states developed strategic plans, industrial policies, and designed roadmaps for upgrading of specific commodity chains and developing effective linkages between different export commodities (Anichini et al. 2013; Castellano and Goizueta 2011; Dabezies 2009; Gabinete Productivo 2010). While there is disagreement among different actors whether the factual and potential value-added already involved in soybean and beef chains should be considered high or low, there is agreement on that it is desirable to incorporate even more value to them.

One of the main tools used in the three countries has been to support private firms through tax exonerations for investments aligned to development criteria such as technology transfer, decentralization, creation of productive clusters, and promotion of biofuels through targeted blends. The three countries have also been working proactively with diversification of export destinations, negotiations for the opening up of new markets, and through certification and quality improvement help the firms to access better-paid markets. States have also given increased priority and funds to Science, Technology, and Innovation (Argentina even created a proper Ministry for the same), and increasing the percentage of GDP spent on education, in order to improve the knowledge content in the production systems. Regardless of specific commodity chain, increased content of knowledge is assumed central for productive upgrading. Moreover, long-term industrial policies have been designed. These policies were particularly strong under *Kirchnerismo* and FA-rule in Argentina and Uruguay (Chapter 5), in a strong historical parallel with the ISI-policies adopted in these countries between the 1930s and the early 1970s (Chapter 2).

Alongside with the neostructuralist aim of "upgrading" the productive structure, all three countries organized in specific private–public "tables" for different agricultural commodity chains with actors from different stages in order to enhance commodity chain governance and joint effort (Gonzaga Belluzo 2015, 115). Regarding the most outstanding export-crop—the soybean—the main idea has in all three countries been to convert it into a product that encapsulates more value-added before exports (Anichini et al. 2013; Castellano and Goizueta 2011; Dabezies 2009; Gabinete Productivo 2010). Results were at best mixed. Argentina exports most of its soybeans as meal and oil, Paraguay exports a mix of meal, oil, and whole beans, whereas Uruguay exclusively exports whole soybeans. It is clear that the extreme cost effective soybean production in Argentina combined with the economic disincentives to export unprocessed soybeans—through higher levies on whole soybeans than on meal and oil—turned the country into the world's biggest soybean oil and meal exporter. However, as Macri took away the tax differential in late 2018, and as China has tax differentials favoring the imports of whole beans (Chapter 3 of this book), the share of whole soybeans in the Argentine exports have increased (Chapter 5 in this book).

The three countries have set up biofuel programs with tax exemptions and passed mandatory blending targets to promote biodiesel as alternative markets for soybeans, as well as for reasons of energy security, climate-related targets, and feedstock availability (Janssen and Rutz 2011).[4] Soy-based biodiesel increased in all three countries, and Argentina became early on a successful soy-based biodiesel producer, representing almost ten percent of the global biodiesel production in 2016 (Rouhany and Montgomery 2019).[5] While processed soybeans (meal, oil, and biodiesel) include more value-added than non-processed soybeans, it is still rather low, compared to the value of the posterior transformation of the soybean in the countries of destination (such as the production of meat, ready-made frozen food, and other food products). Promotion of production of pork, chicken, and feedlot farming has resulted in slight increases, particularly in Argentina, but the overall plans to add value to soybeans have not been very successful in any of the three countries (Barrios et al. 2010; Bértola 2015; Paolino 2015, 39–41).

Regarding incorporating more technology into the soybean chain by moving into upstream stages, all three countries use technology developed and patented by transnational companies (the soybean GM events and patents of agrochemicals). The strong international intellectual property rights regime, with long-lasting patents, and the extremely high costs involved in research and development of new biotechnological events and agrochemical formulas, have not allowed for the states or firms of the Southern Cone to participate successfully in these stages of the commodity chain. However, Argentina has some private and public sector initiatives doing biotechnological R&D on several crops that could eventually in the future become successful. All three countries are engaged in seed adaption and seed improvement. Moreover, Argentina and Uruguay export some high-quality soybean seeds, tapping into their strong national traditions on research for seed improvement. However, the overall profit generated in the soybean chain across the region is not coming from work or domestic technology, but from nature (fertile soils and good climate), external technology (GM, machines, crushing plants) and external demand.

One reason to the relatively poor results of the strategies to add value to soybeans is that whole soybean is one of the most "freely" traded agricultural products in the world, while transformed soybeans into higher value-added products are more regulated and protected. Both Asia and Europe want to have cheap soybeans as input for their own (protected) animal industries.[6] Since soybeans are both ubiquitous and rather invisible in the agrofood system (for end-consumers), animal farmers in the advanced economies can use soybeans as cheap input while still market their meat products as "locally" produced. In this way, these actors have an interest in maintaining the "commodity" character of soybeans with almost no possibility for quality (and thus price) differentiation.[7] Moreover, the handful of multinational actors dominating the international soybean trade (ADM, Cargill, Bunge, Louis Dreyfus—and recently the Chinese state-owned COFCO), also own processing plants and animal farming industries in the countries of destination. Accordingly, they have no economic incentives to develop more added value activities in the countries of the Southern Cone. The extreme corporate concentration thus creates impediments for the states in Latin America to succeed with their strategy to "upgrade". The traders also keep down storage and transport cost by treating the soybean as an interchangeable commodity, without any concerns for where it comes from (traceability) or quality differentiation. As outlined in Chapter 3 of this book, the geographical adaptability of the soybean—enhanced by the biotechnology—together with decreasing costs and increased capacities for storage and container shipping, have allowed it to become the ultimate commodity; a standardized product that can be sourced from anywhere, or nowhere, for the handful of transnational grain traders to make profits on arbitrage. This decreases substantially the bargaining power of the producer countries. "You can arbitrage the soybean sourcing from any origin," as illustratively expressed by Bunge.[8]

Since the international soybean market provides small opportunities for upgrading, the countries have, as mentioned, intended to increase value by transforming soybeans to meat, and in general to upgrade their traditional meat production. The main tools used by the states are designed to improve the quality and international reputation of their meat by improving sanitary standards, introducing new breeds, funding public research, marketing national production, and negotiating new market access. One public policy that oftentimes has been depicted as successful in this respect, is the traceability program of all livestock in Uruguay (Chapter 5). Through this public effort, the country can access specific quotas and well-paid niche-markets for beef, such as the EU 481 grain fed beef quota—with a zero-rated tariff. However, the highly valued EU 481 grain fed quota is set at only 45,000 t/yr, and it is shared between Australia, New Zealand, Uruguay, Argentina, Canada, and the US on a first-come first-served basis. It is thus quickly filled. Outside of the small specific quotas, the beef tariffs are high. Meat—particularly beef—is among the most regulated and protected trade items in the world, in stark contrast to the soybean. Animal farmers of the advanced economies lobby (often quite successfully) for restricted import quotas and/or high tariffs on imported meat (Chapter 3 of this book), at the same time as they receive different types of public subsidies.

Latin America is thus hardly the only region in the world that intends to design policies to capture more value-added activities domestically. This is not a new phenomenon. High protectionism of meat markets has constrained exports from the Southern Cone since the 1920s (Chapter 2 of this book). While the 1995 *Agreement on Agriculture* of the newly formed WTO was a significant step toward trade liberalization, agriculture is still the most protected sector and liberalization has been full of exceptions. The current rules of the game of the international food system, with the strong protectionism of the advanced economies combined with undue influence of large hyper-concentrated transnational corporation, thus make it significantly hard for the neostructuralist "upgrading" strategies to pay-off. However, major shifts seem to be appearing on the horizon. Access to the European beef market will probably increase significantly in the coming years, as Mercosur and the EU after 20 years of negotiation closed a trade agreement in principle, on June 29, 2019. The trade deal will, over time, result in almost complete elimination of import duties on both sides. Before the deal enters force, however, the parties need to proceed to legal revision, to produce the greater agreement text, and to submit the agreement for approval by the Council and the European Parliament, as well as by the four Mercosur governments. No timeline has been given (as of July 2019). Besides increased access to the European beef market, sometimes in the relative near future, the rapidly rising beef demand in China (from very low levels) has already created new opportunities for the beef from the Southern Cone. Since the new millennium, Chinese national beef trade policy has started to liberalize and in record time, as the Asian food giant has no chance in meeting up snowballing demand with only domestic supply. China has become the most important beef importer in the world, revolutionizing the international market conditions of the same.

China's ravenous appetite for agrofood products has provided the natural resource rich Latin America with a new main market, which has allowed the region to become less dependent on its traditional export destinations. However, the high reliance on the Chinese market also raised new concerns. High export concentration on few products and countries implies a high vulnerability to deceleration of demand in those countries, which can bring instability in government income and difficulties in macroeconomic management. The Southern Cone exports could suffer severe consequences from abrupt demand shifts in China (for example, in May 2019 China is suffering a severe outbreak of Swine Fever, which has decreased demand on soybeans in an instant). At the same time, the Asian tiger is well known for its capacity to implement drastic policy shifts, in addition to an overall policy design that aims to domestically capture most of the benefit involved in any productive chain (Svampa and Slipak 2015).[9] Moreover, the trade relation between China and the region is based mainly on the exports of a handful of commodities from Latin America and imports of a diversified basket of manufactured goods of growing technological content from China (Gonzaga Belluzo 2015, 107). In this way, the rapid growth in China's manufactured exports has out-competed many domestic manufactures, while the weight of natural commodities in the export structure has increased (Hoekman 2017, 34). In this way, Latin American exports to China are heavily reliant on natural resources, with a relatively low value-added content, while imports are the other way around (López et al. 2008; Armony and Velásque 2016; Ray and Gallagher 2016).[10] This is often referred to as a "*reprimatización*" of the trade structure, which is quite the opposite to the neostructuralist ideal of upgrading (Bárcena Ibarra and Prado 2015, 19).

Thus, regulations of the international food trade system (selective protectionism and other policies to capture value-added domestically in the advanced economies, and strong international property right protection), combined with the business strategies of the big transnational corporations, put severe constraints to the explicit aim of the Southern Cone countries to use their agrarian-based exports as springboards for long-term development and upgrading. Besides these limitations emerging from the international arena, there are important domestic risks and vulnerabilities involved in the quest for "upgrading". One heated debate is about who should bear the building-up costs for the previous investments required for upgrading; the private actors or the state. For example, as mentioned in Chapter 5, there is an on-going discussion in Paraguay if it would be worthwhile to follow Uruguay's sophisticated and expensive individual traceability system in order to access better-paid markets. The hesitation is due to the significant uncertainty about whether future markets will pay for the costs of detailed information about the whole production process behind a piece of meat. It is also due to the high opportunity costs involved. It is not obvious to prioritize constrained public resources to expensive investments that ultimately will increase the competitiveness of a sector that is dominated by big landowners and foreign agribusiness firms. Besides opportunity costs, the rising amount of public–private arenas for coordination and articulation of policies for upgrading has raised concerns over risks of corruption, clientelism, and co-option of public resources by private actors (Natural Resource Management Specialist IICA Paraguay; Director of Research Promotion CLACSO; Advisory Director DINAMA-MVOTMA, Asunción, Buenos Aires and Montevideo, February–March, 2017). There is a risk that the state incentivizes rent-seeking behavior among the private firms and/or benefits singular firms. For example, as the Uruguayan state (through the state-owned company ALUR) wanted to increase domestic biodiesel production (with meal as a highly valuable "subproduct" from crushing of oil-seeds) for value-added incorporation to the oilseeds production, it made a deal guaranteeing long-term purchases at a fixed price with the only big player in processing and vegetable oil production; COUSA. In this way, COUSA could buy new crushing plants from Denmark without facing any risk of not being able to cash-in for the investment (Director of COUSA; Director of ALUR, Montevideo, 2010). There are expressed fears that states are generally bad in "picking the winners" and can thus end up with huge costs for very little benefit. Also, when the state makes expensive investments to promote upgrading (in technology, in infrastructure), it may risk end up increasing the returns of the private companies and pay for investments that the business sector would otherwise pay for itself. These arguments against state interventionism and industrial policies are not new, they reflect the assumptions of classical trade theory and echo the critique expressed against ISI in the 1970s and 1980s (Chapter 2 of this book).

Besides the discussion about what the states should use their fiscal resources to in order to promote long-term development, there is a parallel pulsating discussion about what activities the states have a legitimate right to extract revenues from. For example, in Argentina, the most effective policy for value-added incorporation in the soybean exports was the differential export taxes, creating incentives for processing through higher levies on unprocessed grains. At the same time, the taxes generated important fiscal revenue to the central state. However, this measure also provoked an extremely critical reaction toward the CFK-government from "the countryside" (as well as recurrent scandals of tax evasion of the big international traders).[11]

In short, there are some important constraints facing the states considering the aim to use the soy and beef chains as springboard for upgrading of the productive structure. Perhaps the regulations, norms, and power relations of the international agrofood system constitute the biggest hinder. The soybean is for example "freely" traded, at the same time as it is hard for the producing countries to upgrade to segments of more added-value. The Southern Cone countries cannot compete with the mega biotech firms developing new GM events and thus holds few patents; the big traders are both vertically integrated and own processing plants in both importing and exporting countries and hence have no incentives to do the processing in the producing countries; the main buyer—China—wants to capture all added value domestically and use differentiated export taxes; there are almost no quality differentiation, since the traders benefit from having the soybean as an interchangeable commodity. The beef encapsulates considerably higher amounts of value-added, but international meat trade is among the most heavily regulated and protected within the international agrofood system. Moreover, the efforts of state led long-term planning of the productive structure for upgrading involve several pitfalls: First, it risks incentivizing rent-seeking behavior within the business sector. Second, it risks making states to bear the costs for investments that would be made anyway by the firms, resulting in higher profit margins of the agribusiness firms, and in inefficient and regressive use of the money from taxpayers. Third, it risks to be very inefficient way of money spent as it can be hard for the states to "pick the winners" and it may invest in incorporating more technology and know-how into commodity chains that still will have no chance in the international market (because of the rules and power relations of the international market). Finally, historical legacies of weak and inefficient industrial sectors and of lack of specific expertise also play a role (Chapter 2 in this book). In relative terms, however, Argentina has always had the strongest industrial base, followed by the relatively successful agrofood industry sector in Uruguay, albeit with rather high costs and low labor productivity. Paraguay, however, never developed a proper domestic industry. Moreover, Argentina and Uruguay rank among the highest in terms of level of education and labor skills, while this is a constraint for Paraguay.

Still, in spite of all these constraints, it is important to remember that the tax differential between whole soybeans and processed soybeans succeeded in turning Argentina to the world's biggest exporter of soybean meal, oil, and biodiesel. Argentina also invested considerably in R&D of the agrarian sector, including biotechnology, which may eventually pay-off and have a potential in international markets. Paraguay has also managed by intentional action to increase the share of processed soybean exports. While Uruguay exclusively exports whole beans, it has been rather successful in accessing well-paid markets for its beef due to its state-imposed traceability program. This (together with marketing and diplomatic efforts) have allowed for Uruguay to access considerably higher prices for its beef than Paraguay. At the same time, production costs are considerably higher in Uruguay than in Paraguay (higher land prices, salaries, and taxes). The next section discusses the states' capacities to regulate agrarian change to increase social inclusion.

### 6.1.2  The Possibilities and Constraints to Achieve Socially Inclusive Development

Latin America—one of the world's most unequal regions—has actually had a rather sustained period of declining rates of inequality and poverty. Between 2000 and 2016, both the incidence of extreme poverty and of total poverty in the region fell. On average, 43% of the reduction in poverty was due to the decline in inequality (FAO 2017, 27). While the fall in inequality may be the result of socially redistributive policies and measurements, the economic growth (Paraguay and Uruguay have had economic growth for the longest sustained time-period ever) is probably largely explained by the favorable external context—high demand of the region's main export products. Not least, the improvement of terms of trade, for which higher soybean prices and soybean export expansion played a significant role, benefitted significantly the economies.

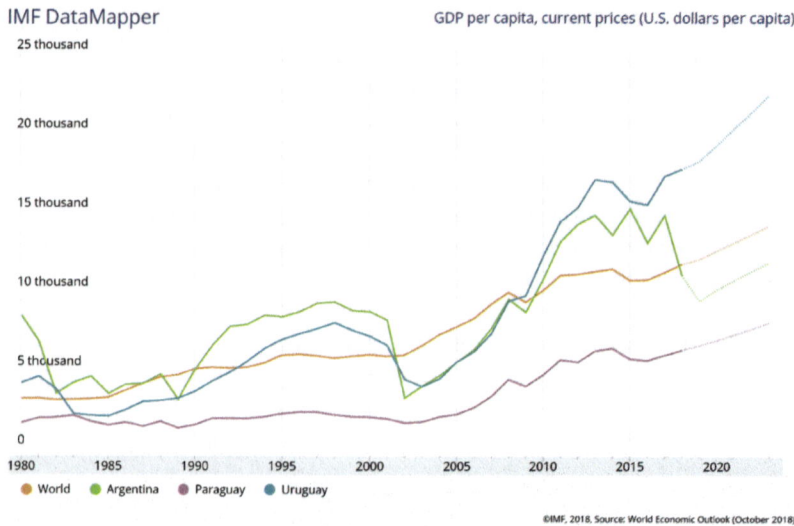

IMF DataMapper                          GDP per capita, current prices (U.S. dollars per capita)

©IMF, 2018. Source: World Economic Outlook (October 2018)

**Chart 6.1**    GDP/c Argentina, Paraguay, Uruguay, and World

Moreover, the importance of the agro-commodity "boom" as a generator of foreign currency through exports is undisputed (Chart 6.1).

As thoroughly discussed in several chapters, however, agrarian change came along with dramatic polarizing shifts in agrarian-related social relations. Large-scale export-oriented agribusiness firms have become increasingly dominating and many family farmers have left activity in all three countries. The actors of input and processing stages of the productive chains (network firms, transnational traders, transnational slaughterhouses, and multinational biotech and seed firms) are even more concentrated and vertically integrated. While staggering inequality of access to land has a long history in the region, concentration and displacement increased further.[12] This process was fueled by the liberalization reforms and state withdrawal from the agrarian sector during the 1990s (Chapter 4 in this book). However, the renewed focus on state intervention, social inclusion and targeted support to family farmers, after the new millennium, did not revert the trend (Chapter 5 in this book). On the contrary, land concentration increased and small family farmers abandoned agricultural activity at a rate that was historically unprecedented. At the same time, it is important to recognize that this pattern could possibly have been even more accentuated still, if it had not been for the specific measures taken to support family farmers and smallholders. All three countries have specific targeted programs including credits and extension services. Organizations representing small farmers are also increasingly participating in the public policy design and decision-making processes through an increased focus on participatory approaches and decentralization. Uruguay has also used a progressive land tax and implemented a historically unprecedented land distribution through INC. Moreover, it uses public procurement to promote family farmers. Paraguay has advanced in its legislation providing indigenous people with relatively strong communal rights to land and new land titling programs with redistribution to small farmers through Indert. Argentina has increased funds for family farming, particularly under *Kirchnerismo* and it prohibited land sales to foreign states. Argentina also has many specific policies at Provincial level, where some have programs of public procurement led by smallholders. While support to family farmers and territorially based rural development were outspoken priorities of *Kirchnerismo*, the Macri administration is described to have largely continued with the same policies, in spite of some budget cuts in response to the snowballing debt (Rural Development Specialist IICA Argentina, Buenos Aires, 3 March 2017). At the same time, private property rights to land is described as almost sacred in Argentina:

> The property right to land is probably the strongest right we have in Argentina. It has been violated only one time in history, when Perón expropriated the land holdings bigger than 5,000 ha and rented out them to sharecroppers. After that, nothing. You see, in the Argentine history there is no agrarian reform. We have destroyed company bonds, we have defaulted state obligations four times; the financial system broke four times!! So all those rights to property turned out to be rather unsafe in the end... So, with this track record the probability that an Argentinean invests in an obligation-bond of his country is not high... By contrast, land is a property where there is almost no jurisprudence of the rights of property ever being violated, with very few exceptions as when the state building new highways, but this is almost nothing in relation to the total size of land. (Former Secretary of MAGyP, Buenos Aires, 2 March 2017)

Besides the strong property rights to land, the inherent large-scale bias of industrial agriculture, particularly in RR soybean production, under current market conditions, constitutes an important stressor acting against successful participation of family and peasant farmers. This model requires very little labor, and thus makes a perfect fit with an agrarian development model of large extensions and almost no people living on the land. The advantages of the agribusiness firms can be listed in the following way: High capital liquidity and state of the art technology, such as satellite monitoring for precision agriculture, and big and more effective machines; "tacit" assets, such as market information and knowledge about how to combine different kinds of resources; preferential treatment from suppliers and buyers, which renders faster response to transport and storage requests, as well as deals at the best possible price (Acosta Reveles 2008; Carriquiry 2015, 30; Errea et al. 2011; Lapegna 2016; Oyhantçabal and Narbondo 2011, 6–7; Teubal 2008).

While particularly soybean production involves important economies of scale, some *pools de siembra* grew so fast that they started to incorporate land of less fertility and hire less professional and experienced agronomists, and thus started to have bad economic results.[13] Notwithstanding these "limits to growth", it is clear that the overall effects of agrarian change have been extremely polarizing. One part of the explanation comes from the abovementioned inherent features of industrial agriculture, and another part of the explanation comes from strong historical legacies of land concertation and lack of rural development. It is nevertheless important to bear in mind the differentiated nature of this pattern. While meat (and wheat in Argentina) trade since the late nineteenth century, from the *Pampas* region of Argentina and Uruguay, created a lot of wealth, it did not follow a settler-like development path as in the US, with a numerous group a highly productive family farmers who created prosperous societies—building up churches, schools, and public services. Instead, the countryside of the Pampas remained sparsely populated, with most of the land concentrated into the hands of big domestic ranchers and their extensive productive systems of little labor generation, and with lacking and/or bad quality public services (Chapter 2 in this book). With the spread of industrial agriculture at the mid-1900s, followed by withdrawal of state support from the 1970s, the small producers that did not manage to keep pace with mechanization and full market integration were wiped out during the second half of the twentieth century (Gras and Hernández 2013, 81–82). When the latest wave of agrarian change, led by soybeans in the Pampas, smallholders had already been wiped out from the area and the family farmers that had managed to hold on were relatively capitalized and big. As the soybean "boom" made land values skyrocket, those who owned land could gain a lot of money from selling, or leasing out, the land.

Many of the sharecroppers, however, became *contratistas* instead (Cepal, CIRAD, and IICA 2014; Manzanal and Schneider 2011; Sabourin et al. 2014; Alianza Internacional de las organizaciones de productores familiares de soya 2008). The agrarian social structure of Eastern Paraguay (Atlantic forest region) and the Northern provinces of Argentina (Argentine Chaco) differs substantially from the Pampas. While public infrastructure and service are of underprovided and of low quality throughout the countryside of the Southern Cone, the social situation is particularly precarious in the poor areas of Northern Argentina and the densely populated Eastern Paraguay. The majorly non-capitalized, largely self-subsistent peasants and indigenous communities of these areas have for the most part not been able to take any benefit from the commodity bonanza (Acosta Reveles 2008, 9; Lapegna 2016; Teubal 2008).

Besides displacement and concentration, foreignization of land is a dominant trait of agrarian change in Paraguay. The first wave of foreign ownership started already in the late nineteenth century, as the Paraguayan state was forced to sell off public land to repay war debts. The second wave started in the 1970s and was led by Brazilian soybean farmers that were rapidly acquiring land and who today dominate completely soybean production in the Eastern region. The Paraguayan Chaco forms part of the same biome as the Argentine Chaco, but is very sparsely populated. The areas have mainly been covered by forests, while also being home to indigenous tribes either living in the forest or engaged in small-scale agriculture, as well as Mennonite communities engaged in farming activities, mainly dairy. However, the big, mostly foreign ranchers, that have expanded in the region since the new millennium are reducing the habitat and displacing native peoples.

Thus, present-day inherent features of the soy and beef chains interact with historical legacies and place-differentiated social-ecological structures. In this equation of multiple stressors, enters also the states, their policies and regulations. Many scholars, activists, and peasant organizations across the region argue that the states are ultimately responsible and should do more to counter-balance the historically formed unequal land structure, and lack of rural development, as well as counter-balance the favoring of agribusiness actors inherent in contemporary agrofood globalization. Particularly the big organizations of small farmers and NGOs adhering to the banners of *Food Sovereignty, Agroecology* and *The Peasant Way* (*Vía Campesina*) throughout the region severely criticizes the states for allowing the neoliberal regulations from the 1980s and

90s to stay largely intact, and for doing too little today to rebalance the relations of force between agribusiness and family farmers. The states are also criticized for talking nice about family farming, while actually favoring and facilitating agribusiness expansion (Ezquerro-Cañete and Fogel 2017; Gudynas 2010; Vergara-Camus and Kay 2017a, b). The states have for example tried to keep down the local currencies (to enhance international competitiveness), invested in export-oriented infrastructure (ports, silos, roads), authorized new GM-events (and in Paraguay and Uruguay accept strong intellectual property rights to breeders) and allowed them to expand "freely" without any consideration of national food security and rural development.

At the same time, all countries have influential century-old producers' organizations—SRA, CRA, ARP, ARU, and FR—faithful to a liberal discourse centered in advocating free trade, strong private property rights, and stability. They argue generally against state interventionism, but from time to time, they have been asking for specific policy interventions in support of the agrarian sectors, such as debt clearance, favorable credits, public investments in infrastructure, and so forth (Chapter 2 in this book). In general, these "traditional" producers' organizations, representing mainly big ranchers, have taken more combative stances against all state intervention and redistributive attempts than the new agribusiness firms that have emerged with the agrarian "boom". Moreover, there are also important legacies in public policies. In Paraguay and in the poorest provinces of Argentina, public policies toward peasant farmers have historically been linked to political campaigns to get voters and/or loyalty. There is a strong historical legacy of "assistentialism", populist short-term gifts or support to poor producers, which have not managed to help people out from poverty. This has created deep-rooted habits and expectations that are described as difficult to change, and the practices have continued to present-day (Director of Social Studies INTA; Paraguay's Deputy Minister of Agriculture, Buenos Aires and Asunción, 2 March and 23 February 2017). This pattern is particularly strong at the level of Provinces in Argentina and Departments in Paraguay, where patronage is used to build up loyal clientelist ties, limit political competition, and strengthen the Governors position, without leading to any structural change. Moreover, in all three countries, the capacity of local private economic interest to do whatever it likes, through direct intervention in public policy design or through bribes and corruption, is described as enhanced by decentralization. This is the flip side of the move toward decentralization. At the same time, the decentralizing reforms have also enhanced opportunities for place-specific solutions and increased participation of local stakeholders in policy formulating and decision-making processes.

While targeted policies to rebalance social relations at the on-farm level have not reverted the trend of loss of family farmers and agribusiness advancement, the states have also intended to redistribute the income from agrarian related activities in more generalized ways. Before going into the (constrained) capacities of the states to do so, it is important to bear in mind that the three countries depart from rather different realities. Paraguay has some of the highest poverty rates in Latin America, while Argentina has less than half the rates of Paraguay, and Uruguay has less than half the poverty rates of Argentina.[14] In terms of income per capita, Uruguay has the highest in the region; slightly higher than Argentina and almost the double than that of Paraguay (GNI/c PPP, 2017: Uruguay—21,870; Argentina—20,250, and Paraguay—12,680). Argentina and Uruguay are the two countries in Latin America with the biggest middle class, encompassing well above half the population, and the lowest income inequality, as measured in Gini.[15] While this differentiated pattern has long historical roots, all three countries improved their social indicators—in terms of real wages, unemployment, public health, education, and housing—considerably between 2002 and 2015 (IBRD 2014, 2015, 2018). However, the social transfers through public programs represent a much larger part of national GDP in Argentina and Uruguay, than in Paraguay.[16] Moreover, fiscal pressure is much higher in Argentina and Uruguay than in Paraguay.

Argentina, under *Kirchnerismo* and even under the most part of the Macri-administration, did in a direct way redistribute some of the wealth generated by the soybean boom through relatively high export taxes. The surplus generated by the export taxes went to many social projects, but they also had an important role in making possible imports (and still have a positive trade-balance, which was of utterly importance for the economic model of *Kirchenrismo*). Moreover, the high export taxes on soybeans made soybean cultivation slightly less profitable. While it continued to represent the most profitable land-use, at least it put a break on the rising land prices. In this way, the export taxes probably represented the most important public policy working against land concentration Argentina. This measure was probably more important than all the cash-transfer programs to small producers. *Kirchnerismo* also held down domestic prices on beef by the use of trade restrictions and tariffs.[17] This was seen to prioritize the urban sector (voters) over the rural sector. At the same time, it is important to bear in mind that the devaluation of the peso aided the exporting sectors, which ultimately implied that at least in this sense, the rural sectors were benefitted over the industrial urban sectors. However, state intervention in the agricultural markets through taxes, tariffs, and quotas generated a severe and long-lasting conflict between the government and the "countryside". All big producer's organizations that historically had had different positions in relation to agrarian policy were united and managed to talk with one voice against the policies of the government. They also criticized *Kirchnerismo* for skyrocketed public spending, transferring money from the hard working and productive parts of the country to lazy people who do not like to work. The rural discontent eventually seems to have played an important role in facilitating the change of government in 2015. As mentioned, "el campo" has good relations with the Macri administration, in spite of the fact that the electoral promise of total abolition of export taxes has not been fulfilled.

The Uruguayan government interfered less directly in the agrarian market, with no taxes or restrictions on agro-exports. Moreover, the overall fiscal pressure of the agrarian sector has been below the average of other sectors (Tambler 2013). However, Uruguay did implement a wider tax reform, mainly taxing income progressively, but also building in progressivity in land taxes (to curb concentration) and allowing family farmers to tribute less than the firms. At the same time, the big agrarian firms can get big tax deductions through the "generous" investment promotion regime, and most agro-commodities are exported from Free Trade Zones. Nonetheless, tax reform coupled with proactive labor politics (reinstitution of tripartite wage councils and collective bargains), yielded positive redistributive effects and reduced inequality. Influential interest groups representing the "countryside" in Uruguay, not least the powerful producers' organizations ARU and FR, have criticized the state for "sucking out" the agrarian sector in order to fuel an ever expanding and ineffective state apparatus and/or to redistribute to lazy people that do not work hard. While the discourse is similar to "el campo" in Argentina, conflicts in Uruguay are less severe. As expressed by a researcher at the Faculty of Agronomy of the University of Buenos Aires, who also works as an extensionist and consultant to crop producers in both Argentina and Uruguay:

Frente Amplio has a better relation with the countryside than what Kirchnerismo had. Things in Uruguay are always more soft. I think that the tax reform helped to create a structure for redistribution without making confrontation necessary. As the tax system in Uruguay is progressive, you pay more tax if you have a lot of profit, and the state gets what it needs from there, and it doesn't matter where the profit is made. (Researcher FAUBA and Agricultural Extensionist, Buenos Aires, 24 February 2017)

Besides the abovementioned tax reform, Uruguay is often described as having a political culture which is less polemic and antagonistic than Argentina, albeit more conflictive and is fond of long on-going discussions than Paraguay. The Director of the Office of Planning and Agrarian Policy, at the Ministry of Livestock, Agriculture and Fisheries, illustratively expressed this view:

In Uruguay, there is a lot of conflict and discussion, but in the end both public and private actors reflect a strong belief in negotiations and the ability to find socially accepted compromises. (Director of OPYPA-MGAP, Montevideo, 6 March 2017)

Public policies are in many respects the outcome of distributional struggles, and distributional struggles in Paraguay have so far mainly strengthened the resource-rich. Paraguay stands out in the region with its low fiscal pressure on agrarian activities and with the lowest income and profit taxes of the region. Reportedly, large foreign owned soy operators are also evading national taxes. Still, each time the state has discussed tax or land reforms for more redistribution, the powerful producers' organizations ARP, APS, and UGP have reacted strongly engaging in demonstrations such as *tractorazos*, blocking roads with tractors and farm equipment (Ezquerro-Cañete and Fogel 2017). As of 2019, when the parliament is, again (sic) discussing to put export taxes on soybeans, the producers' organizations are very active in all arenas (in the press, doing lobby, and in the streets) trying to curb the initiative. At the same time, the organizations of peasants are also mobilizing and increasingly voicing their claims on fair land reform and increased social justice. Paraguay suffers from violent confrontations between peasant farmers and *Brasiguayos* cultivating soybeans in Eastern Paraguay, as well as between indigenous group and big Uruguayan, Brazilian, and Argentine livestock farmers in the Chaco forests. These conflicts nurture from the extreme inequality, by the gloomy horizon for the rural poor to be able to live well from agricultural production, and the limited absorbing capacity of decent jobs in the city.

In sum, there are some important constraints facing the states considering the aim to increase social inclusion, and particularly to strengthen the family farmers. While all three countries have improved social inclusion in terms of poverty alleviation and decreased income inequality, they have not reverted the long-term trend of exclusion of family farmers. Quite the contrary, the pressures from contemporary agrofood globalization—with the important elements of economies of scale and technology substituting for labor—coupled with historical legacies of land concentration and underdeveloped rural areas have resulted in an accelerated pace of displacement processes and advancement of agribusiness firms. Moreover, the influential organized agro-exporting interest groups in all three countries resist any attempt of the state to rebalance relations of force between agribusiness and smallholders. While the Uruguayan state has managed to negotiate with different groups in ways that have avoided tensions to burst out in conflicts, Argentina had a strong backlash in redistributive measures of agrarian wealth as the influential agrarian organizations showed that they did not accept export taxes. Argentina has also suffered violent confrontations in the Northern Provinces and forceful land evictions. Notwithstanding economic progress, Paraguay remains among Latin America's poorest and most unequal countries. Social tensions often erupt in violent conflicts and agrarian wealth continue to benefit the hands of a few. The state has some redistributive initiatives, but severe constraints in the state's administrative, legislative, and funding capacity yield meager results. Paraguay's long history of authoritarian rule, with high control of political adversaries and low priority given to public education, have resulted in weak internal forces for radically more inclusive policies and more redistribution. Lugo managed to put social concerns higher up on the agenda but his political orientation, or any other progressive political orientation, have never achieved majority of the seats in the parliament. While poverty has been reduced the past decade, more than a quarter of the population remains below the poverty line—the majority in the countryside.

### 6.1.3 The Possibilities and Constraints to Achieve Sustainable Development

The rapid agrarian transformation in Argentina, Paraguay, and Uruguay during the past decades has provoked severe social-ecological concerns. As demands for soy and beef have expanded at an accelerating pace, natural and social resources and ecosystems are under increasing pressure throughout the region. The full impact is difficult or even impossible to predict given the complexity of these processes, including nonlinearities, indirect feedbacks and tipping points, as well as the speed at which they occur. While articulated slightly differently over the territory, all states recognize that agropastoral expansion and intensification are causing severe biodiversity loss, contamination of water degradation of land and other negative environmental impacts—which in the end can threaten vital ecosystem services for well-being. Accordingly, environmental regulation is acknowledged as a central state responsibility and ecosystem conservation is a topic of growing concern of all the governments.

Yet, pressures on natural resources are higher than ever before. None of these countries radically challenges the natural resource dependent commodity booms or questions the idea that their country will continue to be important world provider of beef, soybeans, and other natural resource intensive products. "To feed the world and its growing population" is a recurrently mentioned trope in all three countries. The large transnational agribusiness corporations such as Cargill, ADM, Bayer-Monsanto and LDC use the same trope.[18] While long-term care of the natural resources is an increasingly often expressed ideal across all government administrations in charge the past decades, the main agrarian policy lines reflect that the overarching aim is to make agriculture and agro-industry more productive, efficient, and internationally competitive. "Sustainable intensification" (the official motto of the *Ministry of Livestock, Agriculture and Fisheries*, MGAP, of Uruguay) synthesizes the main message of all agrarian development plans and policies in the region; a belief in the possibility to produce more food without any adverse environmental impact and without the conversion of additional non-agricultural land. At the same time, when sustainability is mentioned in official texts, for the most part no particular method or technology to achieve this end is indicated. One rather generalized belief in all three countries, is that the negative environmental aspects of agrarian change are the result of its novelty and lack of knowledge. Accordingly, most problems are expected to eventually disappear by themselves as experience and knowledge become more systematized. Time is thus the most important medicine to cure "unsustainability", but eventual remaining negative externalities, because of persistent market failures, will need to be addressed by the state. In this way, many critics argue that the while "sustainability" is always mentioned in official discourses about agrarian development, it is often mere *greenwashing* as the path de facto taken is "business-as-usual", albeit with minor modifications. The sociologist Maristella Svampa argues that there is an overall acceptance, among all Latin American governments, of a development model centered in exports of natural resources intensive products, which she has labeled a "Commodity-Consensus" (Svampa 2013).

Ecological and social movements all over Latin America and critical scholars have severely criticized the states across the region for doing too little to stop "extractivism" and to increase true and long-term sustainability. They see "intensification" and "sustainability" as inherently irreconcilable. They point at all the recognized ecological concerns that agrarian change has already brought and argue that continuous land-use intensification under current conditions will lead to ever increased pressures from fertilizers, pesticides, waste, and green-house gas emissions, spearheaded by the transnational slaughterhouses, chemical, seed, and trading companies. New philosophical orientations, under the banners of *Buen Vivir* (*Sumak Kawsay*), *Common Property*, *Nature's Rights*, *Agroecology*, and *Post-Extractivism*, suggest radical new ways of human (society)-nature interaction. They all advocate diverse place-based development models, grounded in long time-frames and a notion of intrinsic values (and even rights) of nature, moving beyond anthropocentrism and commodification of nature (Gudynas 2009, 2011, 2017; Svampa 2013). This way of conceptualizing nature—as an integrated whole, impossible to reduce into a commodity—is of course completely incompatible with the current development model. As already mentioned in Chapter 1 of this book, these radical counterpoints express an interesting critique of the mainstream system, but the initiatives are typically too small and fragmented (still) to have impact on the larger system.

Notwithstanding that all governments across the region and across political colors have showed that they are ready to accept significant environmental "costs", it is also important to acknowledge that they have in general increased public spending on projects aiming for sustainable resource management and conservation, as well as enhancing their administrative and regulatory capacities in the environmental area. As explored in depth in Chapter 5, the states have been working in developing national baseline standards, enhancing environmental coordination, interdisciplinary approaches, participatory processes, and territorial planning. They have also diffused standards for "good agricultural practices", promoted systematic data gathering, created national systems of protected areas and imposed new legislation to reduce environmental harm from agrarian change. It is also important to keep in mind that different bodies of the state apparatus, from the congress enacting legislation, to government departments, secretariats, divisions and offices, and public governing bodies at local level, are involved in different ways in formulating and implementing regulations and policies. Power-relations and traditions vary between these different parts of the state apparatus. For example, the Ministries of Agriculture in Argentina, Paraguay, and Uruguay (Agroindustry, MAG, and MGAP) are all reflecting majority "productivist" policies, based on conventional knowledge claims from agronomists and economists. These ministries have generally considerably more influence over public policy and larger budgets than the relatively new Ministries in charge of the Environment and Sustainable Development (MAyDS, MADES, and MVOTMA), which instead often favor policies built on insights from biology, limnology, and ecology, with less consideration of short-term economic gains. While there are important similarities among the three countries regarding "sustainability" at a discursive level, and while all three countries can still be described as mainly operating within a "productivist" paradigm, there exist rather significant differences in priorities and policy tools used, as well as in capacity to assure successful implementation and enforcement.

*Flagship Environmental Regulative Responses in Argentina, Paraguay, and Uruguay*

One of the greatest concerns caused by the recent agropastoral expansion is the rapid deforestation the past decades, particularly in the Gran Chaco region, involving Northern Argentina and Western Paraguay. The fragmentation if the Atlantic forest in Eastern Paraguay has also received significant attention, although massive deforestation in this area started already in the 1970s and today only small fractions are left. In both Argentina and Paraguay, important environmental policy responses to curb deforestation has emerged. The flagship argentine regulation is the 2007 *Native Forest Protection Law* (Law 26.331), and the same in Paraguay is the 2004 *Land Conversion Moratorium for the Atlantic Forest* (the "Zero Deforestation Law"). Both are estimated to have contributed to reduced deforestation. Both are nevertheless also found to have had difficulties in enforcing complete compliance, with many cases of illegal logging. Besides illegal deforestation, the decentralized character of the argentine law, where the Provinces are responsible for categorization of the forests and their territorial planning (stipulating areas where deforestation is prohibited—red, very restricted—yellow, and possibly—green), resulted in a high degree of variance both regarding categorization and interpretation of the categories. Some Provinces marked a high share of its native forests red, but others did not. Some Provinces allowed for almost no deforestation in yellow areas, but others did. While the moratorium of the Atlantic forest in Paraguay does not open up for any ambiguities—all logging is forbidden—it is only a temporary regulation. New laws have extended the period of moratorium many times (No. 2524/04, 3139/06, 3663/08, 5045/12 and 5288/18), and it is currently running until 2020. However, the faith of the remains of the Atlantic forest after 2020 is highly uncertain.

While far from functioning perfect, the new laws have still had significant effect in their respective area of influence. However, they are also found to have created important leakage effects, leading to increasing rates of deforestation in the Paraguayan Chaco (de Waroux et al. 2017). As ranchers have lost land to soybeans in the Pampas region and have constrained possibilities to expand in the argentine Chaco and the Paraguayan Atlantic forest region (due to the argentine forestry law and the Paraguayan zero deforestation moratorium) they have instead expanded in the Paraguayan Chaco. As mentioned, the Paraguayan Chaco is currently one of the places in the world with the fastest rates of native forest loss. Paraguayan legislation (the 1973 *forestry law* - 422/73), forces landowners of forestland to "spare" at least 25% of forest. However, noncompliance is widespread. As explored in Chapter 5, corruption, low fines, legal loopholes, and lack of public monitoring and inspections explain poor law fulfillment. Besides prohibition, the Paraguayan state has installed large protected areas in the Paraguayan Chaco and it has created economic incentives for afforestation, restoration, and conservation. However, these initiatives face the same chronic problems as the majority of Paraguay's public efforts for environmental protection; lack of budget, corruption at all levels, unclear ownership of land, and ambiguous or contradictory legal land rights. In the wake of climate change becoming an increasingly important concern in the international agenda, and since forests are recognized important in their role for carbon storage (as well as climate and water regulators and biodiversity reservoirs), many IOs are also concerned with the rapid forest loss in Paraguay. They fund several projects involving market-based tools and mechanisms to promote biodiversity and sustainable management (Sustainable Development Specialist UNDP Paraguay, Asunción, 20 February 2017). One of the most important recent initiatives comes from REDD+, as part of the UNFCCC with support from UNDP, which is providing funds for the Paraguayan state to be able to finance payment schemes to landholders for their provisioning of ecosystem services, and thus stop logging. This initiative faces the same problems of unclear property rights to land, corruption, and extreme (and contested) land concentration, as many of the other initiatives. It is, however, too early to evaluate any impact yet.

Decentralization reforms in Paraguay and Argentina seem to have significantly weakened state capacity to monitor agrarian change and implement conservation policies. Both countries decentralized their *National Forest Institutes* (INFONA in Paraguay, IFONA in Argentina) during the 1990s, which resulted in significantly reduced capacity to consider the whole territory. Moreover, loyal bonds and agribusiness interest represented within the bodies of local political leaders have often led to the stipulated aims of forest conservation have lost the battle against possible economic returns. While both countries have experienced similar problems, the levels of corruption and lack of resources are much more severe in Paraguay than in Argentina. The argentine state has generally significantly stronger capacity to enforce regulation than the Paraguayan state. However, the decentralized character of the forestry regulation and the important variation in public capacity among the Provinces—where the Provinces of the most deforestation of native forests are among the poorest and with weakest capacities—in Federal Argentina, have resulted in similar problems of enforcement as in Paraguay.

Uruguay's flagship environmental regulation in response to concerns over land degradation associated with the new practice of continuous cropping (Chapter 5 in this book) is the *Soil Use and Management Plans*, which implies that crop producers need to make obligatory plans for rotations. These plans make sure that erosion remains under a stipulated tolerated level, and force all crop producers to get authorization from the state for how the plan, use and manage their lands. The relative success of these plans, with very high compliance, may depend on several factors. One is probably that Uruguay since long already has a rather strict legal regulation for protection of the soils and water, through the 1981 *Law of Use and Conservation of water and soils* (No. 15.239). This gave the state authority to limit the power of the property holders in the name of soil conservation and erosion prevention. This law had been passed by the military regime, without much discussion, but it built upon the ambitious scientific work of the agrarian commission under CIDE in the 1960–1970s (Chapter 2 in this book). It had nevertheless never been properly implemented before 2013. Another contributing factor to the high compliance was probably that the farmers' community accepted it, even though not unanimously so, and it received a lot of criticism in the beginning. This is attributed to the rather long process of broad stakeholder participation for shared knowledge construction and consensus making before the plans were made obligatory.[19] Probably, the strong historical tradition of state intervention, institutionalized collective negotiations, and predominating belief in "rational" and scientific planning (Chapter 2 in this book), also contributed to the acceptance.

While Uruguay has a long history of concerns over soil degradation and erosion, there have been similar concerns in Argentina and Paraguay. Observers in Argentina and Paraguay argued, however, that it would be impossible to implement similar legal tools to force producers to manage the soils in ways that the state stipulates in their countries. The main reason stressed was that the producers would not accept it; they would perceive it as a violation of the private property rights to land. This was illustratively explained by a researcher at FAUBA that has been working with crop producers in both countries:

> The idea of strong private property rights to land; the idea that 'in my land I can do whatever I want' and 'who are you to tell me what to do on my land', is very strong here [in Argentina]. In Uruguay that changed slightly with the soils law, it generated some acceptance of the soils as a public utility or a social good, but it also generated a lot of resistance among the neoliberals. But, it is working. The producers see that it is a good thing and accept that they need to have a plan and that an agronomist needs to accept the plan. In Argentina, that would be impossible to implement. The law in protection of the native forests is the closest thing Argentina have where the state clearly says that there are at least some limitations to private property rights, and getting there generated a lot of resistance and the law is still not well-complied... but it is still an important step. (Researcher FAUBA and Agricultural Extensionist, Buenos Aires, 24 February 2017)

As illustrated in above quote, and in other places of this book, the private property rights to land are interpreted as yielding exclusive rights to the owners to do "whatever" on it. While the notion of strong private property rights to land is rather strong also in Uruguay, it has a longer history, since *Batllismo*, of the state doing things against the short-term economic interest of landowners (such as taxing them quite heavily). Moreover, Uruguay has lower levels of corruption and higher levels of trust than both Argentina and Uruguay, which provides a huge advantage for all types of public initiatives regarding regulations and policies.

However, tiny Uruguay also faces important constraints regarding ensuring sustainability. If continuing with the illustrative example of the obligatory crop plans, many critical voices argue that the set "tolerable" level of erosion is too high, and not sustainable in the long-term. While ecological organizations criticize the plans for being too "tame", the combined effects of falling international soybean prices and the obligatory crop rotations probably made the shareholders of the up until then biggest crop producer in the country, *El Tejar*, to in 2014 all of a sudden leave Uruguay completely. This is an illustrative reminder of the fast moves of the new transnational agribusiness firms in contemporary agrofood globalization. They invest and expand as long as short-term economic margins are profitable and shareholders happy. The minute market conditions change so that margins are lower than expectations, or expected opportunity costs, independently if the reasons are exogenous or endogenous, these firms may leave, creating high volatility.

*Sustainable Pathways Beyond Public Regulation*

Many argue that public regulation alone, relying on tools such as laws, taxes, and fines, in line with some traditional logic of "command-and-control", cannot solve the sustainability problems. One line of argumentation goes that governing through strict laws and prohibitions only causes a lot of conflict and resistance from the private actors of the agrarian sector. Rules need to receive acceptance from landowners and producers in order to be effective (Plant and Animal Health Specialist IICA Paraguay; Consultant and Former President INTA; Director of Agricultural Affairs CARI, Asunción and Buenos Aires, 22, 28 February and 2 March 2017). Moreover, there is a widespread fear that strict regulations may reduce international competitiveness and/or that agribusiness actors and investors abandon the country. Therefore, regulations need to be formulated together with the private actors to secure compliance and avoid both costly and ineffective monitoring systems and the risk of conflict and confrontation.[20] This is one of the often stressed reasons to the increasing use of broad participatory processes for policy design and regulations (Chapter 5 in this book). At the same time, critics argue that these processes blur the line between different development visions and ultimately make it difficult to impose stricter regulation that acts against the agribusiness interest, albeit potentially necessary for "true" sustainability.

Besides participatory policy design, many argue that the state should search for more win-win solutions, showing the private sector that producing in sustainable ways—not deforesting, not over-using pesticides—is the most profitable thing to do. This approach is also adopted by the three states, who often in collaboration with IOs and producers' and agribusiness' organizations work with "capacity building" and diffusion of information. Accordingly, the joint public–private initiatives across the countries have been involved in developing and publishing guides for "good agricultural practices" and in doing campaigns for triple rinsing of jugs and drums to free them from pesticide residue that otherwise may leak into the waterways.[21] These initiatives rest on the belief that landowners and/or producing firms can be made to voluntary change land-use decisions and forms of production into more sustainable directions if they are thoroughly informed about the true costs and benefits involved in different decisions, and internalize them into their (cost/benefit) analysis. Some ecological costs, such as erosion, translates into economic costs in the form of lower productivity fast. Other ecological costs, however, such as loss of biological diversity or pollution of water, are slower, more indirect and complex in their interaction with other stressors, making them more difficult for producers to detect. It is in this way considered to be a pedagogical challenge to make producers see and value these things (Researcher FAUBA and Agricultural Extensionist; Director of ANII, Researcher and former Public Official, Montevideo, Buenos Aires and Stockholm, 2017 and 2018).[22] There is also a widespread expectation of international markets becoming increasingly demanding regarding environmental sustainability, which will translate into economic benefits for the private actors that can show that they produce without causing environmental harm. This will ultimately incentive all actors of the complete productive system into more sustainable forms of production. This is the main philosophy behind many new public–private arenas and programs, where the states support transitions toward "greener" practices.

Others are however more skeptical about the possibilities to change ways of producing by mere capacity building and information, or by expecting markets to become more demanding. While some niche markets indeed ask for more information and some consumers are concerned over the long shadows of their consumption, the overall agrofood market is not moving in that direction, and particularly not for soybeans. China is as mentioned the biggest buyer of both soy and beef and its role will be increasingly important. The quality protocols that China has imposed on the exporting countries are more concerned with no bugs in the cargo, than with environmental practices in production. This will hardly gear the system toward more environmentally friendly practices. Several informants also express that negative "feedbacks" from the ecosystem are often lost in the soybean production system, because an important part, around half, is estimated to be under not owned, but leased, land. This was expressed in the following way by one Argentine consultant with a long trajectory within the agrarian sector:

> A form of doing agriculture was installed with almost 50% under short-term leasing, this creates a logic of short-term gains; the one who leases the land wants to take out as much as possible from it in order to make a profit well and above the rent paid for the land. It becomes impossible to cultivate something that is not well paid on the market. Often there is no choice but to cultivate soybeans. (Consultant and Former President INTA, Buenos Aires, 28 February 2017)

There are also researchers and members of NGOs stressing a mismatch scales considering both time and space, which makes it profitable to adopt unsustainable practices (Researcher and former Public Official, Stockholm, 10 September 2018). It can be economically beneficial to produce in unsustainable ways since the "costs" are not exclusively falling on the farmer but also on landscape or even global level (such as pollution of water and air, risks of flooding downstream, increased risk for weed resistance and increased concentration of greenhouse gases in the atmosphere.), as well as on the future. Thus, the forms of production—including land-use, weed and pesticide management, fertilization—on a particular piece of privately owned land have consequences far beyond the boundaries of that land, since it connects to, and forms part of, a wider ecosystem. The logical conclusion of this line of argumentation is that land cannot be considered exclusively as private property but is always in some respect a public good. Accordingly, land-use and cover is a public concern not to be decided exclusively upon by the property-holders. The advocates of a view on land and water as social goods see optimistically upon the enhanced capacity to control and ensure compliance due to new technological tools that collect and make available huge amounts of data. Some actors highlight however, that this technology can also be used to increase social control of the workers and turn the state into some kind of "Big Brother", and potentially threaten personal integrity.

## 6.2    Differentiated State Autonomy and Capacity

Above section has offered an analysis of the opportunities and constraints facing Argentina, Paraguay, and Uruguay when trying to make agrarian change more "developmental", socially inclusive, and environmentally sustainable. What does this say about differentiated state autonomy and capacity of these countries? As mentioned in Chapter 1, *state autonomy* can be defined as the ability of the state to control its territory and act freely and independently from outside pressures and interference. This involves the capacity to control its borders, have monopoly of the means of legitimate violence and the ability to handle different external factors, such as transnational agribusinesses, foreign states, and international organizations, and prevent them from having overly considerable influence. Foreign debts, poverty, international trade agreements, co-option, and corruption can severely limit state autonomy. *State capacity* can be defined as the ability to formulate, perform, implement and enforce effective, legitimate and coherent rules and decisions in line with established national development targets. This involves government effectiveness, rule of law, and the capacity to generate on-the-ground impacts. It also involves the capacity to align targets, manage field enforcement, make an independent analysis and agenda, encourage participation of affected actors, and systematically monitor and evaluate the effects, including indirect effects. While limited statehood is the rule, there is a wide variation of how well (or poorly) these three countries fulfill the ideals of state autonomy and capacity.

It can be concluded that the Paraguayan state fails in many different respects to live up to the definitions of state autonomy and capacity. Recurrent and systematic violence from landowners, peasants, guerillas, para-military groups, and drug syndicates, gives testimony of that the monopoly on violence of the state is challenged from many fronts.

The disability of the state to regulate land ownership in ways that are clear and accepted as legitimate by the majority of the population, provides constant fuel to this violent fire. Vast amounts of the Paraguayan territory are described as law-less land, without any state presence at all. These places represent "areas of limited statehood". In addition, the chronic scarcity of public resources results in severe constraints in the state's ability to plan, monitor, and to generate on-the-ground impacts. This creates deficiencies in the government's capacity to regulate the private sector and to manage real and perceived social-ecological risks. Rules and regulations are often violated, and the state only occasionally manages to fulfill complete enforcement. Moreover, the predominant role of IOs in agrarian policy formulation and implementation indicate that the state lacks capacity to make an independent analysis and agenda as well as handle external factors in a truly independent way. In addition, the agribusiness firms' ability to make agrarian policies aligned with their interest through lobby, co-option strategies (of the Colorado Party) and by "paying their way" to almost complete dominance over land, agro processing industries, further shows the limits of statehood. Finally, the widespread corruption, which seems to have become a naturalized component of all parts of society, further weaken state capacity. Close to corruption, comes a widespread culture within the Colorado-party of populist *asistentialismo*, introduced by Stroessner. This involves ineffective paternalist programs where the state gives certain "gifts" to poor smallholders in order to secure electoral support, while not structurally changing the situation of the poor. Thus, widespread corruption, political instability, co-option by strong pressure groups, chronic budget constraints, populist measures, and lack of enforcement mechanisms pose significant limitations to state autonomy and capacity.

Compared to Paraguay, both the Argentine and the Uruguayan states are clearly stronger, more autonomous, and with greater capabilities. At the same time, Federal Argentina involves many different realities and the poorer Provinces in the North share many of the vulnerabilities of the Paraguayan state, including unclear and illegitimate land titles, violent struggles over land, co-option by foreign agribusiness, and widespread corruption. Argentina, as Paraguay, also fails in providing indigenous people the rights that Argentine, Paraguayan, and international law afford them. At the central government level, one of the main weaknesses is a polarized and antagonistic political culture, with very low levels of trust. There is for example a culture of not sharing information between different governments, which significantly constraints state capacity to long-term planning, cumulative knowledge creation, and in the end policy effectiveness. Moreover, this culture of noncooperation and mistrust seems to have spread to encompass also different branches or offices of the same political force, or different sections of the same organization. Another Argentine problem that weakens the state capacity is the strong historically formed divide between urban and rural areas with mutual distrust and disrespect. The *conflicto del campo* that emerged under *Kirchnerismo*, illustrates this polemic divide, which significantly constrains the capacity of the state to make the agrarian sector contribute to wider societal development. This conflict also showed the strong political influence that the social actors linked to agro-exports still has in Argentina, and their ability to receive support also from social actors of the countryside that are not receiving benefits from the soybean "boom" and could be expected to articulate other interests. At the same time, the government administration of Macri (2015–2019) showed that the central government could continue to tax soybeans almost as before, without entering in a conflict with the "countryside" as long as the taxes were discursively condemned (while kept) and integrated a general agribusiness-friendly discourse.

In the international arena, Argentina has shown in many respects to be much stronger and independent than its small neighbors have. Argentina has negotiated and sometimes put tough rules on foreign agribusiness firms and states alike. This is partly because Argentina is more than 6 times bigger than that of Paraguay and more than 15 times bigger than that of Uruguay. The 2017 GDP was USD 637B in Argentina, while only 40B in Paraguay and 56B in Uruguay (with the highest GDP/c).[23] Self-conscious about the value of its vast fertile soils and its importance as world agrofood producer, Argentina acts with more leverage, enhancing its bargain power and advantage in negotiating agreements and other conditions significantly. For example, it had a small trade war with China about import taxes on soybeans and textiles, and it put up a fight against Monsanto regarding farmers' right to save proper seeds and won it. However, as Argentina got the biggest loan in IMF's history, in 2018, it is (again) subjected to both conditionality and more subtle influence.[24] The Uruguayan state has a clear monopoly of the means of legitimate violence and is reported to have lower levels of corruption than its neighbors do. Its small size, with constrained resources compared to some of the mega transnational actors operating within it, still put a question mark to whether it can act completely independently from outside pressures.

To secure sustainable pathways of agrarian change requires strong national and regional public regulation and institutional capacity for social-ecological governance. The capacity of the three Southern Cone states to regulate and control the territory has nevertheless evolved in complex, uneven, and contradictory ways under this phase of intensified globalization. As thoroughly showed throughout this book, the features of the current international trade system, and inherent features of the soy and beef chains, put several constraints on what the states can do in order to make agrarian change more developmental, socially inclusive, and sustainable. At the same time, the comparative analysis shows that there is a significant degree of difference in statehood between Argentina, Paraguay, and Uruguay. While they all face constraints, Argentina and even more Uruguay, have from time to time manage to align agrarian change with, and make it contribute to, their broader national development agenda. This shows clear historical parallels to the first globalization wave, or the first *food regime*, where these countries also managed to get more benefits from the commodity "boom" than the other countries in the region. The capacities for poor Paraguay with its weak institutions to do the same is significantly less. Notwithstanding the differences between the countries, it is clear that the main "rules of the game" of the international agrofood system, thoroughly outlined in Chapter 3 and discussed elsewhere in this book, are not formulated by any of these countries. In addition, the recently announced trade deal between the EU and Mercosur will probably substantially increase demand (and price) on agrarian-based products from this region. In the context of the stalled Doha round and globally decelerating food demand growth, the mega trade agreement potentially creates some new development opportunities for the Southern Cones states, but it also further reduces their ability to promote alternative development paths (by for example protecting domestic industries through high tariffs) and it further strengthens the power of the agribusiness sectors vis-à-vis other sectors of the economy.

One general difference to previous food regimes is the pace and scope of change, and the economies of scale that new technological packages and management forms allow. Another important difference, however, is that new technologies have emerged to collect and make available huge amounts of data. With the new information technology, the state can control in a much more efficient and cheap way than before. The technology can bring the state with information about who is cultivating what, where, and how much it yields—and tools to control tax payments, agrochemical use, and livestock movements down to each parcel of land. The former secretary of MAGyP expressed his hopes regarding the new technology in the following way:

> The state does not yet possess the ability to analyze all the information in an efficient way, but that is a matter of time. In any case, I believe that the states have a huge opportunity with the new technology to control things that before were uncontrollable, and inclusively to control Monsanto. (Former Secretary MAGyP, Buenos Aires, 2 March 2017)

New technologies thus bring both new opportunities and constraints for state capacity and autonomy. Another historically "new" aspect of contemporary globalization is the important role of the IOs. While the international arena may put some constraints to state autonomy, many social movements form the region have begun appealing to international human rights frameworks, invoking ethnicity or identity, in order to gain greater legitimacy for their struggles for collective rights to land. This is particularly the case of Paraguay and Northern Argentina. Thus, IOs can also provide leverage to underprivileged domestic groups. These cases illustrate how the capacities of the states to regulate agrarian change in line with established national development targets under globalization—while perhaps increasingly necessary—is not a straightforward task. To conclude, while state autonomy and capacity are constrained in all three countries to varying degrees, stronger and more coordinated regulative efforts could lead to more sustainable pathways for agrarian change in Latin America.

## Notes

1. Carl Sagan, 1980. Cosmos: Episode 2, "One Voice in the Cosmic Fugue".
2. The declaration was an initiative of the *Organization for Economic Cooperation and Development* (OECD) for increased sovereignty of receiving nations.
3. See the annual reports and websites of the national offices in Argentina, Paraguay, and Uruguay of IICA, UNDP, FAO, World Bank, IADB, GTZ, UNEP, and IFAD. After a change in government, the IOs typically meet with the new administration to discuss and set up a new so-called framework agreement, with priority lines and programs for the government period, but there are clearly mainly continuities in the cooperation portfolio.

4. Important legislation in this respect: Argentina (Decree 1396; Decree 109–120; Resolution 129; Laws 26093/2006; 26,190/2006; 27,191/2016- Programa energías renovables para mercados rurales special fund to grant subsidies to private concessionaires), Paraguay (Law 2,748/2005; Decree 2,998/2015; Law 5,444/2015; Resolution 507/2017) Uruguay (Law 17564; Law 18195/2007; 2008 Energy Policy 2005–2030)

5. Argentina is the leading supplier of imported biodiesel for the EU and the U.S. EU imposed a heavy anti-dumping import tax on Argentinian bio-diesel, but removed the tax in 2016 (Rouhany and Montgomery 2019). Argentina's export tax on biodiesel was around half the tax on soy meal and soy oil. Argentina started promoting bio-fuels in 2008, through a new legal framework, and could already for 2010 address a 5% blend target of bio-diesel, since both production and industrial capacity was already there. Paraguay followed a similar path and advanced fast up to a 5% blend y (IICA 2010).

6. While the value-added of crushing is rather low, China has still applied differentiated import taxes between whole beans and soybean meal and flour, in order to serve the domestic crushing industries.

7. One exception to this rule is the 2018 Uruguayan certified GM-free soybean production for exports to China for food and not feed (China does not allow GM soybeans for direct human consumption), which allows for some quality differentiation.

8. Reuters, "ADM, Bunge say they can navigate U.S. -China trade tensions", 16 May 2018 www.reuters.com/article/us-bunge-ceo/adm-bunge-say-they-can-navigate-us-china-trade-tensions-idUSKCN1IH1SW. Accessed 10 July 2018.

9. China plays a particularly unique and extraordinary role as the main buyer of Uruguayan and Argentine agrofood exports. While an important part of the agrarian based exports from landlocked Paraguay also ends up in China, it is typically re-exported via Argentina, Brazil and Uruguay, and is therefore not visible in the trade statistics (CAS 2018, 51–54), due to the lack of diplomatic ties between the countries caused by Paraguay's still recognition of Taiwan as an independent state.

10. However, the assumptions of classical Latin American structuralism of long-term declining terms of trade of agricultural products and of inherent low value-added have been increasingly questioned. The terms of trade for agrofood commodities, particularly soybeans, were favorable for a relatively long period. Prices of soybeans were in general rising between 2001 and 2012—a super cycle—except for the 2008 financial crisis, followed by a rapid recovery. Since then, however, prices have declined significantly. See Macrotrends, "Soybean Prices—45 Year Historical

Chart" https://www.macrotrends.net/2531/soybean-prices-histori-cal-chart-data. Accessed 23 January 2019.

11. El Cronista, "Otra sanción de la AFIP a Bunge deberá pagar más impuestos para exportar", 10 May 2011, https://www.cronista.com/economiapolit-ica/Otra-sancion-de-la-AFIP-a-Bunge-debera-pagar-mas-impuestos-para-exportar-20110510-0065.html. Accessed 22 January 2019.

12. In this way, the size of for example the Argentine average farm in 1947 was 368 ha, while in 2002 it was 524 ha. This can be compared with the average farm in EU in 2002; 40 ha, and in the U.S.; 200 ha (Giarracca 2008, 20).

13. "The old proverb, 'livestock fattens under the gaze of its master', still holds true, the owners need to be close and constantly monitor the crops: it is hard to delegate agricultural work", concluded the Director of CICPES-INTA and former secretary of SAGPyA, Buenos Aires, 2 March 2017.

14. The percentage of the population earning less than USD 5.50 (2011 PPP) per day in 2016: In Argentina—7.8%; in Paraguay—20.6%; in Uruguay—3.7%. Data, Poverty, the World Bank Group, https://data.worldbank.org/topic/poverty?locations=UY-AR-PY. Accessed 22 December 2018.

15. IDB. 2019. "ESTADÍSTICAS DE POBREZA Y DESIGUALDAD DE INGRESOS EN ALC" https://www.iadb.org/es/investigacion-y-datos/pobreza%2C7526.html. Accessed 26 February 2019.

16. IDB. 2019. "Transferencias sociales" https://test-iadb.pantheonsite.io/es/investigacion-y-datos//transferencias-sociales,7531.html. Accessed 26 February 2019.

17. While Argentina's export taxes and quotas in a clearer way opposed export-oriented agrarian interest, they also responded to the fact that income taxes in Argentina are not progressive—and they belong to the provinces, not the central government.

18. LDC: "Since 1851, we've played a vital role in nourishing the world's population." ADM: "For more than a century, the people of Company (ADM) have transformed crops into products that serve the vital needs of a growing world." Cargill: "We help people thrive. Cargill is working to nourish the world." Bayer: "We exist to help people thrive - Advancing health and nutrition is what we do best and care about most." From the official company websites. Accessed 2 May 2019.

19. The plans have later been extended to include dairy farmers in the water basin of Santa Lucia to hinder erosion and agrochemicals to end up in the drinking water.

20. One of the best-known private initiatives to increase the "sustainabil-ity" of the soybean chain is the *Round Table on Responsible Soy* (RTRS), active in all three countries. Within these "tables", in Argentina, Paraguay, and Uruguay, almost all big soybean producers, processors and traders

participate together with organizations such as WWF, FAO, and UNDP. They have developed a voluntary certification scheme, which is supposed to guarantee legal compliance, labor responsibility and social-environmental sustainability, all along the soybean supply/value chain (General coordinator of green production and sustainable commodities at UNDP, in Asunción, 20th of February 2017). One of the criteria for certification is that the soy is not produced in forestland. RTRS have nevertheless received a lot of critique, and most of the national civil society organizations in the region argue that it is nothing but greenwashing. For example, for the Roundtable's 2006 conference in Paraguay, Vía Campesina–Paraguay organized a protest together with many other Paraguayan NGOs and social movements (García-López and Arizpe 2010). The criteria for certification is argued to be in perfect harmony with the standard ways of producing soybeans (GM, no tillage, glyphosate, pesticides) and not demand any sustainable practices at all. When the Uruguayan oilseeds table (MTO) heard about the initiative, it became a member immediately without having to change any practices, since all firms already lived up to the criteria (The secretariat of MTO, in Montevideo, in 2010). Defenders of RTRS claim that it least may slow down deforestation, but critics argue that certified soybeans may still contribute to deforestation, since when soybean production expands in other areas it pushes other agrarian activities, not least cattle grazing, into forest area.

21. On the webpage for ACSOJA it is possible to read all the posts on "good agricultural practices" and how the organization has been working with this approach during the past years, often in collaboration with public entities  http://www.acsoja.org.ar/?s=buenas+practicas+agr%C3%ADD-colas. Accessed 10 September 2018. MTO has together with lecturers from the faculty of agronomy (FAGRO-Udelar) published and actualized a guide for good agricultural practices: http://mto.org.uy/actividad/guia-de-buenas-practicas-agricolas-para-sistemas-con-agricultura-de-se-cano. Accessed 23 February 2019.

22. In this way, unsustainable practices are largely seen as a problem of information deficit, since landowners and farmers benefit from healthy ecosystems as they produce more and generate less (fewer?) negative externalities. The idea is to show how rotations can reduce costs and boost benefits in the medium-run, despite not providing any high value in the market.

23. The World Bank, https://data.worldbank.org/indicator/NY.GDP.MKTP.CD?locations=PY-AR-UY. Accessed 6 March 2019.

24. For example, while the IMF secured social expenditures as a percentage of GDP, Argentina had to cut wages and the number of state workers by 0.5% of GDP to meet IMF loan conditions.

## REFERENCES

### UNCATEGORIZED REFERENCES

Acosta Reveles, Irma L. 2008. "Capitalismo agrario y sojización en la pampa Argentina. Las razones del desalojo laboral." *Lavboratorio, Cambio Estructural y Desigualdad Social (CEyDS) / Facultad de Ciencias Soicales, UBA.* 10 (22): 8–12.

Alianza Internacional de las organizaciones de productores familiares de soya. 2008. *Declaración de Asunción.* IV Encuentro Internacional de Pequeños Productores de Soya y la Sociedad Civil, 6, 7, y 8 de Febrero del 2008. Asunción: Tekokatu, Ser, Probioma and Solidaridad.

Anichini, F., Borda, D., Masi, F., Ramirez, J., Servin, B., and Setrini G. 2013. *Estudio de Potencialidad de Desarrollo de las Cadenas de Valor.* Edited by Centro de Analisis y Difusion de la Economia Paraguaya. Asunción: Agencia de Cooperación Internacional del Japón (JICA).

Armony, Ariel C., and Nicolás G. Velásque. 2016. A Honeymoon with China? Public Perceptions in Latin America and Brazil. *Revista Tempo do Mundo (RTM)* 2 (2): 17.

Bárcena Ibarra, Alicia, and Antonio Prado. 2015. "Introducción." In *Neoestructuralismo y corrientes heterodoxas en América Latina y el Caribe a inicios del siglo XXI*, edited by Alicia Bárcena Ibarra and Antonio Prado. Santiago de Chile: CEPAL.

Barrios, Juan José, Nestor Gandelman, and Gustavo Michelin. 2010. "Analysis of Several Productive Development Policies in Uruguay". IDB Working Paper Series. Montevideo: Universidad ORT Uruguay.

Bértola, Luis. 2015. "Patrones de desarrollo y Estado de Bienestar en América Latina." In *Neoestructuralismo y corrientes heterodoxas en América Latina y el Caribe a inicios del siglo XXI*, 261–295. CEPAL.

Carriquiry, Florencia. 2015. "El papel del agro en el desarrollo económico nacional." In *El desarrollo agropecuario y agroindustrial de Uruguay: Reflexiones en el 50 aniversario de la Oficina de Programación y Política Agropecuaria (OPYPA-MGAP)*, edited by Unidad de Comunicación Organizacional y Difusión, 26–35. Montevideo: OPYPA-MGAP.

CAS. 2018. *Anuario de Comercio Exterior de base agraria de los países CAS 2013–2017.* Edited by M. Ackermann and L. Gorga. Montevideo: Consejo Agropecuario del Sur.

Castellano, Andrés, and Mercedes Elida Goizueta. 2011. *Agregado de valor en la cadena de la soja: alternativa de upgrading para productores primarios.* EEA Balcarce Área de Economía y Sociología Rural. Balcarce: INTA.

Cepal, CIRAD, and IICA. 2014. *Políticas públicas y agriculturas familiares en América Latina y el Caribe: Balance, desafíos y perspectivas.* Edited by Eric Sabourin, Mario Samper and Octavio Sotomayor Echenique. Santiago: CEPAL-ECLAC.

CLAEH, FIDA Mercosur. 2015. *Institucionalidad responsable por la agricultura familiar y las políticas públicas diferenciadas en el Mercosur. Marco de actuación de la REAF.* Montevideo, Uruguay: Programa FIDA Mercosur CLAEH.

Dabezies, Martín. 2009. "Área Cadenas Agroindustriales: Informe final de la consultoría sobre Cadenas Agroindustriales en el marco del Plan Estratégico Nacional en Ciencia, Tecnología e Innovación." In *Plan Estratégico Nacional de Ciencia, Tecnología e Innovación (PENCTI),* edited by Equipo Operativo Gabinete Ministerial de la Innovación. Montevideo: PENCTI.

de Waroux, Yann le Polain, Rachael D. Garrett, Jordan Graesser, Christoph Nolte, Christopher White, and Eric F. Lambin. 2017. "The Restructuring of South American Soy and Beef Production and Trade Under Changing Environmental Regulations." *World Development* 121: 188–202. https://doi.org/10.1016/j.worlddev.2017.05.034.

Errea, Eduardo, Juan Peyrou, Joaquín Secco, and Gonzalo Souto. 2011. *Transformaciones en el agro uruguayo - Nuevas instituciones y modelos de orgnaización empresarial.* Edited by Dámaso Antonio Larrañaga Universidad Católica del Uruguay, *Universidad Católica.* Montevideo: Facutad de Ciencias Empresariales, Programa de Agronegocios.

Ezquerro-Cañete, Arturo, and Ramón Fogel. 2017. A coup foretold: Fernando Lugo and the lost promise of agrarian reform in Paraguay. *Journal of Agrarian Change* 17 (2): 279–295.

FAO. 2017. "Soybean Prices, Economic Growth and Poverty in Argentina and Brazil". In *Background paper to the UNCTAD-FAO Commodities and Development Report 2017 Commodity Markets, Economic Growth and Development.* Rome: Food and Agriculture Organization of the United Nations.

Gabinete Productivo. 2010. *Medidas para el desarrollo de las Cadenas de Valor.* Edited by MIEM. Montevideo (Uruguay): Mastergraf.

García-López, Gustavo A., and Nancy Arizpe. 2010. Participatory Processes in the Soy Conflicts in Paraguay and Argentina. *Ecological Economics* 70 (2): 196–206. https://doi.org/10.1016/j.ecolecon.2010.06.013.

Giarracca, Norma. 2008. "La Argentina y la democratización de la tierra." *Lavboratorio, Cambio Estructural y Desigualdad Social (CEyDS) / Facultad de Ciencias Soicales, UBA.* 10 (22): 18–21.

Gonzaga Belluzo, Luis. 2015. "La reciente internacionalización del régimen del capital." In *Neoestructuralismo y corrientes heterodoxas en América Latina y el Caribe a inicios del siglo XXI.,* edited by Alicia Bárcena and Antonio Prado, 112–125. Santiago de Chile: CEPAL.

Gras, Carla, and Valera Hernández. 2013. *El agro como negocio: producción, sociedad y territorios en la globalización.* Buenos Aires: Editorial Biblos.

Gudynas, Eduardo. 2009. "Diez tesis urgentes sobre el nuevo extractivismo." *Extractivismo, política y sociedad* 187.

Gudynas, Eduardo. 2010. La ecología política del progresismo sudamericano: los límites del progreso y la renovación verde de la izquierda. *Sin Permiso* 8: 149.

Gudynas, Eduardo. 2011. Buen Vivir: Germinando alternativas al desarrollo. *América Latina en movimiento* 462: 1–20.

Gudynas, Eduardo. 2017. Los ambientalismos frente a los extractivismos. *Nueva sociedad* 268: 110–121.

Hoekman, Bernard. 2017. "Trade and the Post-2015 Development Agenda." In *Win-Win: How international Trade Can Help Meet the Sustainable Development Goals*, edited by Matthias Helble and Ben Shepherd, 32–57. Asian Development Bank Institute.

IBRD, IFC, MIGA. 2014. *Country Partnership Framework for the Argentine Republic for the period 2015–2018*. Report Number: 131971.

IBRD, IFC, MIGA. 2015. *Country Partnership Framework for the Oriental Republic of Uruguay for the Period 2016–2020*. Report Number: 97063.

IBRD, IFC, MIGA. 2018. *Country Partnership Framework for the Republic of Paraguay for the Period 2019–2023*. Report Number: 131046.

IICA. 2010. Atlas de la agroenergía y los biocombustibles en las Américas: II Biodiesel. IICA, San José: Instituto Interamericano de Cooperación para la Agricultura.

Janssen, Rainer, and Dominik Damian Rutz. 2011. Sustainability of Biofuels in Latin America: Risks and Opportunities. *Energy Policy* 39 (10): 5717–5725.

Lapegna, Pablo. 2016. Genetically Modified Soybeans, Agrochemical Exposure, and Everyday Forms of Peasant Collaboration in Argentina. *The Journal of Peasant Studies* 43 (2): 517–536. https://doi.org/10.1080/03066150.2015.1041519.

López, Andrés, Daniela Ramos, and Cecilia Simkievich. 2008. "The Impact of China's Global Economic Expansion on Latin America—A Study". Working Paper No. 1. Buenos Aires: Centro de Investigaciones para la Transformación (CENIT).

Manzanal, Mabel, and Sergio Schneider. 2011. "Agricultura Familiar y Políticas de Desarrollo Rural en Argentina y Brasil (análisis comparativo, 1990–2010)." *Revista Interdisciplinaria de Estudios Agrarios* 34 (CIEA, FCE, UBA,):35–71.

Oyhantçabal, Gabriel, and Ignacio Narbondo. 2011. *Radiografía del agronegocio sojero - Descripción de los principales actores y los impactos socio-económicos en Uruguay. Datos actualizados a 2010*. 2nd ed. (first 2008). Montevideo: Redes-AT.

Paolino, Carlos. 2015. "La política públca y el apoyo al sector agropecuario." In *El desarrollo agropecuario y agroindustrial de Uruguay: Reflexiones en el 50 aniversario de la Oficina de Programación y Política Agropecuaria (OPYPA-MGAP)*, edited by Unidad de Comunicación Organizacional y Difusión and Diego Campoy, 37–45. Montevideo: MGAP.

Ray, Rebecca, and Kevin P. Gallagher. 2016. "China in Latin America: Environment and Development Dimensions." *Revista Tempo do Mundo (RTM)* 2 (2): 131–154.

Rouhany, Mahbod, and Hugh Montgomery. 2019. "Global Biodiesel Production: The State of the Art and Impact on Climate Change." In *Biodiesel*, 1–14. Springer.

Sabourin, Eric, Mario Samper, Jean François Le Coq, Gilles Massardier, and Octavio Sotomayor. 2014. El surgimiento de políticas públicas para la agricultura familiar en América Latina: trayectorias, tendencias y perspectivas. *Cadernos de Ciência & Tecnologia* 31 (2): 189–226.

Svampa, Maristella Noemi. 2013. "Consenso de los Commodities y lenguajes de valoración en América Latina." *Nueva Sociedad* 244 (4).

Svampa, Maristella, and Ariel M Slipak. 2015. "China en América Latina: Del consenso de los commodities al consenso de Beijing." *Revista Ensambles* 2(3): 34–63.

Tambler, Adrián. 2013. "Recaudación y presión fiscal en el agro." In *Anuario 2013*, edited by Opypa. Montevideo: Opypa - MGAP.

Teubal, Miguel. 2008. "Soja y agronegocios en la Argentina: la crisis del modelo." *Lavboratorio, Cambio Estructural y Desigualdad Social (CEyDS) / Facultad de Ciencias Soicales, UBA*. 10 (22): 5–8.

Vergara-Camus, Leandro, and Cristobal Kay. 2017a. "New Agrarian Democracies: The Pink Tide's Lost Opportunity." *Socialist Register* 54 (54): 211–230.

Vergara-Camus, Leandro, and Cristobal Kay. 2017b. "Agribusiness, Peasants, Left-Wing Governments, and the State in Latin America: An Overview and Theoretical Reflections." *Journal of Agrarian Change* (17): 239–257.

# Appendix A

## *A.1. A Note on Sources*

The empirical material scrutinized herein is broad and comes from multiple sources. For current soy area and yields, pasture area, cattle heads, input prices, transport costs, agrochemical imports, export prices, and trade flows over the past thirty years are obtained from FAOSTAT (http://faostat.fao.org/), the national departments of Agriculture (called department of agroindustry in Argentinasince 2016, MAG in Paraguay, and MGAP in Uruguay), the national departments of Environment (MAyDS in Argentina, SEAMin Paraguay and MVOTMA in Uruguay) and national export agencies, as well as from reports from the Foreign Agricultural Service (FAS) of the United States Department on Agriculture (USDA) and the Economic Commission for Latin America and the Caribbean (the Spanish acronym CEPAL is used throughout this book). In addition, data from the national institutes of statistics (INDEC in Argentina, DGEEC in Paraguay and INE in Uruguay) are used for a variety of socio-economic indicators. Data on conservation areas comes from the World Database on Protected Areas (http://protectedplanet. net/). For the history of soybeans, the extensive bibliographies, and the sources and literature that they in turn refer to, from Soyinfo Center have provided extremely rich and all-encompassing information related to soybeans and soy products (http://www.soyinfocenter.com/).

© The Editor(s) (if applicable) and The Author(s), under exclusive          375
license to Springer Nature Switzerland AG 2020
M. Baraibar Norberg, *The Political Economy of Agrarian Change
in Latin America*, Governance, Development, and Social Inclusion in
Latin America, https://doi.org/10.1007/978-3-030-24586-3

When possible, data from different sources have been triangulated for increased reliability. The actual regulations are mainly found in laws, decrees and acts obtained from the digital official government journals and websites: in Argentina *Boletín Oficial* (www.boletinoficial.gob.ar), in Paraguay *Gaceta Oficial* (www.gacetaoficial.gov.py) and in Uruguay *Impresiones y Publicaciones Oficiales, IMPO* (www.impo.com.uy).

An extensive literature review of previous research facilitated the navigation through the vast ocean of legislative material. This navigation was also facilitated by important personal guidance on research material from several researchers enrolled in national universities and/or national research institutes.[1] Other informants, such as public officials, politicians, the staff at international development organizations and NGOs, have facilitated the attainment of different reports and white papers as well as indicated new sources for investigation. In addition, white and grey papers and reports from government agencies, international organizations, agribusiness firms, political parties and factions, NGOs and social movements, as well as debates in national media, have been analyzed to grasp the interplay of different social forces behind regulations, and to separate what is contested from what is not.

A key and unique material collected for this book are the transcripts of more than thirty in-depth interviews with public officials, politicians, international organizations, and NGOs, made by the author in Argentina, Paraguay, and Uruguay during 2017–2018. The majority of the interviews were made under personal meetings at the respondents' workplace or at cafés, but four have been made over Skype or WhatsApp. All respondents have been involved in different regulatory processes, but they have participated from different positions (ranging from experts, to lobbyists and legislators) and at different times. The interviews were designed to grasp both the social forces involved in the formulation, shaping and (re)negotiation of the regulatory framework, and the differentiated perceptions of actual implementation and enforcement. The respondents' answers provided insights well beyond the information provided in written sources alone. The interviews gave particularly important insights both to the center and the backstage of policy formulation and implementation processes, yielding new explanatory elements to the analysis of discrepancies between set targets and actual impacts. In addition to the interviews from 2017 to 2018, over fifty transcribed interviews from 2009 to 2011, from a previous research project of the author (Baraibar, 2014) have been used as complementary sources. The "older" transcript has mainly been used to provide background information, but in a few cases, they have been directly referred to in this book, and have hence, in these cases, been included in the reference list.

### A.1.1. Interviews
Paraguay
1. **Sustainable Development Specialist UNDP Paraguay**: Coordinator of the project "Green and inclusive Economy" (Proyecto Economía verde e inclusiva), working with the Japanese Cooperation Agency (JICA), the Secretariat of Environment (SEAM) and its divisions of territorial planning, environmental management and Environmental information, as well as with local municipalities to develop public policies for a more sustainable and inclusive economic system, at the United Nations Development Programme (UNDP) Paraguay, Asunción, 20 February 2017
2. **Public Policy Specialist UNDP Paraguay**: Coordinator of the project "Democratic Governance" in partnership with the Ministry of Finance; working with the road map to implement Agenda 2030 and the sustainable development goals in Paraguay, Asunción, 20 February 2017
3. **Senior Technical Advisor UNDP Paraguay:** Project Coordinator of "National Platform for Sustainable Commodities", working with joint public–private articulation for biodiversity protection and Good Agricultural Practices, at UNDPParaguay; Author of reports on the national development of protected wildlife areas and watershed management, Asunción, 20 February 2017
4. **Director of DCEA-MAG:**The Division of Agrarian Statistics and Censuses, Department of Agriculture and Livestock, Asunción, 21 February 2017
5. **Former Minister of MAG**: President of the Paraguayan Chamber Pro Agro (CAPROA); President of the Seed Chamber; Vice President of the Chamber China-Paraguay; Minister of Agriculture and Livestock of Paraguay (1989–2001), Asunción, 21 February 2017
6. **Representative IICA in Paraguay**: The Inter-American Institute for Cooperation on Agriculture (IICA) in Paraguay; Ph.D. in Agrarian Sciences (USA) and Ph.D. in Sustainable Development and Public Policies (Chile); Author of articles and report on family farming and market integration— Asunción, 22 February 2017

7. **Natural Resource Management Specialist IICA Paraguay**: working with REDD+ Paraguay and the Secretariat of Environment (SEAM). Author of books and reports on "good agricultural practices", climate adaptation and mitigation, and market analysis for small producers, Asunción, 22 February 2017

8. **Innovation Specialist IICA Paraguay**: Former Public official at different divisions of the Minister of Agriculture and Livestock (MAG); working with policies for a national biotechnology framework, development plans to promote agriculture and agroecology, the strategic plan of IPTA, the creation of a division under MAG for organic farming (DEAg/MAG), the creation of a multisectorial technological table on biotechnology as well as a table on plant health, Asunción, 22 February 2017

9. **Plant and Animal Health Specialist IICA Paraguay**: working with joint public–private partnership for the strengthening of family farmerstogether with FAO and the National Service of Plant and Seed Quality and Health (SENAVE). Food Safetyprojects with MAG. Former director of planning office for control of agricultural, livestock, and forestry products at MAG. Long experience in the area of public strategic planning related to production, animal health, and management of rural development projects. Has been international consultant at various international organizations and authored various publications. Asunción, 22 February 2017

10. **Paraguay's Deputy Minister of Agriculture**; Minister of Agriculture and livestock (MAG); National Coordinator of REAF—the Specialized Meeting on Family Agriculture of Mercosur of Mercosur; Lecturer and researcher, Department of family farming, Faculty of Agronomy, University of Asunción, Asunción, 23 February 2017

11. **Public Policy Officer WWF**: World Wildlife Foundation (WWF) Paraguay, working with forest policies, social organization and lobby, Asunción, 23 February 2017

Argentina:
12. **Researcher FAUBA and Agricultural Extensionist:** Associate Professor at the Department of Ecology, Faculty of Agronomy, University of Buenos Aires (FAUBA); the Research Institute on Physiological and Ecological aspects of agriculture, the Argentine

National Research Council (IFEVA-CONICET); Consultant for grain producers in Argentina and Uruguay, Buenos Aires, 24 February 2017

13. **Consultant and Former President INTA**: Consultant for international organizations (the World Bank, the IDB, the EU, FAOand IICA, among others). Private consultant for firms in sectors of agroindustry, agriculture, livestock and forestry (El Tejar and many other firms in Argentina and Uruguay and other countries in the region and Europe); Director of the forestry firm Madersec S.A. and member of the Consorcio Forestal del Río Uruguay; Researcher at the Department of Agribusiness and Food, Faculty of Agronomy, University of Buenos Aires (FAUBA), and associated to other universities; Agricultural producer. Former President of the National Institute of Agricultural Technology (INTA); Former President of Slow Food Argentina; Former National Coordinator of the Programme of Technical Cooperation with FAO; Former National Director of Agricultural Production, the Secretariat of Agriculture, Cattle Farming, Fishing and Food (SAGPyA); Former Advisor to the Secretary of Agriculture, Livestock and Fisheries; Former Head of the cabinet of the Secretary of Commerce, Industry and Mining; Former National Director of Planification and Evaluation of the Secretary of Science and Technology; Head of the cabinet of the Secretary of Science and Technology; Author of books, articles and reports on Agro-industrial Commodity Chains, Competitivity and Markets, Buenos Aires, 28 February 2017

14. **Director of Agricultural Affairs CARI**: The Argentine Council of International Relations (CARI); Director of the CEO Group;; Special Adviser to the General Director of FAO for strategic planning. Previous held positions and work: Coordinator of the Economic Program at the National Institute for Agricultural Technology (INTA); Vice Minister of Agricultural Economy; General Director of the Inter-American Institute for Co-operation on Agriculture (IICA), President of the Governing Board at the International Food Policy Research Institute (IFPRI); member of the International Service for Agricultural Research (ISNAR); Member of the Governing Board of the University Improvement Fund (FOMEC, Ministry of Education and Culture); Piñeiro has provided consulting services to the World Bank, IDB,

FAO, IFAD, Inter-American Foundation, CGIAR and the Ford Foundation among others. Author of books on Development, Agricultural Policy and Economic Integration, Board member of two stock companies; agricultural producer, Buenos Aires, 1 March 2017

15. **Former Secretary of MAGyP**: Director of the Center of Research in Political, Economic and Social Sciences (CICPES) of the National Institute of Agricultural Technology (INTA); Secretary of Agriculture, Livestock and Fisheries, SAGPyA (2013–2015); General Director of Technological Agrarian Innovations of INTA –INTeA S.A. (2013); National Director of Information, Communication and Quality Systems of INTA (2010–2013); Former President of the Argentinean Association for Agrarian Economy (2010), Buenos Aires, 2 March 2017

16. **Director of Social Studies, INTA**: Director of the Institute of Social Studies of the Center for Research of Political, Economic and Social Sciences (CICPES) of the National Institute of Agricultural Technology (INTA); Researcher of Agrarian Economy at the Institute of Economy, INTA (2013–2016); Coordinator of the National Survey of Agrarian Producers (EPA) for 2013, INTA. Researcher at the Sub-secretary of Economic Planning (MECON), of the Secretariat of Political Economy and Development Planning, Buenos Aires, 2 March 2017

17. **Rural Development Specialist IICA Argentina:** Management of Territories, Agriculture and Rural Well-being, Inter-American Institute for Cooperation on Agriculture (IICA) at the office in Argentina, Buenos Aires, 3 March 2017

18. **Director of Research Promotion CLACSO**: The Latin American Council of Social Sciences; Researcher Pedagogy. Buenos Aires, 3 March 2017

19. **Researcher and former Public Official**: Department of Rural Development, Faculty of Agronomy, University of Missiones; Department of Economy, Development and Crop planning Faculty of Agronomy, University of Buenos Aires (FAUBA); Advisor for Productive Processes, at the Secretary of Family Agriculture, Ministry of Agro Industry (2016–2017); Advisor at the Unit of Economic and Social Studies, of the Secretariat of Agriculture, Livestock and fisheries of the Ministry of Agro Industry (2014–2016); Project Coordinator at the NGO Pereyra (agroecology and environmental pedagogy), Stockholm, 10 September 2018

Uruguay

20. **CEO El Tejar**: Taflilar Uruguay; President of the oil-seeds table (MTO); agronomist and consultant. Young, 23 February 2008
21. **Chief of sales ADP**: Agronegocios del Plata, Dolores 13 February 2008
22. **President FRU**: The Rural Federation of Uruguay; Rancher; Veterinary; University lecturer, Udelar. Montevideo, 3 March 2009
23. **Director Schandy Shipping and Logistics**: Montevideo 19 February 2009
24. **President CNFR**: The National Commission for Rural Development (CNFR); Producer, Montevideo, 5 March 2009
25. **Agronomist and family farmer**: Member of the grain cooperative CADOL, Dolores, 5 March 2010
26. **President of INASE**: The National Institute of Seeds; Agronomist; politician of the Socialist Party-FA, Montevideo, 6 March 2010
27. **Director of ALUR**: Alcohols of Uruguay, a state-controlled enterprise jointly owned by the national oil companies of Uruguay (ANCAP, 90%) and Venezuela (PDVSA 10%), Montevideo, March 2010
28. **Director of Cousa**: Compañía Oleaginosa Uruguay, S.A., the largest supplier of soybean oil in Uruguay, Montevideo, March 2010
29. **CEO Marfrig Beef of the Southern Cone** (Argentina, Paraguay, Uruguay and Southern Brazil). Marfrig is the second biggest meat company of the world, Montevideo, 7 March 2010
30. **Director of the Program for Production and Environmental Sustainability, INIA**: The National Institute of Agricultural Research (INIA) of Uruguay; Board Member at International Soil Tillage Research Organization (ISTRO); Lecturer at the post-graduate program in Agrarian Sciences, Faculty of Agronomy (FAGRO), University of the Republic (Udelar), Skype, 3 March 2017
31. **Researcher Political Science**: Faculty of Social Sciences (FCS), University of the Republic (Udelar); Advisory Board and Executive Team Member of the South American Institute for Resilience and Sustainability Studies (SARAS); Previously held positions: Director of the Latin American Faculty of Social Sciences (FLACSO), Uruguay (2009–2012), Montevideo, 6 March 2017

32. **Former Minister of MGAP**: Politician the Movement of Popular Participation (MPP), Frente Amplio (FA), Senate deputy legislator; Consultant in Forestry projects; University lecturer at FAGRO-Udelar; Former President of the National Institute of Colonization (INC) 2010–2013; Former Minister of Livestock, Agriculture and Fishery (2009–2010); Former Vice-Minister of Livestock, Agriculture and Fishery (2008–2009); Former National Forestry Director of MGAP (2005–2008), Montevideo, 6 March 2017

33. **Advisory Director DINAMA-MVOTMA**: The National Office of the Environment (DINAMA) at the Department of Housing, Territorial Planning and Environment (MVOTMA); Political Advisor (IR-FA) to the National Director of Territorial Planning (DINOT-MVOTMA); representative of MVOTMA in the Conference of the Parties to the United Nations Convention to Combat Desertification; Author of books and articles about the soybean expansion in Uruguay and its collateral effects, Montevideo, 6 March 2017

34. **Director of ANII**: The National Agency of Research and Innovation (ANII); Professor in Farming Systems, Production Ecology, Department of Vegetal Production, Faculty of Agronomy (FAGRO), University of the Republic (Udelar), Montevideo 7 March 2017

35. **Senior Researcher INIA**: Department of Nutrition in Intensive Dairy and Livestock Systems, the National Institute of Agricultural Research (INIA) La Estanzuela, Skype, 8 March 2017

36. **Director of OPYPA-MGAP**: The Office of Planning and Agrarian Policy, at the Ministry of Livestock, Agriculture and Fisheries (MGAP); Former tax-specialist at OPYPA; Agronomist, Montevideo, 8 March 2017

37. **Director of SARAS**: The South American Institute for Resilience and Sustainability Studies; Limnologist researcher, Eastern Regional University Centre (CURE), University of the Republic (Udelar); Member of the water-basin management group of Santa Lucia; Punta del Este, 10 March 2017

38. **Former researcher at CONEAT**: The National Commission for the Agronomic Investigation (CONEAT); Independent Consultant and researcher, working for the development of a

Strategic development plan for use and management of the soils-for the Ministry of Agriculture (MGAP) and other public and private institutions on Natural Resources and Development; Director of Future Generations University; Executive Director of the International Network on Migration and Development (Mexico 2010–2012); Director of the Rockefeller Foundation in Mexico (1991–2009); Researcher on Rural Development at Texas A&M University (1981–1991); Researcher and Developer of Uruguayan cartographic and soilsystem, CONEAT (1966–1972), Punta del Este, 11 March 2017

39. **Coordinator CEUTA**: The Uruguayan Center for Applied Technologies (CEUTA)—a nongovernmental Organization, working with Agroecology, Montevideo, 12 March 2017

40. **Dean FCS-Udelar**: Faculty of Social Sciences (FCS), University of the Republic (Udelar); Professor in Sociology, Author of books and articles about farmer concentration and foreignization of land, agrarian history, social relations in agriculture, Montevideo, 13 March 2017

41. **Researcher FAGRO**: Division of Environmental Science, department of water and soil, Faculty of Agronomy, FAGRO-Udelar; Course leader at the social-ecological NGO CEUTA. Stockholm, 23 June 2018

## NOTE

1. I have received suggestions on material from researchers at the University of Buenos Aires (UBA), the Latin American Council of Social Sciences (CLACSO), the National University of Misiones (UNaM), the National University of La Plata (UNLP), the National University of Asunción (UNA), the University of the Republic in Uruguay (UDELAR), the National Institute for Agricultural Technology (INTA) in Argentina, and the National Institute of Agrarian Investigation (INIA) in Uruguay.

# Appendix B

The export profiles of Argentina, Paraguay and Uruguay, 2002 and 2017 (Charts B.1–B.6). Note that areas are proportional to the export value in dollars per commodity. Source: Observatory of Economic Complexity, data from Comtrade FAO statistics. Interactive versions are available (open access) at atlas.media.mit.edu.

Argentina exported for a total value of 25.3B in 2002 and for 53B, more than the double, in 2017. The relative share of the soybean complex (soybeans, soybean oil and soybean oil cake) increased from almost 22% in 2002 (5.57B), to more than 26% in 2017 of total export value (13.78B). The relative share of bovine meatexports doubled between 2002 and 2017. The value of bovine meat exports increased from 0.29B in 2002, to 1.25B in 2017—a rise of more than four times the 2002-value (Charts B.1 and B.2).

Paraguay exported for a total value of 1.13B in 2002, and for 6.45B in 2017—a remarkable rise of almost six times. The export value of the soybean complex increased by six, but the relative share of the soybean complex (soybeans, soybean oil and soybean oil cake) was rather stable, increasing slightly from 44% in 2002 (0.5B), to more than 46% in 2017 (2.99B). Thus, the value of all soy products in 2017 were worth more than double the value of total Paraguayan exports in 2002. The relative share of bovine meatexports increased by almost three times. In 2017, the absolute value of bovine exports was almost fifteen times higher its 2002-value (0.07B in 2002, rose to 1.03B in 2017) (Charts B.3 and B.4).

© The Editor(s) (if applicable) and The Author(s), under exclusive license to Springer Nature Switzerland AG 2020
M. Baraibar Norberg, *The Political Economy of Agrarian Change in Latin America*, Governance, Development, and Social Inclusion in Latin America, https://doi.org/10.1007/978-3-030-24586-3

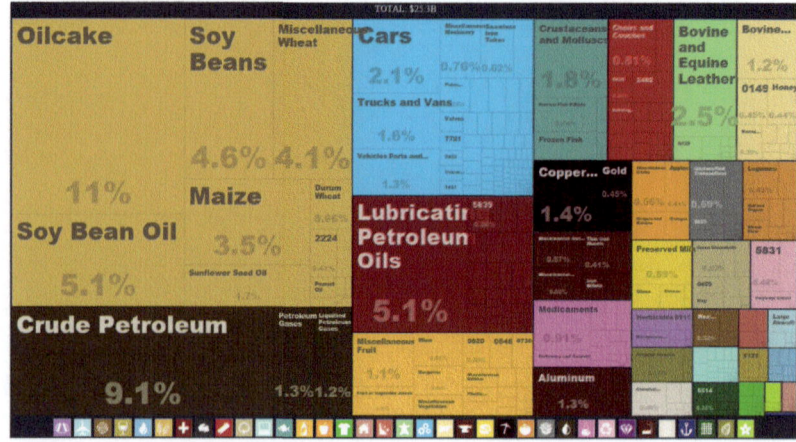

**Chart B.1**    Argentina 2002—total export value USD 25.3 billion

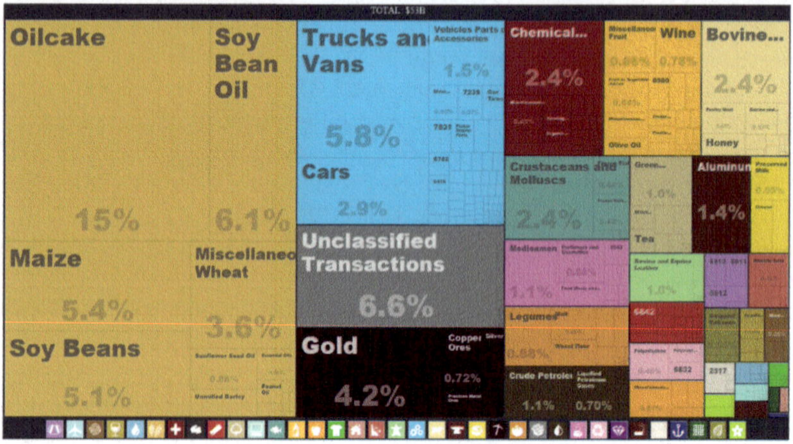

**Chart B.2**    Argentina 2017—total export value USD 53 billion

Uruguay exported for a total value of 1.97B in 2002, and for 7.32B in 2017—a dramatic rise of almost four times. The relative share of the soybean complex increased from 2% in 2002 to more than 13% in 2017. The absolute value of soy products exported in 2017 (0.98B) was almost 29 times more the 2002-value (0.034B). The relative share

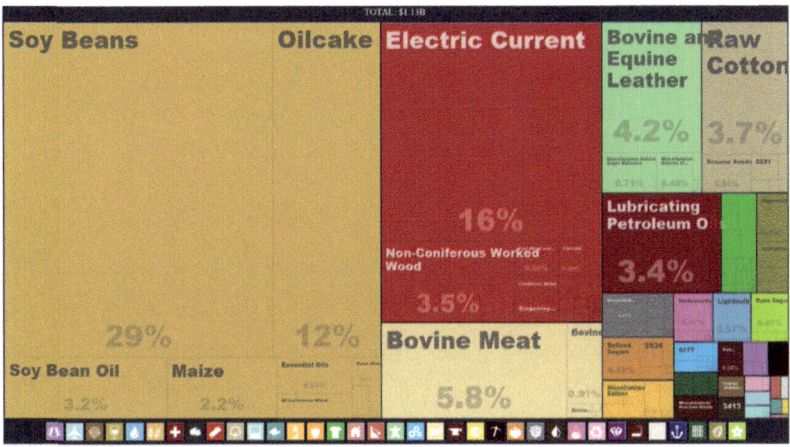

**Chart B.3**    Paraguay 2002—total export value USD 1.13 billion

**Chart B.4**    Paraguay 2017—total export value USD 6.45 billion

of bovine meatexports increased from 12 to 21% of total export value. The absolute value of bovine meat exported in 2017 was 1.52B, with live cattle adding up an extra 0.27B, was almost seven times more the absolute value of bovine meat exports in 2002 of 0.23B (Charts B.5 and B.6).

**Chart B.5**    Uruguay 2002—total export value USD 1.97 billion

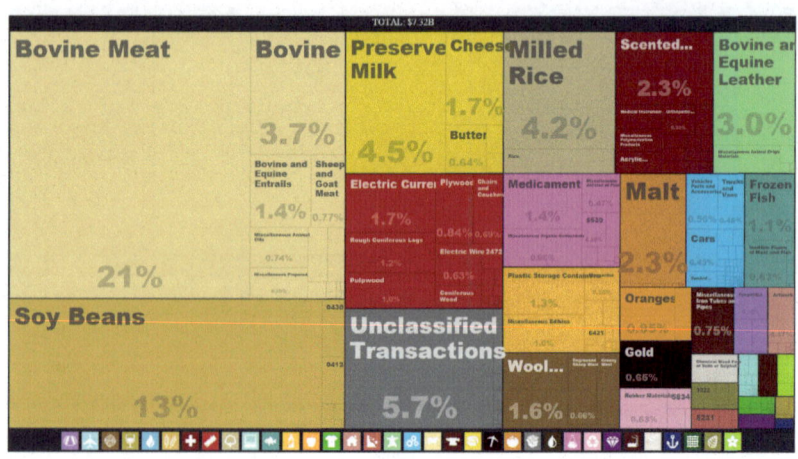

**Chart B.6**    Uruguay 2017—total export value USD 7.32 billion

# INDEX

**A**

Agrarian and Popular Movement (MAP), 247

agrarian census, 91, 97, 102, 176, 241, 242, 249, 269, 270, 304, 309

agrarian reform. *See* land reform

Agrarian Statute, 92, 93, 183

Agreement of Agriculture (AoA), 41, 125, 169, 339

agribusiness, 2, 3, 9, 10, 14, 18–26, 31, 33, 34, 37, 39, 78, 94, 125, 134, 140, 142, 146–148, 157, 165, 171, 175, 184, 185, 190, 192–194, 210, 213, 214, 220, 225, 226, 231, 233, 234, 239–241, 246–248, 251, 254, 256, 263, 266, 268, 269, 271, 276, 277, 279, 280, 282, 283, 286, 296, 298, 299, 302, 327, 329, 334, 335, 341, 342, 344, 346–349, 352, 353, 357, 359, 360, 362–364, 376, 379

Agricultural Credit Agency (CAH), 252

agricultural land, 2, 19, 123, 185, 188, 241–243, 273, 353

agricultural products, 79, 107, 124, 129, 131, 132, 152, 154, 186, 217, 273, 338, 367

agrochemicals, 10, 17, 19, 30, 35, 41, 77, 78, 119, 134, 135, 139, 140, 142, 143, 147, 154, 156, 229, 230, 249, 254, 287–290, 295, 302, 312, 337, 366, 368, 375

agroecology, 24, 279, 296, 354, 378, 380, 383

agrofood globalization, 3–6, 22, 24–26, 28, 32, 34–36, 42, 149, 327–329, 335, 347, 352, 359

agrofood system, 22, 25, 26, 28, 29, 35, 57, 119, 139, 145–147, 166, 169, 195, 327, 329, 338, 342

Alliance of Progress, 92

Alto Paraná (Paraguay), 16, 39, 40, 107, 183

© The Editor(s) (if applicable) and The Author(s), under exclusive license to Springer Nature Switzerland AG 2020
M. Baraibar Norberg, *The Political Economy of Agrarian Change in Latin America*, Governance, Development, and Social Inclusion in Latin America, https://doi.org/10.1007/978-3-030-24586-3

Amazon, 10, 37, 38, 41
American Antitrust Institute (AAI),
    155
Archer Daniels Midland (ADM), 141,
    156, 157, 199, 253, 263, 338,
    353, 367, 368
Argentina, 2–7, 10, 11, 13, 14,
    16, 21, 25–30, 32, 34–40, 58,
    59, 62–66, 68–74, 76, 80–83,
    85, 87–91, 95–103, 105, 118,
    124–126, 128, 129, 132, 137,
    138, 145, 148, 149, 151,
    152, 154, 156, 158, 165–169,
    171, 172, 176, 177, 184,
    185, 189–191, 193–197, 200,
    209–212, 214–218, 220, 225,
    229–231, 235–237, 246, 247,
    249, 251, 254, 255, 265–269,
    273, 274, 286, 288, 290, 295,
    297–303, 327–339, 342–353,
    355, 357–359, 362–369, 375,
    376, 378–381, 385, 386
Argentine Agrarian Federation (FAA),
    74, 84, 175, 220, 223, 224
Argentine Association of Direct
    Seeding Producers (AAPRESID),
    229
Argentine Council for International
    Relations (CARI), 106, 170, 218,
    230, 233, 359, 379
Argentine Institute for the Promotion
    of Trade (IAPI), 83
Argentine Rural Association (SRA),
    64, 65, 71, 74, 83, 84, 88, 100,
    101, 170, 175, 219, 220, 223,
    235, 348
Argentine Rural Confederations
    (CRA), 84, 88, 170, 175, 220,
    223, 348
Asia, 29, 30, 80, 117, 119, 122, 129,
    133, 134, 327, 338
Association of Producers of Paraguay
    (APS), 252, 263, 305, 351

Association of Seed Producers of
    Paraguay (APROSEMP), 305
Association of the Soybean Chain
    (ACSOJA), 229, 301, 369
Asunción treaty, 186
Atlantic forest, 5, 10, 11, 16–18, 37,
    71, 92, 94, 98, 166, 167, 178,
    179, 184, 185, 198, 209, 239,
    240, 242, 243, 245, 254, 255,
    261, 304, 347, 355, 356
authoritarianism, 81, 91, 179


B
Bair, Jennifer, 22, 29, 118, 157
balance of trade, 86, 186, 218
Barrán, José Pedro, 66–68, 101, 289
BASF, 139, 140, 155
Batlle y Ordoñez, José, 61, 67, 75
Batllismo, 67, 68, 81, 102, 359
Bayer, 139–141
Beef, 1, 3–8, 10, 13, 15–17, 19, 22,
    25, 28–30, 32, 34, 35, 37–39,
    60, 63, 85, 87, 95, 98, 100, 101,
    118–120, 122, 126, 128, 129,
    132–134, 143, 145, 147–150,
    152, 153, 158, 165, 166, 172,
    173, 178, 187, 188, 190, 191,
    194–196, 199, 209, 216, 218,
    219, 226, 230, 233, 239–242,
    253, 260–265, 267, 268, 278,
    284, 285, 288, 289, 298–300,
    310, 311, 327, 336, 339, 340,
    342, 343, 347, 349, 353, 361,
    365
Benitez, Abdo, 251, 262, 298, 313
biodiesel, 123, 145, 152, 175, 217,
    286, 301, 337, 341, 343, 367
biodiversity, 2, 39, 61, 77, 80, 95,
    185, 209, 226, 246, 263, 264,
    288, 290, 298, 305, 332, 353,
    357, 377
biological diversity, 14, 19, 287, 360

biotechnology, 22, 30, 35, 119, 135, 137, 140, 143, 146, 280, 307, 338, 343, 378
Bolivia, 2, 10, 11, 75, 103, 125, 151, 152, 154, 191, 299
Bovine Spongiform Encephalopathy (BSE), 152
Brasiguayos, 19, 40, 179, 241, 246, 247, 249, 250, 252, 299, 351
Brazil, 2, 4, 10, 11, 21, 26, 37, 40, 41, 69, 71, 80, 94, 98, 103, 107, 124–126, 128, 132, 133, 138, 145, 148, 150–152, 154, 156–158, 165, 169, 171, 177, 179, 191, 197, 198, 212, 213, 247, 251, 253, 255, 265, 266, 272, 300, 367, 381
breeds, 63, 135, 138, 310, 339
Bretton-Woods system, 76, 89, 99, 328
Britain, 26, 58, 59, 72, 75, 82, 100, 103, 120
Bulmer-Thomas, Victor, 59–65, 67–70, 72, 73, 75, 78, 79, 81, 82, 84, 85, 87–90, 97, 100–107, 167, 196
Bunge, 30, 141–143, 156, 199, 253, 338, 367

C
Campesino, 16, 18, 71, 92, 93, 178, 184, 195, 223, 243, 246, 250, 251, 269
canola, 137
capital accumulation, 61, 66, 70, 71, 97
capitalism, 24, 28, 35, 57, 59, 60, 78, 79, 96, 100, 117, 148, 166, 179
Cargill, 141–143, 145, 157, 199, 253, 263, 338, 353, 368
Cartes, Horacio, 251, 253, 261, 298

Catholic Church, 64, 68, 92
cattle ranching, 257, 260, 301
Chaco, 13, 15, 16, 37–39, 41, 71, 75, 92, 95, 98, 103, 107, 176, 185, 226, 227, 239, 240, 242, 244–246, 249, 255–258, 260–262, 301, 304, 305, 307, 308, 347, 352, 356
Chaco War, 73, 75, 86
Chamber of Grain and Oilseed Exporters and Traders (CAPECO -Paraguay), 252, 305
ChemChina, 140, 142, 156
Chicago Board of Trade (CBOT), 99, 144
chicken, 122, 145, 151, 171, 217, 337
China, 4, 119, 120, 124, 126, 129–131, 133, 134, 137, 140, 147, 150–153, 155, 165, 183, 217, 231, 268, 269, 285, 299, 308, 327, 337, 340, 342, 361, 364, 367
China National Cereals, Oils and Foodstuffs Corporation (COFCO), 142, 156, 338
Christian Agrarian Leagues (LAC), 92, 96
civil society, 180, 226, 231, 250, 274, 290, 296, 369
Clapp, Jennifer, 22, 23, 27, 29, 77, 78, 124, 125, 135, 141, 143–147, 149, 150, 165, 168
colonialism, 57
colonos, 276
Colorado Party (Paraguay), 75, 86, 93, 179, 186, 248, 262, 298, 363
commodity boom, 8, 27, 37, 195, 212, 280, 353
commodity chain, 29, 30, 41, 121, 143, 144, 157, 195, 210, 232, 297, 327, 336, 337, 342

commodity chain approach, 29, 41, 42, 118
commodity consensus, 214
comparative advantage, 27, 78, 89, 99, 125, 186, 195, 330
concentration, 3, 8–10, 14, 19, 30, 61, 119, 135, 136, 140, 141, 145, 146, 148, 188, 195, 209, 241, 243, 249, 252, 271–273, 277, 278, 286, 289, 338, 340, 344, 347, 350, 361, 383
contamination, 20, 209, 246, 289, 290, 294, 353
continuous cropping, 11, 13, 17, 173, 230, 254, 287, 292, 293, 357
contractual relations, 8
contratistas, 8, 14, 270, 271, 347
Coordinator of Rural Producers' Organizations of Paraguay (CAP), 252, 305
corporatism, 81, 275
corruption, 19, 23, 33, 89, 96, 145, 179, 180, 182, 195, 213, 219, 241, 246, 248, 249, 252, 253, 258, 261, 262, 265, 308, 341, 348, 356, 357, 359, 362–364
cotton, 17, 18, 69, 70, 94, 124, 137, 139, 142, 176, 178, 181–184, 195, 242, 243, 311
credits, 59, 79, 92, 103, 104, 124, 167, 169, 174, 175, 179, 182, 194, 195, 211, 214, 215, 219, 224, 249, 252, 260, 275, 276, 333, 334, 345, 348
cropland, 1, 2, 7, 8, 13, 28, 118, 123, 165, 192, 270, 292, 293, 327
crops, 2, 6–8, 11, 13, 14, 17, 19, 20, 36–39, 65–67, 70, 79, 84, 86, 90, 94, 95, 97, 100, 102, 105, 119, 120, 123, 134, 135, 137, 138, 150, 172, 174–176, 178, 187, 191, 192, 196–198, 211, 226, 230, 231, 234, 242,
254, 261, 263, 268, 270–273, 286–289, 293, 294, 338, 350, 357–359
crushing, 10, 121, 122, 129, 138, 143, 151, 152, 171, 225, 246, 338, 341, 367

**D**
dairy, 7, 28, 39, 95, 107, 120–123, 129, 138, 145, 173, 185, 188, 189, 230, 246, 276, 289, 294, 299, 347, 368
debt crisis, 27, 76, 90, 95, 177, 186
decentralization, 168–170, 178, 180, 214, 257, 267, 274, 275, 277, 278, 282, 290, 297, 332, 333, 336, 345, 348, 357
deforestation, 7, 15–17, 20, 38, 39, 41, 93, 95, 178, 184, 191, 226, 227, 229, 240, 244, 245, 254–257, 259–262, 264, 268, 301, 355–357, 369
denationalization, 70
Department of Justice (DOJ), 139
deregulation, 29, 89, 99, 125, 144, 146, 168–170, 174, 178, 186, 188, 193, 327
desarrollismo/desarrollista, 79, 80, 82, 86, 98
devaluation, 73, 177, 183, 191, 195, 216, 217, 350
developmental/developmentalist, 36, 68, 85, 117, 210–212, 267, 281, 284, 297, 328–330, 335, 362, 365
development strategies, 27, 76, 79, 88, 124, 186, 193
displacement, 2, 9, 18, 21, 173, 175, 225, 229, 243, 245–248, 269, 277, 334, 344, 347, 352
Division of Agrarian Statistics (DIEA), 269, 270, 272, 286, 311, 377

Division of Agricultural Services
(DGSA), 41, 295, 312
division of labor, 28, 29, 59, 61, 78
Division of Plant and Seed Health
(SENAVE), 41, 250, 262, 378
Dow, 140
drilling, 14, 270
drinking water, 20, 136, 289, 368
DuPont, 140

E
Eastern Paraguay, 94, 239, 242, 243,
250, 255, 304, 347, 351, 355
economic crisis, 59, 63, 89, 98, 177,
191, 196, 330
economic development, 78, 90, 98,
101, 211, 282, 283, 330, 334
economic growth, 21, 62, 63, 87,
117, 118, 165, 197, 215, 216,
231, 232, 239, 240, 251, 253,
261, 327, 330, 343
economic policy, 69, 102, 189, 232
ecosystem services, 11, 13, 20, 38,
230, 238, 257, 259, 333, 334,
353, 357
El Tejar, 8, 272, 273, 286, 287, 289,
359, 379, 381
enforcement mechanism, 23, 34, 363
Environmental Protection Agency
(EPA), 140
erosion, 2, 7, 11, 13, 20, 38, 59, 66,
76, 85, 86, 105, 133, 172, 185,
187, 191, 194, 230, 254, 257,
287–295, 311, 357–360, 368
Europe, 26, 30, 58–60, 64, 72, 75,
77, 101, 103, 119, 120, 122,
131, 137, 138, 150, 338, 379
European Food Safety Authority
(EFSA), 154
European Union (EU), 104, 129,
132, 136–138, 140, 147,

152–154, 249, 264, 288, 311,
339, 367, 368, 379
exchange rates, 78, 81, 82, 86, 89,
103, 171, 179, 214, 232
export diversification, 64, 90, 188
export-oriented, 26, 27, 35, 59–62,
78, 96, 99, 170, 177, 196, 210,
214, 231, 297, 344, 348, 368
exports, 1, 4, 21, 27, 36, 37, 60–64,
66, 69–75, 79, 81–83, 85,
87–91, 94, 95, 97, 98, 100, 101,
103–105, 107, 124, 126–129,
131, 141, 143, 149–154, 158,
166, 168, 170–173, 176–178,
181, 183, 185–191, 193–195,
199, 200, 212, 214–220, 222,
231, 233, 236, 239, 240, 251,
262, 267, 268, 280–283, 288,
296, 298–301, 307, 308, 311,
330, 331, 334, 336–344, 354,
367, 375, 385, 387
export specialization, 4, 60
export structure, 20, 65, 211, 340
export taxes, 11, 72, 79, 81, 83,
102, 106, 107, 150, 168, 170,
171, 175–177, 186, 191, 193,
194, 197, 198, 217, 219–222,
231–237, 249, 251, 252, 266,
277, 286, 298, 300, 301, 303,
335, 342, 349–352, 367, 368
export value, 36, 37, 64, 126, 129,
134, 195, 199, 239, 267,
385–388
export volume, 82, 85
extensive ranching, 7, 289
extractivism, 225, 235, 354

F
Faculty of Agronomy of UBA
(FAUBA), 11, 13, 228, 350, 358,
378–380

Faculty of Agronomy of the University of the Republic (FAGRO-Udelar), 272
family farmers, 2, 9, 21, 65, 175, 193, 194, 213–216, 224, 225, 238, 239, 241, 242, 247, 249, 268–271, 274, 275, 278, 279, 296, 298, 301, 309, 331, 333, 344–346, 348–350, 352, 378
Federation of Cooperatives of Agrarian Production (FECOPROD), 263, 305
feed, 1, 7–9, 18, 28, 34, 36, 39, 118, 119, 121–123, 126, 134, 137, 138, 143–145, 150, 152, 157, 165, 172, 175, 218, 243, 285, 289, 297, 327, 353, 367
feedlot, 8, 78, 122, 123, 150, 219, 265, 285, 337
fencing, 63, 66, 71, 271
Fernández de Kirchner, Cristina (CFK), 212, 215, 219, 231, 236, 301, 302
fertilizers, 7, 8, 13, 17, 18, 77, 79, 85, 104, 142, 143, 187, 287, 289, 354
financialization, 23, 30, 117, 146, 273, 330, 334
Finch, Henry, 66–68, 75
First Command of the Capital (PCC), 250
First World War, 58, 59, 72
fiscal deficit, 89, 186, 221, 236
fiscal land, 69, 70
fiscal revenue, 64, 73, 104, 106, 177, 189, 220, 231, 233, 252, 300, 303, 305, 342
Food and Agriculture Organization (FAO), 1, 7, 22, 24, 29, 36–38, 103, 118, 122, 123, 125, 126, 128, 129, 131, 134, 146, 147, 149, 150, 165, 170, 172,

198–200, 214, 216–218, 220, 222, 225, 232, 233, 242, 264, 299, 300, 302, 304, 343, 366, 369, 378, 379, 385
food crops, 21, 176, 195
Food Regimes (FR), 6, 22, 28, 30, 35, 41, 42, 57–60, 72–78, 89, 96, 98, 99, 146–149, 166, 365
food safety, 19, 143, 290, 378
food security, 21, 76, 77, 89, 103, 118, 130, 131, 142, 148, 213, 233, 331, 333, 348
food sovereignty, 2, 24, 148, 224, 225, 279, 296, 347
foreigners, 67, 69, 70, 182, 195, 247, 273
foreign investment (including foreign direct investment), 21, 27, 72, 100, 168, 171, 179, 188–190, 194, 211, 214, 239, 266, 267, 273, 281, 334
foreignization, 21, 193, 224, 246, 247, 252, 268, 272, 273, 276, 279, 347, 383
forest, 2, 4, 7, 9–11, 14–17, 20, 37–39, 41, 57, 70, 71, 90, 92–95, 100, 118, 170, 176, 178, 180, 184, 185, 188, 189, 223, 225–229, 231, 240–242, 244, 245, 249, 254–262, 264, 267, 268, 273, 301, 304, 307, 347, 352, 356–358, 369, 378
Forest Law (422/73), 16, 93, 228, 256–259, 261, 356
forestry, 24, 93, 107, 188, 189, 227, 267, 268, 273, 289, 295, 299, 307, 311, 357, 379, 382
Formosa (Argentina), 15, 176, 228
forms of production, 2, 5, 6, 9, 19, 24, 30, 31, 34–36, 42, 196, 209, 216, 254, 271, 297, 298, 360, 361

free trade zones (FTZs), 188, 191, 192, 194, 266, 279, 282, 286, 350

freight prices, 63

Frente Amplio (FA), 106, 190, 193, 212, 266, 331, 351, 382

Frente Guasu, 251

Friedmann, Harriet, 28, 58, 59, 61, 76, 77, 89, 99, 104, 148

Front for Victory (FPV), 212, 227, 301

**G**

General Agreement on Tariffs and Trade (GATT), 76, 89

General Office of Rural Development (DGDR-MGAP), 274, 309

genetically modified (GM), 6, 7, 125, 135–137, 139, 147, 155, 172, 225, 231, 243, 296

Gereffi, Gary, 22, 157

Global Commodity Chain (GCC), 3, 6, 22, 29, 30, 35, 41, 42, 118, 144, 145, 147, 149

glufosinate, 136, 139, 154

glyphosate, 6, 38, 135, 136, 140, 141, 146, 147, 154, 191, 222, 229, 230, 286–288, 311, 369

gold standard, 59, 72, 75, 102

good agricultural practices, 17, 293, 355, 360, 369, 377, 378

Gran Chaco, 2, 5, 7, 10, 14–16, 20, 166, 176, 197, 209, 226, 355

Great Depression, 27, 58, 73, 75, 76, 78, 98

green revolution, 8, 77, 79, 80, 104, 134, 135, 172

growth rates, 62, 79, 122, 133, 138, 189, 215, 216, 266, 308, 330

growth strategies, 62

**H**

habitat, 15, 41, 95, 178, 226, 229, 244, 256, 288, 347

harvest, 1, 8, 13, 65, 101, 122, 126, 144, 174, 181, 187, 192, 198, 236, 246, 270, 271, 301, 308

harvester, 7, 21, 174

herbicides, 7, 13, 38, 77, 136, 137, 140, 154, 230

Herbicide-tolerant (HT), 135, 136, 236

hides, 26, 60, 62, 63, 67, 69, 70, 72, 101, 258

Hilton quota, 132, 218, 285

hormones, 134, 138, 150

**I**

imperial preference, 72, 73

imports, 79, 81, 86, 90, 129, 131, 152, 169, 186, 217, 219, 231, 232, 237, 282, 339, 364, 367

import substitution industrialization (ISI), 76, 78, 83, 88, 89, 342

income, 4, 21, 63, 66, 68, 79, 81, 82, 91, 97, 101, 104, 142, 171, 187, 188, 198, 217, 219, 232, 235, 247, 251, 253, 260, 266, 270, 277–279, 282, 285, 287, 302, 303, 308–310, 340, 349–352, 368

indebted state, 26

indigenous people, 64, 185, 244, 251, 259, 345, 363

industrial capitalism, 60

industrialization, 27, 33, 58, 63, 65, 68, 76–81, 83, 84, 87, 98, 146, 177, 186, 194, 280

inequality rates, 4, 68

infamous decade, 73

inflation, 75, 87, 89, 90, 98, 104, 171, 176, 196, 216, 218, 232, 310

infrastructure, 3, 10, 16, 21, 60, 69, 71, 84, 94, 95, 103, 124, 141–144, 147, 170, 175, 183, 185, 192, 198, 214, 217, 222, 224, 226, 231, 237, 253, 261–263, 273, 279, 283, 286, 311, 334, 341, 347, 348
insecticide, 19, 77, 140, 287, 288, 311
Intellectual property rights (IPRs), 30, 135, 146, 165, 190, 194, 246, 263, 286, 337, 348
intensification, 2, 3, 6, 14, 19, 29, 30, 32, 34, 107, 173, 196, 209, 219, 240, 286, 289, 290, 295, 353, 354
Inter-American Development Bank (IADB), 95, 168, 180, 181, 299, 366
Inter-American Institute for Cooperation on Agriculture (IICA), 1, 7, 11, 14, 24, 128, 175, 180, 192, 214, 224, 237, 258, 299, 302, 307, 332–334, 347, 366, 367, 377, 379, 380
Inter-Cooperative Agrarian Federation (CONINAGRO), 220, 223
interest group, 3, 67, 72, 81, 86, 166, 275, 350, 352
International Fund for Agricultural Development (IFAD), 175, 224, 237, 299, 304, 309, 366, 380
International Labour Organization (ILO), 244
International Monetary Fund (IMF), 168, 169, 177, 233, 235, 300, 308, 364, 369
International Organizations (IOs), 3, 17, 23, 24, 31, 33, 107, 168, 175, 180, 210, 211, 213, 237, 238, 255, 329, 332, 362, 376, 378, 379

international prices, 17, 67, 74, 83, 85, 99, 170, 176, 330
International Products Corporation (IPC), 73
International Service Acquisition (ISAAA), 38, 135, 137, 154
International Soybean Growers Alliance (ISGA), 136, 138, 155
international trade, 33, 41, 72, 99, 126, 134, 211, 362, 365
interventionist/interventionism, 64, 72, 78, 81, 85, 86, 89, 98, 175, 177, 193, 195, 196, 215, 223, 266, 277, 279, 281, 342, 348
inward-orientation, 87, 88
irrigation, 7, 10, 77, 85, 103, 134, 289, 295
Itapúa (Paraguay), 16, 39, 40, 107, 183

J
JBS, 10, 145, 157, 158, 265

K
Kirchner, Néstor, 212, 215–218, 224
Kirchnerismo, 212, 215, 216, 223, 225, 229, 231, 232, 298, 331, 332, 335, 336, 345, 349–351, 364
Korean War, 85, 87

L
laissez-faire, 70, 72, 86, 215, 229, 231, 290, 335
land concentration, 2, 9, 19, 65, 80, 166, 193, 216, 242, 260, 268, 276, 277, 344, 349, 352, 357
land occupation, 21, 184, 225, 247, 248, 252

landowners, 14, 15, 66, 69, 70, 72, 80, 83, 84, 86, 92, 93, 97, 101, 197, 221, 228, 231, 234, 241, 247, 248, 256, 257, 259–261, 270, 291, 303, 304, 308, 341, 356, 359, 360, 362, 369
land reform, 84, 86, 88, 96, 97, 104, 181, 198, 225, 231, 241, 248, 249, 276, 305, 334, 351
land rents, 14, 66, 80, 175, 245, 270
land structure, 8, 18, 35, 64, 67, 68, 80, 84, 88, 92, 97, 98, 173, 178, 187, 195, 239, 347
land tax, 106, 188, 269, 277, 279, 345, 350
Land Use and Land Cover Change (LULCC), 1–3, 16, 118, 226
land-use intensification, 2, 5, 17, 35, 118, 185, 211, 240, 254, 262, 268, 289, 354
land values, 11, 13, 14, 17, 21, 66, 175, 184, 191, 245, 278, 286, 346
latifundio, 19, 66, 80, 92, 93, 120, 254, 271
Latin America, 1–5, 8, 14, 22, 23, 25–27, 29, 34–36, 40, 42, 57, 59–63, 72, 73, 75, 76, 78, 79, 81, 87, 91, 96, 100, 101, 104, 118, 124–126, 128, 129, 131, 133, 136, 137, 147, 149, 150, 165–167, 169, 185, 196, 213, 266, 273, 327, 328, 330, 338–340, 343, 349, 352, 354, 366
Latin American Coordinator for Rural Organizations (CLOC), 40, 305
Latin American structuralism, 79, 211, 330, 367
law of water and soil conservation (15.239), 86
law proposal, 86, 227
liberalization, 22, 30, 89, 99, 117, 125, 144, 146, 149, 165, 167, 169, 174, 185, 186, 195, 211, 232, 266, 330, 339, 344
Littoral (Uruguay), 38, 90, 105, 268, 270, 287, 288
livestock, 1, 2, 4, 6, 9, 13–16, 18, 28, 59, 64, 66–68, 84, 85, 87, 90, 95, 97, 118, 119, 121–123, 138, 143, 145, 165, 175, 176, 178, 216, 229, 239, 240, 243, 244, 254, 256, 269, 284, 286–290, 293, 295, 339, 352, 366
livestock sector, 9, 39, 40, 63, 66, 67, 71, 74, 87, 100, 107, 126, 176, 184, 186, 188, 191, 216, 219, 232, 254, 308
loan conditionality, 23, 33, 168
London Corn Trade Association (LCTA), 99
Louise Dreyfus Commodities (LDC), 199, 263, 353, 368
Lugo, Fernando, 248–251, 253, 298, 331, 352

**M**
Macri, Mauricio, 216, 232–237, 298, 302, 337, 345, 350, 364
maize, 37, 65, 82, 104, 124, 128, 137, 139, 219, 222, 249, 296, 300, 301
manufactures, 59, 60, 78, 79, 89, 97, 98, 100, 154, 280, 312, 340
Marfrig, 8–10, 271, 381
McMichael, Philip, 22–24, 28, 30, 41, 57, 58, 60, 76–78, 89, 99, 144, 146, 147, 166, 168
meat, 3, 8–10, 26, 28, 30, 35, 38, 42, 60, 62–65, 67, 70–75, 83, 86, 89, 95, 98, 100, 107, 118–123, 125, 126, 128, 129, 131–134, 137, 139, 143–150, 153, 154, 157, 165, 185, 186, 189,

218–220, 222, 233, 240, 264,
265, 267, 284, 285, 289, 297,
311, 327, 337–339, 341, 342,
346, 381, 385, 387
meat consumption, 123, 147, 149
meat packers, 63, 86, 95, 100, 103,
188, 218
mechanization, 84, 107, 182, 346
Menem, Carlos, 168–171, 189, 197,
215, 332
Mercosur, 132, 148, 153, 154, 165,
169, 186, 187, 197, 212, 213,
251, 274, 311, 331
Mergers and acquisitions (M&A), 134,
136, 139–141
migration, 40, 88, 183
military government, 90
minifundio, 66, 67, 96, 243
Ministry of Agriculture and Livestock
(MAG), 107, 178, 180, 181, 183,
184, 238, 239, 241–243, 245,
257, 261–263, 303, 304, 332,
333, 355, 375, 377, 378
Ministry of Environment and
Sustainable Development
(MAyDS), 227, 228, 232, 303,
355, 375
Ministry of Housing, Land
Management and Environment
(MVOTMA), 187, 214, 289,
294–296, 333, 334, 355, 375,
382
Ministry of Livestock, Agriculture,
Forestry and Fishery (MGAP),
16, 198, 199, 267, 269, 270,
274–277, 280–282, 285, 286,
289–297, 309, 311, 312, 332,
333, 353, 355, 375, 382, 383
Ministry of Livestock, Agriculture and
Fishery (MAGyP), 216–219, 223,
224, 233, 235, 345, 366, 380

mixed systems, 7, 11, 90, 187, 287
modernization, 17, 66, 71, 77, 79, 98,
124, 167, 175, 178, 188, 232,
289
monoculture, 21, 61, 79, 100, 148,
246
Monsanto, 6, 135, 136, 138–140,
154–156, 172, 221, 225, 231,
235, 246, 253, 281, 302, 305,
364, 366
moratorium, 16, 227, 255, 261, 307,
356
Mujica, José "Pepe", 212
multilateral agreements, 23, 41


N
Nahum, Benjamin, 66–68, 101, 289
National Agrarian Council (CAN), 74,
84, 274, 304
National Agrarian Research Institute
(INIA), 189, 199, 271, 274, 280,
292, 293, 296, 381, 382
National Autonomist Party (PAN),
101
National Commission for Rural
Development (CNFR), 273, 278,
381
National Commission for the
Agronomic Investigation
(CONEAT), 85, 86, 291, 292,
382, 383
National Development Bank (BNF),
252
National Directorate of the
Environment (DINAMA), 296,
312
National Directorate of Waters
(DINAGUA), 294
National Forest Institute (IFONA),
170, 226, 357

National Forest Institute (INFONA), 15, 20, 38, 244, 255, 256, 259–263, 303, 307, 308, 357

National Forum for Family Farming (FONAF), 223, 224, 301

National Inter-sectorial Coordinator (CNI), 252

nationalism, 24, 41, 69, 81, 91

nationalization, 78, 82, 86, 105

National Meat Institute (INAC), 264, 274, 284

National Peasant Federation (FNC), 18, 243, 247

National Service for Animal Quality and Health (SENACSA), 256, 265

National Service of Sanitation and Food Quality (SENASA), 40, 229

National Society of Agriculture (SNA), 93, 102

National System of Protected Areas of Paraguay (SINASIP), 258, 259

National System of Protected Areas (SNAP), 311

native communities, 9, 16, 242, 244, 252

natural resources, 2, 14, 23, 26, 36, 57, 59–61, 80, 85, 98, 100, 212, 238, 240, 243, 261, 266–268, 290, 291, 293, 298, 304, 328–330, 340, 353, 354

neobatllismo, 80–82, 84, 89, 104

neoliberalism/neoliberalization/neo-liberal, 24, 26–28, 36, 76, 90, 91, 99, 106, 125, 166, 169, 175, 177–180, 190, 193, 195, 196, 209–212, 215, 225, 237, 266, 275, 291, 297, 298, 327, 328, 330–332, 334, 347, 358

neostructuralism, 36, 210, 211, 215, 330–332

network firms, 8, 174, 191, 344

nineteenth century, 10, 26, 58, 61, 67, 97, 99, 102, 103, 120, 124, 168, 346, 347

nitrogen, 13, 38, 187, 289

Non-Governmental Organizations (NGOs), 21, 33, 40, 138, 225–227, 229, 246, 255, 262, 269, 273, 279, 302, 305, 335, 347, 361, 369, 376, 380, 383

nontariff barriers, 59, 88, 168

no-tillage, 6, 10, 11, 136, 172, 173, 191, 230, 287, 292

**O**

Office for Control of Agricultural Business (ONCCA), 219

Office of Agrarian Censuses and Statistics (DCEA), 183, 245

Organization for Economic Co-operation and Development (OECD), 1, 29, 118, 123, 125, 129, 131, 134, 147, 149, 165, 303, 366

Organization for Land Struggle (OLT), 247

outward-orientation, 90

**P**

Pampas, 5, 10, 11, 13–16, 18, 20, 38, 65, 90, 166, 171–176, 191, 193, 197, 209, 220, 229, 230, 234, 245, 255, 270, 286, 288, 301, 311, 346, 347, 356

Paraguay, 2–7, 10, 11, 13, 15–18, 21, 25–30, 32, 34–41, 58, 59, 62, 69, 70, 72, 73, 75, 76, 80, 81, 86, 87, 91, 93–99, 102, 103, 107, 118, 124–126, 128, 132, 137, 138, 142, 145, 148, 149, 151–154, 156, 158, 165–169, 177–185, 191, 193–198,

209–212, 214, 237–243, 245,
246, 250, 251, 253–255,
257–261, 263–265, 267–269,
286, 290, 295, 297–299,
303, 305–307, 327–337, 341,
343–345, 347–349, 351–353,
355–358, 362–369, 375–378,
381, 385
Paraguayan Association of Producers
of Soybeans, Cereals, and
Oilseeds, 248
Paraguayan People's Army (EPP),
250, 252, 253, 305
Paraguay's Deputy Minister of
Agriculture, 16, 18, 181, 182,
238, 240, 245, 256, 260, 261,
348, 378
participatory processes, 275, 290,
292–294, 297, 332, 355, 360
pastureland, 1, 7, 9, 13–15, 28, 38,
95, 118, 165, 175, 219, 240,
242, 262, 286, 288, 327
patent, 135, 136, 138, 141, 146, 337,
342
paternalistic/paternalism, 91, 181,
182, 221, 225, 363
Patriotic Alliance for Change, 212
peasants, 9, 16, 18, 20, 21, 23, 26, 39,
71, 80, 92, 94, 96, 99, 148, 178,
179, 184, 195, 223, 225, 226,
244, 246–250, 252, 258, 269,
299, 305, 347, 348, 351, 362
Permanent Meeting of Family
Agriculture (REAF), 198, 213,
215, 274, 331, 378
peronismo, 80, 81, 89
Perón, Juan Domingo, 80, 83, 170,
215, 345
Pesticides, 7, 13, 38, 77, 104, 135,
139–141, 154, 225, 243, 262,
287, 295, 312, 354, 360, 361,
369

Pink Tide, 36, 166, 210, 212, 216,
297–299, 331, 334
Pollution, 2, 20, 194, 240, 254, 262,
268, 288, 304, 332, 333, 360,
361
pools de siembra, 8, 14, 174, 191,
277, 346
populism, 81, 182
pork, 107, 121, 122, 129, 131, 145,
151, 171, 217, 218, 300, 337
port, 3, 100, 142, 143, 192, 199, 225
post-Washington consensus, 210, 213,
237, 238, 275, 297, 330, 332,
334
poverty, 4, 33, 37, 80, 88, 90, 168,
177, 183, 196, 213–215, 224,
232, 235, 238, 239, 253, 260,
266, 297, 308, 330, 333, 343,
348, 349, 352, 362, 368
power relations, 3, 25, 27–31, 35,
134, 329, 330, 342, 343
primary products, 59–63, 78, 81, 91,
103, 104, 129
private property rights, 27, 63, 64,
66, 89, 168, 175, 192, 214, 231,
244, 246, 266, 279, 281, 291,
294, 327, 345, 348, 358, 359
privatization, 70, 99, 167, 168, 170,
186, 190, 193, 199, 245, 330
processing, 6, 9, 10, 22, 30, 35,
63, 65, 70, 71, 73, 78, 86, 94,
97, 107, 119, 120, 124, 134,
142–146, 151, 157, 166, 180,
188, 192, 217, 219, 225, 253,
267, 338, 341, 342, 344, 363
producers' organizations, 31, 68, 84,
88, 93, 97, 103, 170, 171, 175,
210, 220, 225, 233, 234, 246,
252, 277, 278, 296, 301, 302,
305, 329, 335, 348, 350, 351
production costs, 62, 144, 174, 187,
189, 192, 279, 343

productivity, 7, 17, 21, 37, 39, 62, 63, 67, 77, 80, 81, 84, 85, 95, 100, 102, 106, 121, 126, 138, 150, 173, 181, 183, 184, 187–189, 194, 232, 243, 263, 267, 284, 290, 297, 300, 343, 360
property rights, 16, 29, 104, 153, 189, 212, 241, 244, 259, 269, 340, 345, 357
protectionism, 73, 79, 186, 339, 340
protective tariffs, 72, 76
public goods, 284, 297, 362
public land, 63, 80, 178, 347
public policy, 68, 192, 214, 237, 238, 260, 262, 290, 332, 339, 345, 348, 349, 355

**R**
railway, 61, 64, 67, 82, 101, 120
ranchers, 9, 13, 14, 16, 64–68, 70, 71, 75, 84, 85, 89, 95, 97, 100–102, 157, 170, 175, 179, 185, 187, 194, 219, 244–246, 248, 255, 270, 287, 309, 346–348, 356, 381
real wages, 59, 83, 90, 105, 196, 266, 349
Reducing Emissions from Deforestation and forest Degradation (REDD), 260
regional integration, 88, 169, 187
regulations, 3, 5, 6, 21, 22, 24, 25, 27, 29, 31, 32, 34–36, 41, 42, 59, 72, 76, 85, 89, 128, 146, 147, 149, 166, 167, 180, 189, 192, 209, 210, 214, 216, 231, 234, 238, 241, 254, 257, 261, 262, 269, 284, 290, 294–298, 302, 328, 329, 333, 335, 340, 342, 347, 355, 359, 360, 363, 376

regulatory framework, 3, 5, 6, 24, 25, 27, 30–32, 42, 196, 210, 255, 329, 335, 376
rentier capitalism, 66
rent-seeking, 23, 341, 342
re-primarization, 211, 269
Research and Development (R&D), 135, 167, 173, 281, 337
Río de la Plata, 61, 97
rotation, 7, 11, 13, 17, 90, 147, 187, 230, 231, 263, 269, 272, 292–294, 357, 359, 369
Round Table on Responsible Soy (RTRS), 264, 368, 369
Roundup Ready (RR), 6, 125, 135, 136
Rural Association of Paraguay (ARP), 71, 93, 102, 179, 248, 263, 348, 351
Rural Association of Uruguay (ARU), 67, 68, 71, 277, 279, 348, 350
rural development, 107, 175, 183, 198, 213, 224, 274, 331, 333, 345–348, 378
Rural Development and Land Institute (INDERT), 183, 198, 249, 305
Rural Federation (FR), 68, 279, 348, 350
Rural Welfare Institute (IBR), 92, 93, 96, 182, 183, 198, 249, 305

**S**
Salta (Argentina), 15, 176, 228
Santiago del Estero (Argentina), 15, 176
Scobie, James, 63–66, 101
Second World War, 58, 76
Secretariat for Agriculture, Livestock, Fisheries and Food (SAGPyA), 171, 175, 177, 197, 224, 368, 379, 380

Secretary of the Environment (SEAM), 15, 20, 238, 239, 244, 250, 255, 256, 258–263, 303, 307, 308, 375, 377, 378

seeds, 10, 30, 35, 85, 119, 134–136, 139–141, 143, 156, 167, 172, 173, 175, 181, 190, 192, 211, 225, 246, 257, 281, 302, 338, 364

service crops, 38

sharecroppers, 9, 14, 65, 84, 101, 175, 187, 194, 269, 270, 309, 345, 347

slaughterhouses, 9, 10, 29, 344, 354

social function of land, 80, 93, 168, 179

social movements, 21, 23, 31, 40, 92, 148, 193, 210, 225, 246, 252, 305, 329, 354, 366, 369, 376

social relations, 2, 3, 5, 6, 8, 20, 24, 30, 31, 34–36, 42, 196, 209, 216, 226, 259, 266, 269, 279, 297, 298, 344, 349, 383

soil, 6, 11, 14, 15, 17–20, 24, 38, 57, 65–67, 85, 86, 90, 105, 106, 123, 135, 136, 146, 167, 172, 181, 187, 191, 194, 221, 223, 226, 230, 233, 234, 240, 246, 257, 260, 262, 263, 267, 269, 276, 287, 289–294, 297, 298, 307, 338, 358, 364, 383

soil degradation, 11, 20, 105, 172, 230, 254, 268, 292, 293, 358

sojization, 13, 20, 175, 194, 217, 223, 286, 288

Southern Agricultural Council (CAS), 4, 7, 38, 131, 149, 158, 216, 267, 299, 300, 367

Southern Common Market (Mercosur), 166, 169

Southern Cone, 3, 8, 9, 62, 72, 97, 120, 126, 149, 167, 196, 271, 337–342, 347, 365

soybean consumption, 129

soybean demand, 129

soybean meal, 30, 94, 107, 121–123, 126, 129, 131, 137, 138, 151, 152, 155, 300, 343, 367

soybean oil, 37, 121, 126, 151, 152, 200, 231, 301, 337, 381, 385

soybean production, 7, 11, 17–19, 90, 121, 124–126, 150–152, 167, 172, 173, 179, 182–184, 190, 199, 217, 221, 223, 230, 231, 234, 242, 254, 264, 270, 271, 288, 311, 337, 345–347, 361, 367, 369

soybean trade, 10, 30, 120, 126, 131, 143, 147, 150, 338

specialization, agriculture, 2, 6, 8, 87, 98, 148

speculation, 87, 106, 143, 144

state autonomy, 6, 23, 25, 32–34, 36, 329, 362, 363, 366

state capacity, 5, 23, 25, 32–34, 180, 299, 329, 335, 357, 362–364, 366

state intervention, 59, 77, 81, 147, 185, 344, 348, 350, 358

Strategic Agrofood and Agroindustry Plan (PEA), 232

Stroessner, Alfredo, 17, 76, 87, 91, 92, 94, 96, 98, 99, 107, 125, 167, 177–180, 182, 183, 195, 211, 241, 247, 313, 332, 363

Stroessnerismo, 91

Sub-Secretariat of Rural Development and Family Farming (SSDRAF), 224, 301

subsidies, 27, 74, 77, 79, 84, 89, 90, 124, 126, 145, 147, 151, 187, 188, 191, 211, 219, 299, 300, 339, 367

sugar, 26, 60, 100, 142

sustainability, 21, 100, 143, 210, 213, 214, 231, 234, 240, 241, 246,

261, 263, 264, 266, 285, 297, 353–355, 359, 360, 368, 369
Sustainable Development, 214, 272, 294, 311, 377
sustainable intensification, 267, 290, 293, 297, 353
Syngenta, 140, 142, 154, 156

**T**
Taiwan, 152, 183, 367
tariff, 59, 61, 73–75, 79, 81, 88, 90, 95, 100, 101, 103, 106, 107, 126, 131, 132, 153, 168–170, 185, 186, 197, 231, 339, 349, 350
taxes, 86, 92, 101, 102, 129, 170, 171, 177, 188, 189, 214, 217, 220, 231, 233–235, 237, 246, 248, 251, 263, 277–279, 303, 331, 334, 342, 343, 350, 351, 359, 364, 367, 368
technological innovation, 60, 63, 135
Technological Laboratory of Uruguay (LATU), 272
Technological Oil-Seeds Table (MTO), 271, 272, 280, 283, 288, 293, 309, 369, 381
Tequila crisis, 176
tobacco, 65, 69, 70, 154
traceability, 123, 264, 265, 284, 285, 297, 338, 339, 341, 343
trade liberalization, 29, 76, 89, 130, 168, 169, 212, 327, 339
Trade-Related Aspects of Intellectual Property Rights (TRIPS), 125, 169
traders, 10, 77, 78, 98, 139, 141, 143–145, 170, 174, 175, 183, 192, 220, 252, 253, 263, 264, 338, 342, 344, 368
transnational corporations, 33, 117, 339, 341

transport, 8, 10, 59, 68, 79, 141, 143, 154, 189, 192, 217, 270, 273, 293, 346
transport costs, 60, 63, 67, 95, 125, 146, 189, 262, 273, 338, 375
twentieth century, 61, 81, 120, 126, 135, 166, 187, 211, 271, 275, 328, 330, 346
twenty-first century, 1, 118, 166, 193, 195, 209–211, 216, 247, 254, 275, 297, 328–330

**U**
Union for the Protection of New Varieties of Plants (UPOV), 104, 172
Union of Producer's Associations (UGP), 249, 252, 351
United Nations Development Programme (UNDP), 181, 214, 263, 264, 302, 332–334, 357, 366, 369, 377
United Nations Economic Commission for Latin America and the Caribbean (CEPAL), 78–80, 86, 88, 211, 267, 302, 304, 330, 331
United Nations Framework Convention on Climate Change (UNFCCC), 259, 303, 357
United States, 27, 28, 103, 104, 139, 151–153, 157, 183
United States Department of Agriculture (USDA), 2, 7, 16, 18, 39, 120–126, 128–134, 137, 147, 149, 151, 152, 254, 256, 268, 285, 311
upgrade/upgrading, 36, 210, 212, 217, 222, 232, 237, 238, 267, 269, 278, 280, 282, 283, 285, 286, 297, 328, 329, 331, 335, 336, 338–342

urbanization, 2, 29, 58, 66, 100, 117,
    118, 133, 226, 243, 248
Uruguay, 2–7, 11, 13, 16, 21, 25–30,
    32, 34–38, 58, 59, 61–64, 66–73,
    75, 76, 80–91, 96–100, 102,
    103, 105, 106, 118, 124–126,
    128, 132, 137, 138, 145, 148,
    149, 151, 152, 154, 156, 158,
    165–169, 185, 186, 188–199,
    209–212, 214, 237, 251, 254,
    255, 264–278, 281, 283–299,
    308–311, 327–339, 341,
    343–346, 348–351, 353, 355,
    357–359, 362, 364–368, 375,
    376, 379, 381–383, 385, 386

V
value added, 30, 198, 281–283, 337,
    342
Value-added Tax (VAT), 129, 152,
    171, 188, 217, 237, 248, 252,
    278, 282, 303
value chain, 142, 212, 225, 275, 280,
    283, 300, 336, 369
Vázquez, Tabaré, 212
vertical integration, 10, 22, 30, 119,
    135, 140, 143, 146, 165

W
War of the Triple Alliance, 26, 69, 97,
    241
Washington Consensus, 35, 91, 99,
    148, 167, 214
water, 7, 14, 19, 20, 67, 85, 117,
    185, 193, 209, 226, 230, 233,
    240, 246, 254, 256, 267, 275,
    288–290, 293–295, 297, 298,
    353, 357, 358, 360–362, 383

water cycle, 20, 289
Western Paraguay, 240, 355
wheat, 11, 64, 72, 79, 84, 94, 99,
    101, 104, 124, 128, 139, 141,
    142, 172, 173, 187, 198, 219,
    220, 222, 230, 251, 299–301,
    303, 346
whole soybeans, 4, 120, 121, 126,
    129–131, 151, 152, 171, 200,
    217, 222, 225, 249, 251, 252,
    300, 301, 337, 338, 343
wool, 26, 62, 65, 72, 73, 85, 87, 89,
    100, 101, 171, 189
World Bank, 37, 95, 134, 168, 175,
    180, 181, 227, 237, 238, 240,
    275, 283, 300, 304, 309, 366,
    368, 369, 379
World Health Organization (WHO),
    140
world production, 121, 123, 124, 126
world system, 3, 22, 23, 35, 42, 57,
    59, 60, 88, 166
World Trade Organization (WTO),
    41, 125, 131, 146, 153, 169, 339
World Wildlife Foundation (WWF),
    185, 225, 255, 259, 264, 303,
    308, 369, 378

Y
yerba maté, 69, 70
yields, 6–8, 13, 36, 66, 74, 77, 80,
    84, 90, 125, 136, 138, 146, 167,
    181, 184, 187, 191, 196, 198,
    254, 257, 263, 287, 352, 365,
    375

Z
Zero deforestation, 261, 356

Printed by Printforce, the Netherlands